T0335614

Thermal Imaging Techniques to Survey and Monitor Animals in the Wild

A Methodology

Thermal Imaging Techniques to Survey and Monitor Animals in the Wild

A Methodology

Kirk J Havens

Edward J Sharp

AMSTERDAM • BOSTON • HEIDELBERG • LONDON
NEW YORK • OXFORD • PARIS • SAN DIEGO
SAN FRANCISCO • SINGAPORE • SYDNEY • TOKYO

Academic Press is an imprint of Elsevier

Academic Press is an imprint of Elsevier
125, London Wall, EC2Y 5AS, UK
525 B Street, Suite 1800, San Diego, CA 92101-4495, USA
225 Wyman Street, Waltham, MA 02451, USA
The Boulevard, Langford Lane, Kidlington, Oxford OX5 1GB, UK

ISBN: 978-0-12-803384-5

British Library Cataloguing-in-Publication Data
A catalogue record for this book is available from the British Library

Library of Congress Cataloging-in-Publication Data
A catalog record for this book is available from the Library of Congress

For information on all Academic Press publications
visit our website at http://store.elsevier.com/

Dedication

To my wife, Karla, who only occasionally raised an eyebrow and rarely questioned the late night trips to "study wildlife." To my son, Kade, who understands the wisdom in questioning everything and to my parents, Bill and Ginny, who gave me the childhood freedom to explore.

Kirk J Havens

Contents

Preface

Over the past few decades there has been a marked increase in areas of remote sensing, including thermal imaging, to study and count wildlife in their natural surroundings. While much of the work with thermal imagers to date has been devoted to testing equipment during surveys, little advancement has actually been achieved. This is primarily due to three basic problems:

1. Early field studies were conducted with cryogenically cooled thermal imagers (photon detectors) with sensitivities an order of magnitude lower than those available today. With few exceptions, the new and improved models of thermal imagers with superior sensitivities and resolution have not been used in the field because of the perceived difficulty in data acquisition and to some extent limited availability and cost. The more recent fieldwork has been for the most part confined to the use of uncooled bolometric cameras that use thermal detectors as opposed to photon detectors.
2. A pervasive misunderstanding of what thermal imagers detect and record and what ultimately constitutes ideal conditions for conducting thermal imaging observations.
3. The promulgation of results that have erroneously compared survey data collected with thermal imaging equipment to that obtained with standard techniques such as spotlighting or visual surveys.

In this volume, we spend considerable effort reviewing the literature and pointing out fallacies that have been built upon as a result of these problems. This book presents a methodology for maximizing the detectability of both vertebrates (homotherms and poikilotherms) and invertebrates during a census or survey when using proper thermal imaging techniques. It also provides details for optimizing the performance of thermal cameras under a wide variety of field conditions. It is intended to guide field biologists in the creation of a window of opportunity (a set of ideal conditions) for data gathering efforts. In fact, when thermal imaging cameras are used properly, under ideal conditions, detectivity approaching 100% can be achieved.

Recent attempts of researchers and field biologists to use thermal imagers to survey, census, and monitor wildlife have in most cases met with limited success and while there are a number of good books that treat the theory and applications of remote sensing and thermal imaging in significant detail for applications in land mapping, construction, manufacturing, building and vehicle inspections, surveillance, and medical procedures and analyses (Barrett and Curtis, 1992; Budzier and Gerlach, 2011; Burney et al., 1988; Holst, 2000; Kaplan, 1999; Kozlowski and Kosonocky, 1995; Kruse et al., 1962; Vollmer and Mollmann, 2010; Williams, 2009; Wolfe and Kruse, 1995), they contain very little on how wildlife biologists should go about using this equipment in the field to survey and monitor wildlife. This book provides detailed information on the theory and performance characteristics of thermal imaging cameras utilizing cooled quantum detectors as the sensitive element and also the popular uncooled microbolometric imagers introduced into the camera market in the past decades, which rely on thermal effects to generate an image. In addition, there are numerous excellent texts devoted to survey design and statistical modeling to aid in the monitoring and determination of wildlife populations (Bookhout, 1996; Borchers et al., 2004; Buckland et al., 1993; Buckland et al., 2001; Caughley, 1977; Conroy and Carroll, 2009; Garton et al., 2012; Krebs, 1989; Pollock et al., 2004; Seber, 1982, 1986; Silvy, 2012; Thompson et al., 1998; Thompson, 2004; Williams et al., 2001), but they do not include the treatment of thermal imaging capabilities to help achieve these tasks. This book is being offered as a bridge between the two technologies and the teachings presented in these excellent volumes so that their combined strengths might be united to improve upon past efforts to assess animal populations and to monitor their behavior.

Even though there has been a technological disconnect since the earliest field experiments, there has still been a considerable amount of work carried out by biologists using thermal imagers to study and monitor wildlife. These studies began in the late 1960s and early 1970s when cryogenically cooled thermal imagers using photon detectors were first used for surveys and field work (Croon et al., 1968; Parker and Driscoll, 1972) and this phenomena continued to grow as thermal imagers became more readily available to field biologists. At the time, these early cameras were acknowledged as being only marginally sensitive for the task of aerial surveying. The more recent introduction of the low-cost uncooled bolometric cameras generated a new wave of experimentation with thermal imagers in the field. The sensitivity and range of bolometric cameras are limited due to the fact that they rely on a thermal process to generate an image. So we see at the start that all thermal imagers are not the same and if they are used in the field they must be used to exploit the strengths of the particular imaging camera so that reliable data can be obtained. There are appropriate uses for imagers utilizing photon detectors where high sensitivity and long ranges are characteristics making them suitable for surveying applications. There are also applications suitable for imagers fitted with thermal detectors that have lower sensitivities and ranges. Their advantages are their availability, cost, and that they are uncooled. Field applications favoring bolometric cameras that do not require long ranges or high sensitivity will also be addressed in this book.

The process of using thermal imagers as a tool to collect field data has been compared with other data collection techniques; however, in nearly all cases the thermal imager was not used correctly and perhaps was even inadequate for the task. This practice has led to a number of misconceptions about the basic use of a thermal imager and the correct interpretation of the results. There is a big distinction between thermal imagers that utilize quantum detectors as the sensitive element and detectors that rely on thermal effects to generate an image. The differences are enormous as far as fieldwork goes for censusing and surveying, particularly on a landscape scale. Unfortunately, a text describing the use of 3–5 and 8–12 μm photon detectors for animal surveys and field studies has not emerged. This is probably due to the fact that 3–5 and 8–14 μm imagers were not widely used since the first field experiments. These experiments used cryogenically cooled units typically borrowed from military installations. These robust units are now becoming available at a reasonable cost and should see increased use by field biologists. An excellent text describing the practical use of pyroelectric and bolometric imagers for a wide range of applications has been written (Vollmer and Mollmann, 2010) and a number of distinctions are pointed out between these imagers and those using photon detectors as the focal plane.

Past work using thermal imagers in the field has mainly been carried out so that comparisons could be made with other data gathering methods. From the outset we see that comparing the results obtained with thermal imagers with that of data collected with other methods such as spotlighting and visual surveys must necessarily be skewed and these efforts, while commendable, do not allow for a fair comparison of the data collection capability of the compared techniques. Thermal cameras are suitable for surveys and counts throughout the 24-h diurnal cycle while other methods are not. These studies by their nature and design mean that the results of data collected with a thermal imager will be compared with data collected using a method that was optimized for the conditions of the survey at hand. For example, consider the comparison of data collected during a visual survey and the data collected via thermal imagery using the same temporal and spatial conditions. Note that the survey must be conducted during daylight hours because the visual spotters need daylight to see the animals of interest. Thermal cameras can also detect the animals of interest during daylight hours but there are concomitant conditions required for the optimization of the thermal survey if it is conducted during daylight hours. These conditions can be met in a relatively easy manner but were not generally addressed during these past comparisons so the results reported were skewed and in some cases grossly inaccurate. We review many of these comparisons and offer alternatives. A variety of statistical methods, such as distance sampling and mark recapture, among others, were used for estimating the abundance of animal populations in these comparisons and the results of these studies were built upon by others. We do not treat these statistical methods here but point out that each of them has strengths and weaknesses (Borchers et al., 2004), depending on the species of the animal being surveyed. All will benefit from data collection methods that produce a detectability (see Chapter 1) that approaches ~100%.

The widespread dissemination of these results is the existing foundation that later work has been built upon and it has led to a confusing and widespread misunderstanding of the capabilities of thermal imaging as a powerful survey tool in these applications. This distribution of erroneous or badly skewed information regarding the performance of thermal imaging for these tasks needs to be rectified and it is one of the major goals of this book to start that process.

The work of Romesburg (1981, p. 293) pointed out the fallacies of building on unreliable knowledge: "Unreliable knowledge is the set of false ideas that are mistaken for knowledge. If we let unreliable knowledge in, then others, accepting these false laws, will build new knowledge on a false foundation." We still overlook important aspects of the scientific inquiry to gain reliable scientific knowledge. All the statistical methods applied to data gathered in the field are better predictors when the count is completely random and the sample is large. It is also known that the general methods used to count animals in the field during a survey are usually biased and yield animal counts less than what is actually there; however, in some cases there will be more counted than are actually there. These statistical losses or gains are presumably accounted for in the statistical formulation being used. The problems arise when the estimated parameters to account for losses or gains in populations, along with other parameters to account for such things as species mingling, group sizes, mortality rates, and sometimes double counting, are folded into the calculations. Even though these parameters are often very good guesses, they all come with systematic and random errors attached and cannot predict valid outcomes except by chance (Romesburg, 1981, p. 309). This is because the more parameters a model contains that are guesses the more they are amplified by their interaction with one another through the calculations, such that the resulting errors can be quite large at the output of the calculations.

It is essential for wildlife management and the preservation of healthy populations that we seek and promulgate reliable knowledge regarding the current status of animals in the wild. Ratti and Garton (1996) advance the important realization put forth by Romesburg by showing that in order for wildlife research to be useful to wildlife managers and their varied programs, it must be founded on high-quality scientific investigations that are in turn based upon carefully designed experiments and methodologies. Limitations to achieving the desired high quality and reliable knowledge must be identified and rectified. We postulate that the single most important thing to do at the present time to mitigate the unreliable knowledge stemming from skewed and distorted animal surveys and counts is to look very carefully at the detectability possible by different counting methodologies.

The components of science required for meaningful and reliable outcomes are mingled together in a relatively complex way. Wildlife managers and field biologists must incorporate biology, chemistry, atmospheric science, physics, and climatology, as well as the behavioral ecology and physiology of the animals surveyed or studied. All must be considered when forming a research plan for a species. The best window of opportunity for collecting data must be determined based on the best science available. To this end, a detailed methodology for using infrared thermal imaging to conduct animal surveys in the field and other studies requiring nondisruptive observation of wildlife in their natural surroundings is developed in this book. We show that ~100% detection can be achieved for surveys if the methodology is formulated to take full advantage of the infrared cameras used for observation and if it is coupled with the details of the behavioral ecology and physiology of the animals being surveyed or studied.

In this book we address the primary difficulty with surveying or censusing animals and demonstrate that it is not the sampling methodology (i.e., distance sampling, aerial transect sampling, quadrat sampling, etc.) or the statistical model being used on the collected data, but rather lies with the detectability that can be achieved with any particular sampling or data collecting technique. This suggests that more work needs to be done on comparing factors that influence the detectability of a species of interest rather than the statistical methods to compensate for the inadequacies of over or undercounting. There are many other details of a research plan that could grossly skew or render the resulting survey invalid (Thompson et al., 1998; Lancia et al., 1996; Krebs, 1989) but the visual observation (or other counting methods) are well-known to be skewed by a number of factors and limit data collection to daylight hours or when the landscapes or transects are artificially illuminated. It is also known that artificial illumination introduces behavioral modifications that can adversely influence the detectability and introduce bias (Focardi et al., 2001). There are various treatments proposed to deal with known biases. They are adjustments to the calculations to deal with under- or overcounting animals during surveys resulting from biased detectability. In this work, we will concentrate on the task of increasing detectability by eliminating bias in the data collection aspect of wildlife monitoring.

Because thermal imaging can be conducted at any time during the diurnal cycle and can be conducted from various aerial or ground-based viewing platforms, it offers a host of configurations to observe animals of interest while using their preferred habitat. If performed correctly, the observations can be conducted from a distance that precludes disturbances to the animals under study, thus reducing the possibilities of skewing the counts or surveys caused by anthropogenic-produced behavioral changes or double counting. Each variable introduced by some recognized uncertainty in the counting or observation techniques used must be accounted for and if it is done statistically the results become more and more questionable. If an uncertainty in the counting technique can be fixed at the field level, the resulting counts are closer in line with the true situation because there is one less layer of data manipulation to perform due to under- or overcounting.

As noted earlier, there is already a significant amount of up-to-date information available on methods for treating collections of field data with various statistical formulations and appropriate assumptions. These mathematical tools allow the evaluation of field data (if correctly collected) so that meaningful estimations of the abundance and/or the density of wildlife populations can be determined. As a result, we do not delve into these methods but rather focus on the details of establishing a technique for correctly collecting data and achieving the highest detectability possible when conducting field work. Applications other than those dealing with wildlife will not be treated here unless we need to make a specific point about some aspect of the workings of a thermal imager or if the application would clarify some aspect of the proposed methodology. Applications such as military, surveillance, police work, fire detection, manufacturing, and building inspection have been well-treated by others and can be found in the references mentioned earlier. The results of many studies of animal behavior, thermoregulation, pathology, and physiology are also reviewed.

In order to appreciate the advantages that thermal imaging has to offer we must recognize that our eyes are sensors that are limited in a number of ways that limit their utility as effective detectors of wildlife in their preferred habitat. Our eyes are confined to the visible region of the spectrum and at low-light levels they do not collect enough data so that our brain is able to form images that are recognizable; however, there are a number of ways that we can easily extend their functional range for our applications. For example, binoculars greatly enhance the probability of observing an object when faced with low-light levels and long viewing ranges. If we can use various technologies and instrumentation to aid our vision by seeing in the dark and seeing at longer ranges, then we need to add these things to our set of observational tools. In short we need to detect objects in order to count them and we need to see them in some fashion to detect them. The acquisition of images in the infrared region of the spectrum can be provided by thermal imagers and as such serve as an aid to our overall visual capability. By utilizing thermal imagers we can create images of very high contrast so that objects of interest are clear and distinct from their backgrounds, allowing us to extend our visual capability into the dark portion of the diurnal cycle. Once this is accomplished, the brain can process the images that the eyes see. In fact, in recent work at Cal Tech and UCLA, researchers found that individual nerve cells fired when subjects were shown photos of well-known personalities. The same individual nerve cell would fire for many different photos of the same personality and a different single nerve cell would fire for many different photos of another personality. Follow-up research suggests that relatively few neurons are involved in representing any given person, place, or concept, which makes the brain extremely efficient at storing and recalling information after receiving visual stimulation.

Without going into a detailed mathematical description of thermal imaging and the complex principles behind the operation of thermal imagers (thermal cameras) we instead introduce basic laws and principles that allow us to set the stage for data collection with thermal imagers. However, field biologists need to have a basic understanding of the physics governing heat transfer processes in the environment (Monteith and Unsworth, 2008) and the effects of local meteorological changes on the performance of a thermal imager. The proper use of a thermal imager requires a basic knowledge of how an imager works, why we see what we see with a thermal imager, and how we can optimize those images for the tasks at hand. Simple "point-and-shoot" infrared imagery for data collection will not work nor will using someone else's "point-and-shoot" imagery in sophisticated statistical calculations. What the imagery actually represents and how it was acquired must be known for it to be useful. While the performance capability of uncooled thermal imagers has improved remarkably over the last decade and the cost of these cameras has become reasonable for most researchers, field biologists must understand how they work, how to use them, and what they are actually recording as imagery. Unfortunately, for the most part, the rapid technological advancement and availability of thermal imagers has outpaced the knowledge and understanding required of the specialists using them in the field (Vollmer and Mollmann, 2010, p. xv). This sad commentary regarding the use of thermal imagers stems, for the most part, from applications associated with monitoring inanimate objects in fixed backgrounds. Our applications, as we have already pointed out, are much more difficult and complex so we need to be particularly careful and thorough in our understanding of a few basic principles regarding thermal imaging and wildlife ecology.

This book is about formulating a methodology to optimize a window of opportunity so that wildlife can be observed and studied in its natural habitat. This requires that biologists and program managers get together and formulate a sound survey design, which assumes that they know the ecology of the species of interest plus all mitigating factors that could possibly distort the outcome of a thermal imaging survey. The methodology presented here is logical and simple yet it demands a detailed understanding and incorporation of critically interlinked disciplines arising from biology, physics, micrometeorology, animal physiology, and common sense. Thermal imaging is a technique that forms images from heat radiating from objects and their backgrounds, so much of the information contained in this book is devoted to managing the interplay of the heat transfer processes of conduction, convection, and radiation between the objects of interest (animals) and their backgrounds to obtain the best thermal images. We will see that creating this window of opportunity is not as restrictive as one might think. Data can be collected from ground- or aerial-based platforms at any time during the diurnal cycle without compromising detectivity, disturbing the animals, or altering their behavior. Even though the methodology used to obtain meaningful data brings together a wide range of criterion and requirements that must be met concomitantly, it boils down to creating a window of opportunity that will allow researchers to conduct surveys with near 100% detectability by properly using thermal imagers as a detection tool.

About the Authors

Kirk J. Havens was born in Vienna, Virginia and received his BS in Biology (1981) and MS in Oceanography (1987) from Old Dominion University and a PhD in Environmental Science and Public Policy (1996) from George Mason University.

He is a Research Associate Professor, Director of the Coastal Watersheds Program, and Asst. Director of the Center for Coastal Resources Management at the Virginia Institute of Marine Science. He also serves as a collaborating partner at the College of William & Mary School of Law, Virginia Coastal Policy Clinic. His research has spanned topics as diverse as hormonal activity in blue crabs to tracking black bears and panthers using helicopters and thermal imaging equipment. His present work involves coastal wetlands ecology, microplastics, marine debris, derelict fishing gear, and adaptive management processes. He hosts the VIMS event "A Healthy Bay for Healthy Kids: Cooking with the First Lady" and the public service program "Chesapeake Bay Watch with Dr Kirk Havens".

He is Chair of the Chesapeake Bay Partnership's Scientific and Technical Advisory Committee. He was originally appointed to STAC by Gov. Warner and was reappointed by Gov. Kaine, Gov. McDonnell, and Gov. McAuliffe. He was also appointed by North Carolina Gov. Perdue to serve on the Executive Policy Board for the North Carolina Albemarle-Pamlico National Estuary Partnership and is presently vice-chair. He serves on the Board of Directors and is past Board Chair of the nonprofit American Canoe Association, the Nation's largest and oldest (est. 1880) organization dedicated to paddlesports with 40,000 members in every state and 38 countries.

Edward J. Sharp was born in Uniontown, Pennsylvania, attended Wheeling College and John Carroll University and received PhD degree from Texas A&M University in 1966. He conducted basic research in the area of applied nonlinear optics at the US Army Night Vision & Electro-Optics Laboratory and the US Army Research Laboratory. Presently, he is working as a consultant on the use of infrared imaging equipment in novel application areas. His major areas of interest include laser crystal physics, thermal imaging materials and devices, electro-optic and nonlinear-optical processes in organic materials, beam-control devices, optical solitons, harmonic generation, optical processing, holographic storage, photorefractive effects in ferroelectric materials, and the study of animal ecology using thermal imaging equipment. He is the author or coauthor of more than 100 technical publications and holds over 15 patents on optical materials and devices. He is a member of the American Optical Society. Recently, he has been working on new methods for using thermal imaging to address issues related to animal ecology and natural resource studies with faculty at the Virginia Institute of Marine Science (VIMS), College of William & Mary.

Acknowledgments

A special thanks to the following people and organizations: David Stanhope and Kory Angstadt, Virginia Institute of Marine Science/Center for Coastal Resources Management/Coastal Watersheds Program; Bryan Watts, College of William & Mary; Richard Pace, Louisiana State University; Deborah Jansen, US Fish & Wildlife/Big Cypress National Reserve; Kenny Miller, US Army Night Vision & Electronic Sensors Directorate; Greg Guirard, US Fish & Wildlife Service; US Fish & Wildlife Great Dismal Swamp Refuge, Virginia Living Museum, Peninsula SPCA, Newport News, VA; and Carl Hershner, Virginia Institute of Marine Science/Center for Coastal Resources Management.

Chapter 1

Introduction

Finding, monitoring, and accurately counting animals in the wild are very complex tasks that have been attempted in a variety of ways and with varying degrees of success. The sheer volume of literature devoted to this topic is staggering and the activity devoted to these tasks is becoming increasingly more important as suitable wildlife habitats shrink due to the ever-increasing demands of humanity. There are new conflicts arising on a daily basis between potential user groups for these lands in urban, rural, and wilderness areas. The recreational, energy, farming, livestock, manufacturing, timber, mining, petroleum, housing, and transportation industries, among others, all make arguments for the best use of these resources. While each group argues for the best management of these resources based on their own perception of value, they do so for the most part lacking accurate counts of the living resources indigenous to these areas.

In the absence of verifiable scientific information on the population status and trends in specific regions and in some cases for specific animals listed under the Endangered Species Act, the resource management issues can be significant. Areas such as game lands, military installations, national forests, and parklands are facing pressures in the form of restrictions or lack thereof, because management decisions are being made based on incomplete or inaccurate field data. These uninformed decisions can be very costly, because unwarranted restrictions placed on the use or development of land for recreation, power production, timber, oil, etc. represents a clear loss of revenue. Likewise, the improper use of a critical habitat places the living resources in the affected area at risk and in some cases threatens them with extinction.

A variety of techniques can be used effectively to manage and recover endangered species; some are identical to techniques used with more abundant species, but many others are specially adapted to the needs of rare species. Special approaches are needed because it is uncommon for most endangered species to have had their habitat requirements defined specifically enough to guide a recovery effort (Scott et al., 1996). The management of endangered species is complicated by their rarity, by legal restrictions intended to protect such species, and by the public and political scrutiny under which endangered species management is conducted.

Lands that have already been set aside and established for particular uses would also benefit from accurate counts, particularly if the animals concerned are listed as threatened or endangered under the Endangered Species Act. For

Thermal Imaging Techniques to Survey and Monitor Animals in the Wild: A Methodology
http://dx.doi.org/10.1016/B978-0-12-803384-5.00001-4

example, it has been noted that the determination of the population status and trends of threatened or endangered species on Department of Defense (DOD) installations are inadequate. As a result, the US Fish and Wildlife Service has developed management practices for these installations that place restrictions on training activities for certain periods of time during the year and on certain areas of the DOD land. Detection and identification of animals on these lands are essential in determining whether these activities can go forward. The Endangered Species Act of 1973 calls for a rare, threatened, or endangered determination and the resulting protective measures that the law provides if the number of individuals within a species is reduced to dangerously low levels, such that the extinction of the species is a real probability. These issues point to the need for simple, accurate, and inexpensive monitoring and survey techniques that can be conducted on the ground or from the air for a variety of habitats.

If field data is timely and accurate, a comprehensive management plan might be formulated that only periodically mandates restrictions or permits certain activities within the boundaries of contested lands. These restrictions and/or special uses may be implemented periodically or only implemented on portions of the land that are deemed suitable based on accurate field data. Most animal surveys are done to aid wildlife managers, particularly managers of public game lands. For example, decisions to control herd size either by increased or decreased harvesting are frequently based on inaccurate or outdated animal counts. The increased demand for the habitat that remains available to game animals has raised the need for population information to a new level. Since the regions of habitat are often fragmented and connected only by narrow corridors the survey information must be of a spatial or temporal nature or both. That is, in many cases the managers need to know how many animals there are, where they are located, and when they are there.

Decisions are made every day about how best to maintain the health and stability of wild animal populations. These decisions are influenced by a number of factors, many of which are the result of anthropogenic-induced changes, whether intentional or not. Such changes may include habitat loss, habitat modification through pollution (light, toxics, noise, etc.), and habitat fragmentation. These changes can lead to highly skewed redistributions and/or population loss or, in some cases, such as white-tail deer, to unsustainable population gains due to a lack of predators and/or hunting. Even so, there are decisions made that can further exacerbate existing problems. In many cases management decisions to alter the population density or distribution of wildlife are determined by economics or politics.

Chadwick (2013) pointed out in a news release that cougars (*Puma concolor*) are now the most common apex predator across one-third of the lower 48 states and that most of the other two-thirds lack any big predatory mammals. Even so, since predation by cougars was deemed responsible for a reduced deer population in South Dakota, hunting permits were issued for 100 cougars out of a total population estimate of 300 even though the decline of elk and deer in South Dakota was actually due mainly to excessive sport hunting. It is ironic that this planned change to reduce the total population of cougars by a third came about because hunters complained to state game commissioners that "there's no game left in the woods." To put this in perspective, consider that the hunters of South Dakota can now shoot cougars so that the deer and elk populations can increase and they too can be hunted. Chadwick (2013) further points out that in Texas, cougars are classified as varmints; you can shoot one almost anywhere at any time. California, on the other hand, has not allowed cougar hunting since 1972 and now has the most cougars of any state. It also has an abundance of deer and one of the lowest rates of cougar conflicts with humans. On the flip side, there are cases where deer numbers are deemed to be too large and sharpshooters are called in to reduce herd size, thereby reducing auto/deer collisions in suburban environments. This emphasizes the need for accurate data for all species involved in a management decision to alter existing population densities for whatever reason.

As mentioned earlier, determining a wildlife population density is not an easy task. To get an idea of the difficulty first consider an animal population that is not wild and is merely spread over twenty acres. The farmer who has twenty cows in a rolling pasture of 20 acres can guess that at any given time he has a population density of 1 cow/acre, but he would have to check to make absolutely sure. He can do a survey or census, which can be done in a number of relatively easy ways. Some choices might be walking the perimeter of his pasture and noting the location and number of cows or he might drive the old pickup truck along the fence line (it is a fenced and closed population at the moment). Note that this might be easy or very difficult since the one or two cows that are not accounted for may be unobservable from the truck or on foot because of the features of the terrain, unless he gets very close to them. He may have to walk or drive the pasture several times to locate all of his cows with certainty. On the other hand, if each of his cows is identifiable with a tag, he could wait at the watering trough and count them as they come to drink. However, if one cow is not thirsty then he has to take a hike in his 20-acre pasture to find the missing cow. Another (albeit far-fetched) option might be to take video of his pasture with a thermal imager and record the animals within the fenced area. This video session could be carried out during the day or night, whichever is convenient for the farmer. Figure 1.1 is provided as a sample of what the thermal imagery might look like for his herd of cows and provides a record for the farmer for future comparisons. Each of the above methods requires effort, takes time, and costs money, but when the farmer is finished with his census he knows how many cows are in his pasture. Based on this information he can make good decisions that are important to him and the health of his cows.

When biologists go into the field to conduct a "survey" or "census" of some animal population (the animals that occupy a particular area at a particular time) the objective is to count all the animals of interest in the immediate area of observation. Simply put, all animals of interest should be detected. Note that a "census" is designed to count all the animals or the complete population so only special cases and relatively small sections of the habitat can be included in the

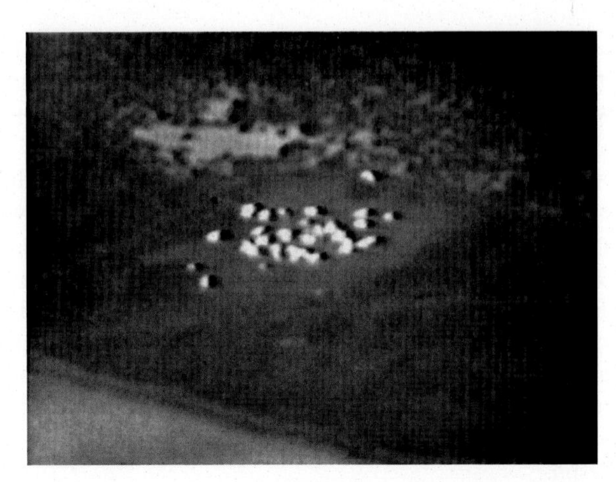

FIGURE 1.1 **A thermal image of a small herd of cows including adults and calves.** It is a single frame extracted from a video that was taken in daylight hours under partly sunny skies.

count. Generally, a census of animals in the wild is not undertaken because of the difficulty with geographic closure. Some examples of where a census might be appropriate could be an island, a section of fenced range, a roosting site for birds, or an ice flow for walrus. If the condition of geographic closure is met and there are no animals moving into (immigration) or out of (emigration) the census area then we will obtain the population of the island, section of fenced range, bird roost, or ice flow. A "survey" on the other hand does not require a complete count of all the animals but only the animals included in the field of view when sampling the animals' habitat. This allows surveys to be taken on a much larger scale to include landscapes such as range lands, deserts, vast expanses of open water, and game lands. A robust population estimate can be made if the survey techniques provide high detectability of the animals of interest within the field-of-view.

The objective of this work is to develop a methodology for the use of thermal imaging techniques in the inventorying and monitoring of a broad range of animals (both homothermic and poikilothermic), including threatened and endangered species. These sampling methodologies can be applied at the landscape scale and are applicable to multiple species. Chapter 2 provides a brief review of population surveys using visual and photographic counting techniques. Chapter 3 covers remote sensing techniques as a tool for counting and monitoring wildlife where the use and the benefits of trip cameras, video recorders, image intensifiers or night vision devices, and radars are reviewed.

The multitude of problems associated with achieving high detection rates in past animal surveys will be examined and a new formulation of techniques for using infrared thermal imaging systems to overcome these problems will be covered in the remaining chapters. Chapter 4 covers the heat transfer processes of conduction, convection, and phase changes. Chapter 5 is devoted to the radiation heat transfer process, which is the basic underlying process responsible for the formation of thermal images. Chapter 6 reviews the emissivity (number ranging from 0 to 1), a ratio that compares the radiating capability of a surface to that of an ideal radiator or "black body" and which depends on a wide range of physical conditions. These chapters provide the details necessary for understanding the physical phenomena that can affect thermal radiation and subsequently influence the quality of imagery that can be formed by a thermal imager.

The current status and availability of thermal imagers, including detailed information on the theory and performance characteristics for cameras utilizing cooled quantum detectors as the sensitive element or uncooled micro bolometric imagers, is covered in Chapter 7. Suggestions are included for the selection of a thermal imaging camera to meet specific applications based on range, sensitivity, resolution, camera availability, and cost. A review of the latest infrared imaging equipment available and its use provides a foundation for those seeking to use the thermal imaging technique for wildlife field studies.

Much like the farmer and his cows, wildlife managers would like to know the animal abundance and/or the population density of the species for which they are responsible. They may also want to determine the sex of individual animals or determine the ratio of adult to juvenile animals within a particular species. To do this they only need to completely count (as did the farmer) all the animals of interest on the landscape of interest. The magnitude of this challenge is truly daunting. The problem of 20 cows confined to a fenced 20-acre pasture has mutated into a much more complex problem. We now need to determine an unknown number of animals of interest that are mixed with several other species of animals of similar size and ecology. The fenced pasture is replaced with a vast landscape of variable terrain and vegetation ranging from bare ground to heavily forested. On this landscape the animals of interest are in a constant state of change both in number (reproduction and death) and location (immigration and emigration) as they seek food and shelter. A census would be impractical; however, we can conduct properly designed surveys that are well planned and executed to determine the number of animals in the area of interest (which can be of varying size, depending on the present interest of the survey). If we can repeatedly detect all target species that are being surveyed at a particular location and time with $\sim$100% detectability, then we can determine an accurate population density for the landscape. The key point here is detectability.

Throughout this book we try to use terminology which is considered common (Krebs, 1989; Lancia et al., 1996; Pierce et al., 2012; Thompson et al., 1998) in the studies and surveys of wildlife. There are a few terms that we want to define for the sake of clarity.

Detectability: The probability of correctly noting the presence of an animal of interest within some specified area and period of time (Thompson et al., 1998). This definition has been advanced by a number of authors and we shall use it here.

Sightability: The probability that an animal within the field-of-search will be seen by an observer.
Observability: The probability of observing (seeing or catching) an animal within the field-of-search.

We note that these definitions are similar and have been used interchangeably in the literature. The definitions of sightability and observability are essentially the same (seeing an *animal* in the field-of search). Since these are not as specific as detectability (seeing an animal of interest within the field-of-search) we elect to use the term detectability in this book.

The techniques provided in this work are capable of being applied at the landscape scale in order to supply inventory and provide monitoring of animals that will produce population levels and demographic data, in addition to confirming species' presence or absence. Both ground-based and aerial-based applications of thermal imaging are presented. The use of thermal imaging significantly improves estimates of animal populations and overcomes the problems that render other techniques inadequate during the detection phase of the surveys. These improvements are sought because typical aerial surveys conducted of animals in a forested habitat or partially forested habitats are strongly skewed as a result of visibility bias. That is, animals are very difficult to detect in their natural habitat with the naked eye due to the fact that quite often the coloration of the animal and its background are very similar. Compounding this obvious camouflage problem is the fact that the amount of skewing is affected by a host of factors such as aircraft speed, altitude, weather conditions, spotter experience (also including fatigue and distractions), animal group size, vehicle access, time of day, and ground cover, among others. It is essential that a method of surveying animal populations be developed that is capable of completely eliminating visibility bias and allows for maximum detectability. Once an adequate survey design has been established this is the first step toward obtaining accurate animal surveys, regardless of the statistical technique used to determine the animal abundance. It allows accurate population estimates to be determined from any number of statistical models (Seber, 1982, 1986; Buckland et al., 1993, 2001; Lancia et al., 1996; Thompson et al., 1998; Borchers et al., 2004; Conroy and Carroll, 2009) and coupled with other parameters, such as birth-death rates and harvesting numbers, should be adequate to determine populations at a given point in time precluding any abnormal losses due to extreme weather conditions or disease.

Counting and monitoring animals in their natural environment is difficult because of the conflicting requirements of finding out as much as possible about the demographics of the population while leaving it undisturbed. Specifically, the lack of control over natural populations coupled with the possibility of nocturnal and reclusive behavior, large group sizes, inaccessible habitats, visibility bias, and comingling of species makes counting animals in the wild a daunting task. Another significant problem involves the monitoring and counting of reintroduced species. Their numbers could be small and they may be widely dispersed and comingled with species of similar size, so finding these animals in the wild would be difficult without radio telemetry or other signaling devices placed on the animals at their release (Havens and Sharp, 1998). However, once the general location of such individuals or groups is established, the monitoring of their activities would be straightforward using thermal imaging methods.

Thermal imaging technology developed by the military has recently found its way into the commercial market place. For example, thermal imaging systems, both handheld and airborne units, are now available with sensitivities more than an order-of-magnitude better than the units used in the early experiments devoted to large mammal surveys (Croon et al., 1968; Parker and Driscoll, 1972). With these improved thermal cameras one can easily detect all faunae that radiate energy as a part of their basic metabolic function (i.e., homotherms) and insects that collectively generate heat within the hive or nesting cavity. The present work will provide the field researcher with the techniques and methodology to locate and identify individual animals or distributions of animals (homotherms and poikilotherms) in their natural habitats. Present methods for inventorying and surveying most species (particularly animals with extended home ranges) such as spotlight counts, mark to recapture, and aerial surveys introduce behavioral variables and viewer bias (LeResche and Rausch, 1974; McCullough et al., 1994). Thermal imaging technology provides a method for obtaining counts of animals with little risk of behavioral or sampling bias. The basic performance parameters and important system considerations for thermal imagers are covered in Chapter 8.

Three levels of information can be extracted from the thermal imagery collected: detection (observation and number of warm objects contained in the thermal image), recognition (a determination if the detected objects in the thermal scene are biotic objects of interest), and identification (what species have been detected). It is important to note that these three levels of information are assumed to be contained within the detectability but in fact refer to completely different levels of knowledge regarding the thermal signatures extracted from the imagery. In prior work we demonstrated that thermal imagery could identify individuals within a species (Havens and Sharp, 1998). Radio-collared panthers (*Puma concolor coryi*) could be distinguished from noncollared panthers from the air due to the unique thermal signature of the collar (cool band across the neck). In many cases it is only necessary to achieve detection with the thermal imagery collected. For bats and birds one needs only the detection phase for accurate and complete counts. For herding animals one may only need detection capability when the species location is known but numbers are not. In other situations, where more than one species of similar size, shape, and numbers may occupy the same habitat, it may be necessary to achieve identification for accurate surveys.

In Chapter 10 we review many past efforts to find, monitor, and count animals in the wild. We also review the results of thermal imaging experiments for monitoring and counting wildlife as described in the literature. Most of these efforts were attempts to compare thermal imaging techniques with some other methodology for surveying or estimating animal abundance. In almost every case thermal imaging proved to be superior even though the use of the thermal imagers was not optimized. Remarkably, in some cases researchers refused to accept the results of their own work that showed better performance using thermal imaging to improve detectability. These works include the use of aerial and ground-based platforms to monitor both vertebrates and invertebrates in terrestrial, aquatic, and air environments. The strengths and weaknesses of the techniques used in those efforts are examined and suggestions are offered for improvement through the use of remote thermal imaging as a technique. We look critically at the past work done during field studies such as surveys and counts as well as experiments that compared the detectability obtained with thermal imaging with other techniques. We illustrate that using a thermal imager correctly is more important than having the most expensive imager.

What exactly is a thermal image and what does one look like? All objects radiate heat and the amount of heat radiated is determined by the condition of the object's surface and by its temperature. Modern thermal imaging cameras are capable of measuring the heat radiating from objects. Since heat transfer by radiation occurs at the speed of light, images of the objects can be formed. One can record thermal images captured by the infrared (IR) camera on video, view the camera display on a monitor, or simply view the objects of interest through a viewfinder as one could with a conventional camcorder. The only difference is that the IR camera senses and displays a spatial distribution of thermal (heat) energy instead of visible light. This allows one to see in total darkness, through smoke, and other low visibility, low contrast situations. These cameras can also be used during daylight hours to see heat generated images when visual observation is inadequate to distinguish a heat emitting object from its background. The imager detects the infrared energy given off by all objects in a particular scene. Since thermal imaging is a technique to form images from heat radiated from objects and their background, much of the information contained in this book is devoted to managing the interplay of the heat transfer processes of conduction, convection, and radiation between the object of interest and its background to obtain the best thermal images possible for a wide range of uses. Chapters 4, 5, and 6 are devoted to a discussion of this interplay and how it can affect the formation and usefulness of thermal images. The details of the properties of a thermal signature (a particular image within a scene) and a discussion of image interpretation are contained in Chapter 9.

As we mentioned earlier, the texts currently available that describe the use of remote sensing, including texts devoted to applications utilizing thermal imagers, do not address the problems associated with monitoring and/or conducting animal surveys. The books devoted to animal counting and surveys do not properly treat the use of thermal imaging to carry out these tasks. Of those listed above the book by Barrett and Curtis (1992, p. 58) is perhaps the

FIGURE 1.2 **Infrared line-scan imagery of land near Mark Yeo, Somerset, UK.** *(Courtesy: Barrett and Curtis, 1992; with kind permission of Springer Science + Business Media)*

most informative regarding the quality of infrared images taken from aircraft. Without having a great deal of understanding about thermal imagers and their capabilities, we are still able to look at the photo presented in their book of a thermal image taken from an aircraft of the rural countryside in England and get a good understanding of the strength that thermal imaging can bring to census and survey work.

The image depicted of the countryside in Somerset, UK (Figure 1.2) was captured with an older model line scanning imager and it shows hundreds of individual farm animals dispersed over a landscape of considerable extent, yet the high contrast imagery leaves the individual animals easily detected and countable. The imager used has a relatively wide field-of-view and, if a fixed portion of this field-of-view were used to survey transects across this landscape, the detectability of these animals could be ~100% with very little deviation. When examining this single photo, keep in mind that this imagery is typically recorded as a video that can be studied frame by frame and can be enhanced to examine particular features of interest. There may be a small percentage of the animals lost in the lee of the hedgerows when comparing the thermal signatures of the animals with their backgrounds (the surface soil that has not been cooled by the prevailing wind through evaporation or convection). This possible source of divergence from perfect detectability could easily be rectified at the field level in a number of ways using an appropriate methodology. The concepts and the effects of heat conduction, convection, and phase changes are covered in Chapter 4.

The results of many studies of animal behavior, thermoregulation, pathology, and physiology are also reviewed. A brief review of thermographic applications in studies of wild animals that included disease diagnosis, thermoregulation, control of reproductive processes, analyses of animal behavior, and detection of animals and estimation of population size was carried out by Cilulko et al. (2013). These studies were conducted with thermal imagers based on thermal detectors such as microbolometers as opposed to photon or quantum detectors typically used for surveys and censusing applications. The main difference between these two types of imagers is discussed in Chapter 7 and their properties and limitations are described.

In Chapter 11 we devote sections to each of the important aspects of an appropriate thermal imaging methodology and its function in the overall convergence of critical information and requirements to create a window of opportunity for data collection. This book is about optimizing that window of opportunity to observe wildlife in its natural habitat. The methodology is logical and simple yet it demands a detailed understanding and incorporation of critically interlinked disciplines arising from biology, physics, meteorology, animal physiology, and common sense. The techniques of remote sensing with a thermal imager and the progression from the detection of thermal signatures to the recognition and identification of species are described. We discuss the multitude of problems associated with achieving high detection rates in past animal surveys and present a new formulation of techniques for using infrared thermal imaging systems to overcome these problems and make it possible to achieve ~100% detectability in the field. The techniques forming the basis of the procedural methodology can be used for ground and aerial-based surveys as well as behavioral studies in the field and are not confined to low-light level situations and, when used during daylight hours, eliminate the problems associated with visibility bias. We conclude with Chapter 12, a short discourse on the latest technological developments directed at miniaturizing thermal imaging cameras and the prospects of flying these cameras with remote piloted vehicles (drones).

Chapter 2

Background

Chapter Outline

OVERVIEW AND BASIC CONCEPTS

A fundamental requirement for the proper management, protection, or preservation of any animal species is an accurate determination of its estimated population. To find and count animals in the wild is a very complex task that has been attempted in a variety of ways and with varying degrees of success. The sheer volume of literature devoted to the topic of estimating animal populations is staggering and the activity devoted to these tasks is becoming increasingly more important as suitable wildlife habitats shrink due to the ever-increasing demands of humanity. Accuracy in accomplishing these tasks is of the utmost importance since the information acquired can be used in decision making to help solve problems regarding the welfare of the animals in the estimated population. This information can also aid in resolving problems perceived by the public, such as over/under harvesting of game animals, losses of habitats due to urbanization, or perhaps public concerns of a suspected wildlife population decrease due to man-made pollutants.

The interest in population dynamics (Johnson, 1996) is becoming a subject of increasing importance as the demand for limited habitats by competing species grows. Information regarding the relationships among species, subspecies, and populations is essential for making informed and timely decisions needed to maintain sound wildlife management practices. A broad but useful definition of population is a group of organisms of the same species living in a particular space at a particular time (Krebs, 1985). In most cases a species is made up of many populations, and a population is only one segment of a species. The exception to this is perhaps a species that is faced with extinction, which is a situation that is becoming more common. Ratti and Garton (1996) point out that the wildlife scientific community usually deals with three types of populations: the *biological population*, the *political population*, and the *research population*. The *biological population* is an aggregation of individuals of the same species

Thermal Imaging Techniques to Survey and Monitor Animals in the Wild: A Methodology
http://dx.doi.org/10.1016/B978-0-12-803384-5.00002-6

that occupies a specific locality, and often the boundaries can be described with accuracy. The *political population* has artificial constraints of political boundaries, often dictated by city, county, state, and federal or international jurisdictions. The *research population* is usually only a portion or segment of the biological population and it is this segment that is sampled to obtain information regarding the relationships among species, subspecies, and populations.

The quality of an estimate is determined by its accuracy, precision, and bias and their relationship to one another and is usually discussed in conjunction with an illustrated target diagram proposed by Overton and Davis (1969); see also Ratti and Garton (1996), Lancia et al. (1996), Conroy and Carroll (2009), and Pierce et al. (2012).

An accurate estimate is one that is both unbiased and precise. It is determined by the average of the squared deviations between the true population size and the population estimate repeated many times.

The precision of an estimate depends mainly on the size of the sample and the closeness of repeated measurements to one another. The difference between the repeated measurements is call the variation and it can be broken out into temporal, spatial, and sampling variations. The temporal and spatial variations refer to changes in the number or distribution of the target species over time and space within the sampling area, which is pretty much in a state of constant flux due to the availability and abundance of food, seasonal changes, predation, weather, fires, and perhaps the presence of humans. The sampling variations can be further divided into two components: one consisting of variation in counts between sampling plots dispersed according to a selected survey design across the particular landscape of interest and the other variation coming from incomplete counts or surveys within individual plots. Siniff and Skoog (1964) conducted an aerial survey for caribou (*Rangifer tarandus*) in central Alaska using sampling plots (quadrats) of 4 square miles. Their entire study area was comprised of six strata based on a pilot study of caribou densities in different regions. The 699 quadrats were divided unequally among the six strata (18 in the smallest and 400 in the largest). The idea here is to pick strata to be as homogenous as possible so that the precision can be improved. If one were able to divide a highly variable population into homogeneous strata such that all measurements within a stratum were equal, the variance of the stratified mean would be zero or there would be no error. There are therefore advantages for using stratified random sampling.

The bias of an estimate defines how far the average value of the estimate is from the true wildlife population. Ratti and Garton (1996) point out that evaluating bias in an estimate is difficult and usually has been done in the past on the basis of the researcher's biological knowledge and intuition. Bias can occur, for example, at the sample plot level from poor placement of plots within the sampling area such that there may be plots that overlap or share borders (leading to double counting). Bias can and frequently does occur at the counting level where two types of errors are possible. If an animal is misidentified, such as a deer for an antelope, where there are mixed species sharing the area being surveyed, it can lead to the situation of seeing an animal that wasn't there and concomitantly missing an animal that was there. The second type of error is simply one of omission or failing to detect animals included in the target species. This latter form of bias is usually simply called visibility bias and is a direct result of imperfect detectability. It has been problematic for most aerial surveys of large mammals and its causes are many. For example, the detectability within a sampling plot can change because of the degree of vegetative cover, the size of individual animals (age related), weather, observer experience, the counting method being used, and so on. There have, however, been a number of surveys and censuses conducted with thermal imagers that were not plagued with the problems arising from visibility bias. These will be discussed in Chapter 10. A word of caution is offered here because even if one properly counts all the animals in sample plots without bias (i.e., the detectability is perfect) there can still be a biased outcome in the estimate of the target population if the plot selection, for example, was already biased.

There are a variety of factors that can influence the way a census or survey should be carried out. Wildlife population monitoring is a complicated task requiring considerable investments in time and money. The practical side of important factors involved in monitoring wildlife populations such as objectives, method selection, and implementation are discussed by Witmer (2005). Before going out and gathering data we first need to decide what the objectives of the survey actually are. What do we want to accomplish and what factors will affect our ability to accomplish the survey? An understanding of the ecology of the target species is required. We need to be prepared for the landscape (hilly, flat, marshy, mountainous, fences, roads, power lines) and how it is vegetated (trees, open grassland, fields, brushy, or a mix of differing cover), the size of the survey area (large or small), if it can be covered in a day or if it will take many weeks or months or even years to survey, and if there will be adequate manpower, funds, and experienced field scientists to accomplish the undertaking. After considering all these things the sampling design and enumeration methodology must be formulated with a goal of achieving an accurate, precise, and unbiased accounting of the target species. Pierce et al. (2012) provide a list of twenty questions that should be carefully considered before any survey is attempted. We suggest that this list be consulted early on in the planning process rather than later or after the fact.

It is extremely important to have a command of the basic principles needed to properly formulate a meaningful and robust program for monitoring animal populations. Additionally, a successful survey design must be coupled with detailed knowledge of the ecology of the species being monitored and the plan must include the determination of exactly what data will be required from the field. There are numerous texts and review papers devoted to helping researchers, field biologists, and resource managers accomplish these tasks (Seber, 1982, 1986; Krebs, 1989; Lancia et al., 1996; Ratti and Garton, 1996;

Bookhout, 1996; Johnson, 1996; Thompson et al., 1998; Williams et al., 2001; Buckland et al., 2001; Thompson, 2004; Borchers et al., 2004; Pierce et al., 2012; Conroy and Carroll, 2009; Garton et al., 2012; Silvy, 2012). We do not intend to address the formulation of survey designs. The reason for this is two-fold. First, the above mentioned texts provide a comprehensive treatment to guide research scientists in formulating appropriate sampling designs, sampling methodologies, and enumeration methods to estimate animal populations. Anything we might add to the information already available would be redundant. Lancia et al. (1996) recommend that "because of the variety and complexity of methods available for estimating animal population size, it is becoming increasingly important to involve a statistician or quantitative population ecologist in the selection and application of a method." Second, we think that it would be better to devote the space in this book to exploring the many possibilities offered by modern thermal imaging cameras for studying animal behavior and physiology (both in the field and the laboratory) and for improving data collection aspects when surveying or censusing animals in the field by improving detectability.

COUNTING METHODS

Since the early 1960s there has been a tremendous effort put forth to correct inaccurate censuses and surveys by offering improved survey designs and sampling techniques. These improvements are sought to primarily deal with bias introduced in counting techniques as well as a host of other problems that stem from an improper survey design; these issues affect accuracy. Numerous research papers and books have devoted considerable space to looking into the problems associated with visibility bias and what to do about it. The adjustments for incomplete detectability when collecting field data to be used in programs for monitoring animal populations have received much of this attention. Generally wildlife managers and wildlife scientists use total or complete counts, incomplete counts, mark to recapture counts, and indirect counts to estimate wildlife populations, and numerous direct and indirect methods have been used in the past. Indirect methods do not directly detect animals but instead detect some quantity or feature that must then be related to the density of animals that produced the detected feature (Buckland et al., 2001; Seber, 1982; Lancia et al., 1996; Stephens et al., 2006; Wilson and Delahay, 2001).

Direct Counting Methods

Most animal counts and surveys are based on the direct count of the animal. Here direct counting refers to actually counting the animals and not some related parameter such as track, dung, or other evidence of an animal's presence.

Note that direct counting can be either complete or incomplete. That is to say a complete count is one where all animals present in a survey plot (regardless of its size) are counted. Note that this is generally not possible except for special cases where the survey area is relatively small and the detectability (e.g., for a visual survey) is perfect. If the survey plot were to be sampled via transects or subsamples then the count would be incomplete at the survey plot level but complete at the subsample level. An estimate of the population can be made in either case.

Complete Counts

There are very few instances or situations where it is possible to conduct complete counts of animals. Most animals have such large home ranges with varying habitats that total counts are impossible except under a very exacting set of conditions. One can obtain total counts of animals on sample plots, where the size and location of the plots are selected according to a predetermined survey design and are within a larger area that is home to the population of interest. If, for example, a spatially well-defined sampling plot or a sampling area where an enclosure confines and delineates the survey area is completely accessible then a total count is possible. Such plots might be quadrats or strip transects. As an example, consider a photograph of a window pane covered with ladybugs in the spring. One could easily count all the bugs on the pane and it would be the population on the pane (at the time the photograph was taken), which is related in some way to the population on the entire window. In other words, it is a complete count of the ladybugs on the pane but an incomplete count of the lady bugs on the window. There are numerous examples of complete counts of large mammals such as deer, moose, and antelope that have been counted from the air on properly selected sample plots. Some examples are provided further.

Drive counts: Sometimes it is possible to use drive counts for animals concentrated in a group or for animals in enclosed populations where the enclosure might be an island, peninsula, fenced pasture, plots of wooded areas bounded by roads, and urban areas such as parks or perhaps segments of riparian zones within cities, etc. Drive counts are conducted to get a complete count of all the animals in the sampling unit. This method usually requires a large group of people that crosses the enclosure in a line, counting all animals that pass in each direction. The distance between the drive crew members is adjusted to ensure that none of the animals will be missed, even those that might not be flushed by the crew. Drive counts are expensive and impractical because of the manpower required not to mention that the animals and their habitat are purposely disturbed. We point out that this approach is seldom used except in experiments where a total count is needed to verify the detectability of a count by some other method that is being tested for use in monitoring programs. Drive counts were used by Stoll et al. (1991) in their work to determine the accuracy of helicopter counts of white-tailed deer in a western Ohio farmland habitat through a comparison with ground counts. The accuracy of the count was confirmed (ground-truthing) by driving the deer from the patches by using both a team of drivers and the helicopter for flushing deer. Using this method in this type of setting they were able to detect 119 of 120 deer in the survey area. They acknowledged that the utility of such a technique, while very accurate, can be applied only to very special types of habitats.

Aerial counts: There is value in using aerial survey techniques to gather data on animals in the wild but there are also recognized problems associated with aerial surveys. An aerial survey can be carried out to get a sample of the population present or as a census to count all the animals present. Generally total counts are expensive, especially if the area to be covered is large and requires a lot of flight time. Furthermore, it is assumed that in total counts no animals were counted twice (double counting) and no animals were missed in the count. These assumptions are difficult to substantiate and therefore leave the accuracy of total counts suspicious at best. Note that when an observer counts animals, errors can be introduced by overcounting or by undercounting. In aerial surveys it is almost always undercounting. Consider the situation of counting a group of animals from a fixed wing aircraft flying at a relatively high altitude. The group of animals will be available for counting (within the field-of view of the observer) for a period of time, which will give the observer a pretty good look. However, the aircraft might be too high to resolve the individual animals in the group for counting purposes. This is a problem of inadequate resolution. Without some sort of optical system to improve the resolution the next obvious step would be to reduce the altitude of the flight to improve the resolution such that individual animals within the group are distinguishable. This means that the aircraft will be passing over the group of animals such that now the time the group is within the field-of-view is considerably reduced so that even though the observer can see the individual animals in the group there is not enough time to count them. Note that the option of flying at low altitudes in a fixed wing plane may not be advisable because of the terrain or is not always possible due to safety regulations. The bias introduced by these counting errors can be mitigated somewhat by using helicopters or drones (remote piloted vehicles or RPVs) for the aerial surveys in place of fixed wing aircraft.

Large mammals such as moose and mule deer, which are spread over wide ranges of forested habitats, are frequently the objects of aerial surveys or counts from helicopters while fast moving fixed wing aircrafts flying at moderate altitudes are used for animals like antelope in the more open habitats. Many efforts to provide complete coverage of sampling units consisting of strips or quadrats have resulted in a concomitant development of procedures to correct for visibility during these surveys. For example, a detectability model based on logistic regression was developed by Samuel et al. (1987) for elk and the model was applied by Steinhorst and Samuel (1989) to correct counts from a sample of quadrats to get an unbiased estimation of population size. There have been many efforts (reviewed below) to count different species of animals from the air (see Table 2.1), including examples of complete counts on a sample of quadrats for mule deer (*Odocoileus hemionus*) by helicopter (Kufeld et al., 1980).

A detailed review of thermal imaging surveys is found in Chapter 10 but we include here the report of two early thermal imaging surveys that were conducted

TABLE 2.1 A Listing of Animal Species That Have Been Observed, Studied, or Counted During Aerial Surveys in the Literature Reviewed in This Chapter.

Large Mammals	References
White-tailed deer (*Odocoileus virginianus*)	Rice and Harder (1977); Beasom et al. (1981); Beasom et al. (1986); DeYoung (1985); Stoll et al. (1991); Koerth et al. (1997); Berringer et al. (1998)
Collared peccary (*Tayassu tajacu*)	Shupe and Beasom (1987); Hone (1988)
Coyote (*Canis latrans*)	Shupe and Beasom (1987); Hone (1988)
Feral pig (*Sus scrofa*)	Shupe and Beasom (1987); Hone (1988)
Moose (*Alces alces*)	Evans et al. (1966); LeResche and Rausch (1974); Gasaway et al. (1985)
Mule deer (*Odocoileus hemionus*)	Gilbert and Grieb (1957); Kufeld et al. (1980)
Mountain goat (*Oreamnos americanus*)	Pauley and Crenshaw (2006)
Antelope (*Antilocapra americana*)	Pojar et al. (1995)
Elk (*Cervus elaphus*)	Samuel et al. (1987); Anderson et al. (1998); Eberhardt et al. (1998)
Red deer (*Cervus elaphus*)	Daniels (2006)
Caribou (*Rangifer tarandus*)	Siniff and Skoog (1964)
Red kangaroo (*Megaleia rula*)	Caughley et al. (1976)
Wildebeest (*Connochaetes taurinus*)	Caughley (1974)
Elephant (*Loxodontia africana*)	Caughley (1974)
Marine mammals	
Sea otter (*Enhydra lutris*)	Eberhardt et al. (1979); Samuel and Pollock (1981)
Manatee (*Trichechus manatus*)	Packard et al. (1985)
Birds	
Emu (*Dromaeus novaehollandiea*)	Caughley and Grice (1982)
Brown pelican (*Pelecanus occidentalis*)	Rodgers et al. (2005)
Double-crested cormorant (*Phalacrocorax auritus*)	Rodgers et al. (2005)
Anhinga (*Anhingha anhinga*)	Rodgers et al. (2005)
Cattle egret (*Bubulcus ibis*)	Rodgers et al. (2005)
Wood stork (*Mycteria americana*)	Rodgers et al. (2005)
Black duck (*Anus rubripes*)	Conroy et al. (1988)
Canada goose (*Branta canadensis*)	Best et al. (1982)

with thermal imagers of inferior performance by today's standards (relatively poor spatial and thermal resolution). They are included here to provide a comparison between visual and thermal imaging counts conducted during the same timeframe (mid-1990s). An important experiment carried out by Wiggers and Beckerman (1993) showed that a scanning 8–12 μm thermal imager mounted on a fixed wing aircraft flying at 160–200 km/h at a minimum of 271–370 m above ground level (agl) could collect imagery with sufficient resolution to discern the morphological characteristics of the penned deer and permitted the accurate determination of the age and sex structure of the deer present. Detectability was good enough to count all deer in the pens, including one with 73% tree canopy.

In another important thermal imaging survey Garner et al. (1995) flew a fixed wing aircraft fitted with a thermal imager to survey moose (*Alces alces*), white-tailed deer (*Odocoileus virginanus*), and wild turkeys (*Meleagris gallopavo*) They were able to view flocks of turkeys from many different oblique angles by circling in the aircraft and they were confident that they achieved 100% detection of all the birds in each flock.

Incomplete Counts

Since complete counts across an entire survey area or over groups of sampling plots are rarely achievable, most monitoring needs rely on incomplete counts. Incomplete counts can arise from an incomplete count at the survey level, incomplete counts over sampling plots, or even complete counts over sampling plots. There are a number of approaches to dealing with incomplete counts or counts where the detectability is imperfect, which in some cases can be large (reports of ~ 50% detectability are common). In an incomplete or partial count not all animals are counted on the survey plots but there are ways to use the counted fraction of the total animals present on the plots to obtain estimates of the population size. These sampling and/or analysis techniques must be considered in conjunction with a survey design that will ultimately provide the samples needed to permit precise and accurate abundance or density estimates for target populations. A brief list of some of these methods and/or techniques include double sampling, multiple observer methods (where the observers can be independent or dependent), distance sampling, marked sample methods, removal methods (catch-per-unit-effort and change-in-ratio), and mark-resight methods. For detailed discussions of these and other methods, see Lancia et al. (1996), Thompson et al. (1998), Williams et al. (2001), and Pierce et al. (2012). The behavioral ecology of the target species, the size of the survey area, what the goals of the study are, and the available resources all have an influence on the method one chooses to use in estimating the abundance.

Wildlife managers are limited to essentially two choices: (1) estimate the true abundance or (2) use indices that relate in some way to the true abundance. In the first case population estimates are usually calculated from incomplete counts (since complete counts are rarely obtainable) and a mathematically derived estimate of population density is calculated as a function of the detectability and sampling strategy. In both the complete and incomplete counts scenarios the main elements in the design of the study are the detectivity and the sampling strategy. The behavioral ecology of the target species such as its movement patterns and its spatial distribution throughout the survey area must be taken into account so that the results, in the form of a population density, can be reliably extrapolated from the counts.

By estimating the fraction counted, the incomplete counts can be converted into estimates of the target population. For example, on an individual sample plot we can relate the number of animals detected, n, (the sample size or number counted) to the population size, N, and if the detectability is perfect then $n = N$. If the detectability is not perfect then we count only a fraction of the animals present or $n = DN$ where D is detectability and its value is $0 \leq D \leq 1$ so there is uncertainty in how many animals there actually are on the sample plot. In this situation we can only estimate the number of animals present as $\tilde{N} = n/D$. It is now obvious that the problem has been reduced to finding an estimate for the detectability.

The second major problem is that it is nearly always impossible to apply a particular survey method to an entire area of interest. If a group of sample plots are selected to represent a fraction of the total area of interest (a = area of sample plots), then the estimate of the number of animals $\tilde{N}$ in the study area can be expressed as $\tilde{N} = N_{SP}/a$ where N_{SP} is the total estimated number of animals in the sample plots. This type of problem is the result of uncertainty in the selection and placement of sampling plots relative to the distribution of animals in the area of interest, but these problems can be mitigated somewhat by utilizing stratified random sampling techniques or double sampling. There are ways to estimate D in conjunction with animal counts; it can be calculated from the sampling process based on statistical theory and used to adjust the abundance estimates; however, the methods whose estimators adjust for $D \leq 1$ are nearly always more costly and take more time than those based on indices. Furthermore, there is no guarantee they will be any better than indices unless all the assumptions required for the method are met.

Indices: Techniques that do not adequately account for values of $D \leq 1$ are generally referred to as index methods. Most of the time index methods are applied because they are relatively inexpensive and easy to use, which makes them attractive for assessing changes or trends in abundance or population density. An index is a measure of some aspect of a population that is assumed to fluctuate with the actual population size. For example, a count of animals, animal track counts, dung and pellet group counts, or call counts from birds are indices of relative abundance or density. It is assumed that changes in these counts over space and time accurately represent actual changes in population numbers. This assumption must be properly tested if it is to be used in monitoring populations. For *relative index* methods (indices of relative abundance or density) to be used for spatial and temporal comparisons they must be subject to some standardization in the counting effort and procedure. Bart et al. (2004) and Pollock et al. (2004) suggested that index methods are often a cost-effective component of valid wildlife monitoring but that double sampling or another procedure that corrects for bias or establishes bounds on bias is essential. For example, aerial surveys of animals frequently use a double-sampling approach in which complete ground counts are conducted for sampling plots representative of the entire sample survey area and the ratio of the mean aerial count on the sampling plots to the mean ground count of the subplots then gives an estimate of *D* or the proportion of animals seen from the air. Bart et al. (2004) further point out that the common assertion that index methods require constant detectability for trend estimation is mathematically incorrect; the requirement is that there is no long-term trend in the detection "ratios" (index result/parameter of interest), a requirement that is probably approximately met by many well-designed index surveys. Evaluation of this issue usually will show that surveys will be inconclusive unless it can be argued convincingly that the index ratio (survey result/population size) has not changed by more than 15–20% during the survey period (Bart et al., 2004).

Lancia et al. (1996) separated indices from all other population estimation methods because they are a special case for which no population estimate is intended. They point out that the application of indices to make inferences about differences in population size principally involves detectability concerns. Many statistics that are considered as indices of relative abundance or density are based on counts of animals seen (road counts and spotlight counts), animals caught (trapping efforts), animals harvested (hunt counts), and animals that have been heard (auditory cues such as howling or bird calls). Other statistics have been used to quantify abundance by relating a count of some physical sign of animal presence (nests, dens, tracks, dung, or pellets).

Roadside and spotlight counts of animals take advantage of light reflected from the tapeta (e.g., Figure 2.1) of animals' eyes at night. McCullough (1982)

FIGURE 2.1 Eye-shine from the tapeta of a white-tailed deer (*Odocoileus virginianus*). *(Photo taken in coastal Virginia in 2011 by Havens and Sharp.)*

found that seldom were more than 50% of deer detected during spotlight surveys. Bucks were typically under-represented, with the highest counts being in July. Fawns were also grossly undercounted and did not approach base counts until 10 months old. Collier et al. (2007) found similar bias for road-based spotlight counts when compared to thermal imaging counts. Spotlights detected only 50.6 % of the deer detected by thermal imagers. Fafarman and DeYoung (1986) found that estimates of the white-tailed deer density in south Texas that were obtained later in the night (after 23:00) by spotlight counting were probably less biased than those obtained from helicopter surveys. Focardi et al. (2001) reported that spotlighting detected only 50.8% of the animals detected by a thermal imager. It should be noted that spotlight surveys are conducted across the United States on private and public lands as a method for deer population monitoring and harvest planning. Estimating corrections for each location, observer, and timeframe are obviously impractical.

Aerial counts: For large mammals, particularly those with extended home ranges, the present methods for inventorying and surveying such as spotlighting (McCullough, 1982; Focardi et al., 2001), roadside counts (Collier et al., 2007), mark-to-recapture methods (McCullough and Hirth, 1988), and aerial surveys can introduce behavioral variables and viewer bias. In the majority of the papers reviewed in this section the radical changes induced in the behavior of the animals being surveyed was called for in the survey protocol. In these efforts the practice of flushing was felt to be necessary to enhance the probability of detection, thereby reducing the visibility bias associated with aerial surveying techniques.

There are a number of factors that can reduce detectability. The phenomenon of visibility bias, which is different from the bias introduced through counting errors associated with aircraft speed and altitude, is the main source of this problem. Visibility bias is linked to factors such as the coloration of the animal relative to its background (camouflage). Furthermore, the animal's size and how much visual obscuration is caused by vegetative cover or light levels (e.g., dusk and dawn) will reduce the visual cues available to the observer, thereby distorting or biasing the observer's ability to detect the animal. These types of bias stem from the inability of the human brain to recognize an animal by observing only a portion of the animal and by not being able to distinguish an animal from its background due to the animal's coloration. The amount of effort put into the search also affects the outcome (the longer and more person-hours devoted to the search the higher the detection probability should be). The result of these factors is that bias is introduced into all visual aerial surveys. Caughley (1974) and Caughley et al. (1976) reviewed the problem of bias in aerial surveys and determined that for fixed wing aircrafts, the accuracy of aerial surveys deteriorates progressively with increasing width of transect, cruising speed, and altitude. They suggested that even with refining the techniques there seemed to be no technical solution. In particular, they noted a direct correlation between strip width and speed and the number of animals observed. Beasom et al. (1981)

suggested that the use of a helicopter theoretically overcomes many of the practical censusing difficulties because both the speed and altitude can be adjusted to obtain improved detectability. A study area of 80,000 ha of brush-covered range lands with a canopy cover of 30–70% at heights to 5 m was selected on 11 ranches in Texas. They examined the influence of strip width on detectability from a helicopter flying at 25–30 km/h at an altitude of 20 m. They reported sighting data from two strip widths defined as inside (0–50 m from the flight line) and outside (50–100 m from the flight line) and found that the detectability was 53% lower in the outside strip (farthest from the aircraft) than the inside strip. The decrease in detectability was attributed to the fact that animals in the outside strip were farther from the flight line with the conclusion that the apparent underestimation of total animals was at least 26% on the average. There are several other factors that may have influenced the detectability. Since the canopy cover was as high as 70% and at heights to 5 m some of the deer may not have been flushed at the extremes of the transect width and with the short observation times (even in the slow moving helicopter) approximately 7–10 m/sec deer could easily be missed.

The accuracy and precision of eight line transect and one strip transect estimators were examined by helicopter aerial survey on the floodplains and surrounding areas of the Mary and Adelaide rivers in the Northern Territory, Australia (Hone, 1988). The survey area was virtually treeless, with large areas of green grasses and herbs to heights of 1–2 m. A known population of feral pig (*Sus scrofa*) carcasses was located in the survey area resulting from a recent application of lethal control measures. Four strip widths (0–25, 26–50, 51–75, and 76–100 m) were delineated by tape on a pole that projected perpendicular to the flight path and positioned just in front of the observer. A total of 51 carcasses were counted and the number counted in each strip-width class declined as the classes were more distant from the flight path. The results of the survey showed that line transect estimators may be useful in helicopter aerial surveys. Conroy et al. (1988) evaluated aerial transect surveys for wintering American black ducks (*Anas rubripes*) and concluded that the estimates of the surveys were biased; however, bias is not a primary concern for a survey whose main purpose is detecting population trends. They suggested that techniques to decrease bias through air-ground comparisons are likely to be expensive and will require more development but further suggested that air-ground comparisons could probably be justified if there was a demonstrable need for an estimate of abundance of the absolute size of the black duck population versus an index.

Ground-truthing surveys were conducted in conjunction with a state wide aerial survey of Florida wading bird colonies to evaluate the efficacy of the aerial technique (Rodgers et al., 2005). Five species, brown pelican (*Pelecanus occidentalis*), double-crested cormorant (*Phalacrocorax auritus*), anhinga (*Anhinga anhinga*), cattle egret (*Bubulcus ibis*), and wood stork (*Mycteria americana*), which are large birds that tend to nest higher in the canopy or are white-plumaged, were most often detected during the aerial survey. Five other species that were smaller and more cryptically colored and typically nested beneath the canopy were not detected in the aerial survey. The aerial detection rate for previously unknown colonies defined as the proportion of active colonies found from ground surveys that were also found from the air was 71%. They caution that the level of variability and probability of species detection should be assessed prior to conducting large-scale inventories for colonial water-birds and cite results from 1999 surveys requiring 30,140 km of transects spaced at 5 km widths that cost $95,320, of which $38,560 was spent on aircraft rental. In that effort the estimated colony detection rates of 56–84% based on the 5 km wide corridors suggest that smaller width corridors may be required during an aerial survey and that reducing the corridor width to 2.5 km would require flying 60,000 km of transects to cover Florida. As a result, the survey could not be accomplished in a single breeding season.

Caughley and Grice (1982) used the mathematics of the mark-recapture model to derive a factor correcting count of emus (*Dromaeus novaehollandiae*) surveyed from the air in Western Australia. The emus were neither marked nor recaptured, the correction factor being derived from the number of emu groups counted independently by two observers scanning the same transect. The analysis suggests that about 68% of emu groups on the transect are counted by a given observer during a standard survey, and those counts must be multiplied by 1.47 before they give a true density of groups.

A number of studies have been conducted that demonstrate the superiority of helicopters over fixed wing aircraft as an observation platform for visual counting during aerial surveys of large mammals. The recognized advantages of low altitude and low flight speed were common to all these efforts. Both of these features aid in the observer's ability to locate and identify animals on the ground (Thompson and Baker, 1981; Shupe and Beasom, 1987). Intuitively, it would also seem that a snow-covered landscape would provide maximum contrast between the animal and its surroundings, making it easier to observe. We can also assume that the detection rates of animals would be higher in open habitats rather than in areas with dense vegetative cover. When all these conditions are met, aerial surveys from helicopters should be capable of producing a detectability approaching 100%. Unfortunately, this is not the case. The detection rates reported for traditional helicopter surveys using visual counting techniques are between 35% and 78% regardless of the habitat, weather conditions, or methodology. DeYoung (1985) tested the accuracy of helicopter surveys of deer in south Texas (traditional ranch counts) where untested claims of 90–95% accuracy are commonly made and where these unadjusted counts are widely used for management purposes. The terrain surveyed by DeYoung (1985) consisted of two separate areas (one surveyed in the winter and the other in the fall), both of which were dominated by vegetative cover comprised of low to medium height brush. Population estimates were obtained from mark-recapture methods described by Rice and Harder (1977). Marked deer were considered recaptured when sighted during a low altitude flight (23 m agl) along 200 m wide transects at 56 km/h, which would flush deer so they could be counted. Traditional counts were also conducted over the same terrain for comparison. The traditional ranch counts averaged 65% of the mark-recapture estimate for the winter count and 36% of the mark-recapture estimate for the fall count. Form this work we can see the importance of accuracy when obtaining population estimates to be used as herd management purposes.

Aerial mark-resight estimates of mountain goat (*Oreamnos americanus*) abundance using paintball marks at Seven Devils and Black Mountain, Idaho, were conducted by Pauley and Crenshaw (2006). Mountain goats were marked with highly visible and persistent (lasting 71 days) bright yellow, orange, and red oil-based paintballs using recreational paintball markers fired from helicopters. The precision of abundance estimates was reasonable with mark samples ≥51% of the estimated abundance. Sighting probabilities, calculated from the proportion of marks observed on resight surveys, ranged between 0.34 and 0.46 across both study areas. These detectabilities are a function of both visibility and a potential change in behavior caused by marking. That is, their primary concerns were that paintball marking could cause a significant change in mountain goat behavior, causing abnormal movement at the presence of helicopters and that highly conspicuous orange marks potentially increased the visibility of marked mountain goats with respect to unmarked mountain goats.

LeResche and Rausch (1974) examined the accuracy and precision of aerial censusing of moose (*Alces alces*) confined to four 1-square-mile enclosures located in a large 1947 burn in Alaska. The enclosures contained representative vegetation of both burned (regenerative) and remnant (mixed birch-spruce-aspen) stands. The number of moose in individual pens ranged from 7 to 43 and a total of 74 counts were flown over each pen during three censusing periods from January 1970 to December 1971. During the flights an observer sat behind the pilot in a Piper PA-18-150 "Supercub" aircraft that was flown at an altitude of 200–300 ft. agl and airspeed of 50–60 mph. Each observer had 15 min of observation time over each pen, or an hour to census the four pens. There were a total of 49 observers who were characterized according to experienced (current), experienced (but not current), and inexperienced. Ten experienced (but not current) observers saw only 46% of the moose they flew over in excellent snow conditions compared to 21 experienced (current) observers who saw 68% of the deer they flew over under excellent snow conditions. The 12 inexperienced observers saw 43% under excellent snow conditions. As would be expected, the count of deer in the pen with the highest percentage of mature habitat was the lowest for the experienced (current) observers. The conclusions drawn from this work are that there are a host of variables responsible for the observed counts including: (1) factors affecting the performance of individual observers such as experience and fatigue, (2) the density of moose and their diurnal behavior patterns, (3) the local physiography such as the features of the terrain and vegetative cover, (4) the weather, including cloud cover, turbulence, and snow cover, and (5) the equipment used for the census work, including the aircraft and pilot.

Kufeld et al. (1980) pointed out that for animal surveys in rugged mountainous terrain a helicopter was better suited for meeting the survey requirements for several reasons. There are obvious dangers associated with flying fixed wing aircrafts in mountainous terrain where low altitudes are required for surveying applications, and even when using helicopters they must be powerful enough to handle erratic wind conditions and large changes in elevation quickly. In their survey for mule deer (*Odocoileus hemionus*) on the Uncompahgre Plateau, Colorado, total counts on sample quadrats and a stratified random sampling design were employed. The survey area was divided into eight strata and each stratum was gridded into 0.6475 km^2 quadrats. Allocation of the quadrats among the eight strata followed optimum allocation procedures whereby the number of quadrats assigned to each stratum was proportional to both stratum area and subjective estimates of relative deer density therein. Strata with high deer densities received a higher sampling intensity because they anticipated that they would have greater variances. They determined that during the 3 years that the census was conducted, the stratified random design reduced variance of the mean number of deer seen per quadrat by 42, 21, and 37%, respectively, from that which would have resulted if treated as a simple random sample. Some of the quadrats had elevation changes as much as 305 m across the quadrat, which a fixed wing aircraft would not be able to negotiate. Sample quadrats were censused by helicopter using three observers and they estimated that the population means were within 20% of the true value with 90% confidence. Lancia et al. (1996) cautioned that these estimates should be considered conservative and used as minimum estimates or used as a constant-proportion index because of the possibility of visibility bias during the quadrat surveys. Kufeld et al. (1980) concluded that the use of a helicopter had several advantages but did not attempt to quantify them in their work. They suggested that when quadrats must be used because of steep terrain a high-powered maneuverable helicopter with good visibility should be used as the viewing platform. They noted that the advantages of helicopters over fixed wing planes are: (1) with a pilot and two observers sitting three abreast there are three observers searching for deer, (2) terrain and cover problems are minimized since the helicopter can turn and ascend/descend rapidly and it can hover over thick patches of vegetation to flush animals, (3) helicopters are capable of operating at air speeds much lower than that of fixed wing aircrafts, which increases observation time, (4) the sound of the rotor blades tend to flush deer, increasing their detectability, (5) forward visibility is superior to fixed wing aircraft, (6) helicopters are less likely to subject observers to forces that are capable of producing disorientation and air-sickness, and (7) a helicopter can fly safely at lower altitudes than fixed wing aircraft.

Pojar et al. (1995) conducted a series of aerial counting experiments from 1981–1987 to estimate pronghorn (*Antilocapra americana*) density and herd structure in Colorado sagebrush (*Artemisia spp.*) steppe (SBS) and shortgrass prairie (SGP) habitats. They used stratified random sampling methods in both the SBS and SGS habitats. Random quadrat and random transects (line and strip) were selected for both the habitats and the cadastral survey of the two habitat areas to be sampled was divided by 2.59 km^2 units which was used for the quadrat size. This selection was in agreement with Seber (1982) who suggested that the quadrat size be as small as possible to maximize precision. For surveying the quadrats they used a two-person helicopter crew (pilot and observer) for the SGS habitat and a three-person crew (adding a navigator) for the SBS habitat. They flew at 60–70 km/h at 15–30 m agl. Strip transects (1.6 km wide) were flown faster (100 km/h) but at the same altitude as the quadrats. In the second experiment they included narrower transects (200 m) and line transects described by Buckland et al. (1993). During these experiments they concurrently tested the observer differences. The final test was for quadrat bias where they used two helicopters during a survey of one stratum of the SBS habitat containing 40 quadrats. One helicopter flew approximately 150 m above the survey crew, which gave a wider field-of-view to detect movement of flushed animals. Population estimates from wide strip transects in SBS were 54% lower than comparable estimates from quadrats. In the SGS habitat the wide strip transect population estimates were 60% less than estimates from narrow strip transects and line transect surveys. The test for observer effects showed that there was no difference in buck to doe and fawn to doe ratios between observers. On samples of 449 animals for the primary observer and 436 for the secondary observer, the buck to doe ratio was 28 and 27, respectively; the fawn to doe ratio was 32 and 35, respectively. The test for quadrat bias was negligible, with both crews counting identical numbers of pronghorn on 20 quadrats and within 4 animals on 12 quadrats and ≥ 5 animals on 8 quadrats. They recommend that quadrat sampling be used since intense searches of 2.59 km^2 quadrats resulted in a large reduction of relative bias when compared to wide strip transect units. The quadrat counts produced results similar to those for narrow strips and line transects but did not have to contend with the inherent subjectivity associated with observers trying to keep the delineation of the strip width constant during the surveys. They were confident that there was no double counting on the quadrats and that overall the quadrat sampling was probably the least biased.

Complete snow cover: A nearly universal conclusion reached in all of the aerial surveys of big game animals was that complete snow cover will provide maximum contrast for aerial surveys and was a prerequisite for improving the accuracy of surveys (Evans et al., 1966; LeResche and Rausch, 1974; Gasaway et al., 1985; Gilbert and Grieb, 1957; Berringer et al., 1998). This is not surprising since the decision to conduct surveys during times of complete or nearly complete snow cover actually combines two conditions that collectively improve the opportunities for detection. A nearly uniform white background provided by the snow cover tends to maximize the visual contrast between the animals and their background, thereby minimizing the problem of visibility bias when viewing animals from the air. At the same time, the tree canopy is considerably reduced during the late fall and winter months in forests dominated by deciduous vegetation so that the obscuration provided by vegetative cover is also reduced. A classification of snow conditions for the detectability of moose during aerial surveys in early winter, when they form larger groups, is given by Gasaway et al. (1986, p. 19).

A number of reports point out that detectability is highest in habitats with snow cover present and little or no vegetation. For example, Ludwig (1981) reported that detectivities ranging from 65 to 76% were obtained during a helicopter survey for deer (*Odocoileus virginianus*) in Minnesota farmland under snow conditions where cattails were present. The detectivity was highest (76%) over timbered habitats with a rolling topography but with no cattail. Rice and Harder (1977) reported detectivities of 51–70% for deer in a 122-ha snow-covered enclosure with brushy habitat in northern Ohio. The deer in the enclosure were censused via a drive count and the density of deer in the pen was very high (127/km^2). Berringer et al. (1998) carefully evaluated a flight protocol for counting deer (*Odocoileus virginianus*) on snow with the use of a helicopter. In their experiment they used marked deer as the known population to survey a 794-ha fenced forested area dominated by mature hardwoods. They flew 200 m transects at an altitude of 60 m at an air speed of 23 knots to obtain an average detectability of 78.5% (range = 72.4–86.9%). This detectability is the highest thus far reported for counting deer on snow, so other factors must be playing a role in preventing the near perfect detection rates that would be expected. There are certainly the human factors of experience in spotting, fatigue, and visibility bias. The latter may be more important than it appears even with complete snow cover. For example, if an animal is partly obscured by vegetative cover or large woody debris the normal profile of the animal presented to the observer is distorted or incomplete and the shape factor is removed from the observer's list of recognition cues. The work of LeResche and Rausch (1974) would seem to support this argument based on the results of experienced and inexperienced spotters counting moose (*Alces alces*) in 1-square mile pens with excellent snow cover.

The work of Stoll et al. (1991) utilizes all the desired features for increasing the detectability and accuracy of helicopter counts. They designed and implemented a deer survey for an intensely farmed region in Western Ohio where winter habitat for white-tailed deer (*Odocoileus virginianus*) was typically restricted to small, isolated, deciduous wood lots characteristic of this region. It was felt that this survey area met the basic criterion for an idealized situation with low deer densities on snow-covered, mostly flat open farm fields characterized by small isolated patches of deer habitat. The habitat patches were intensively scanned from a helicopter along 100–300 m wide transects using low speed (30–70 km/h), low altitude (45–60 m), and slow circling until the observers were confident of their count. If animals were flushed from the habitat patch during the count they were easily spotted. The accuracy of the count was confirmed (ground-truthing) by driving the deer from the patches using drivers and the helicopter. Using this method in this type of setting, they were able to detect 119 of 120 deer in the survey area. They acknowledged that the utility of such a technique, while very accurate, can be applied only to very special types of habitats. In most surveys, counting the animals without disturbing them, especially threatened or endangered species, is desirable since disruptions in their activities can be stressful to the animals, and disrupting them to the point of flushing them can easily result in duplicate counting. When these surveys were carried out the practice of flushing animals (particularly large ungulates) was typically part of a prescribed protocol for most aerial surveys. Note that even though many workers insist that snow cover is required for high detectability, the work of DeYoung (1985) and Beasom et al. (1986) obtained comparable detectability without snow cover for deer in Texas brush habitats.

Models and techniques are being developed to improve estimates of animal populations that are being biased in some way during the detection phase of the survey. For example, new statistical methods have been used to improve estimates of elk (*Cervus elaphus*) populations acquired in aerial surveys (Eberhardt et al., 1998). Samuel et al. (1987) developed a detectability model to account for visibility bias in aerial surveys of elk from helicopters during winter months in Idaho. The method attempted to standardize survey conditions that are controllable such as the air speed, number and experience level of observers, and aircraft type. They then developed correction factors, through logistic regression, for factors influencing the detectability such as group size and vegetative cover. The main advantage of using detectability models is that once an appropriate model has been selected then subsequent surveys will not be as labor intensive or as costly. Thompson et al. (1998) point out that unfortunately, the conditions under which the model was developed must be closely mirrored or the model will yield spurious estimates of detectability, which will lead to biased estimates of abundance. The problem may be even worse if the model is applied to a different area, that is, an area where detection factors such as vegetative cover will likely be different than the original area. If the original area is substantially changed after the model was developed (maturing trees, fire, timbering activity) the model will likewise be deficient.

Detectability models were also developed and tested for elk surveys conducted during the summer months by Anderson et al. (1998). These efforts were undertaken because of problems associated with detecting smaller groups of elk on both summer and winter ranges. A deterioration of the detectability conditions stemming from changes in the weather, observer skill, and time of day were treated by Beavers and Ramsey (1998) for surveys employing line transect sampling. Visibility bias was also found to be a problem when conducting field trials using line transect methods and distance sampling applied to the estimation of desert tortoise (*Gopherus agassizzi*) abundance (Anderson et al., 2001). The importance of observer experience in finding desert tortoises was pointed out by Freilich and LaRue (1998). A correction of visibility bias in aerial surveys where animals occur in groups was carried out by Samuel and Pollock (1981). They applied a proposed sighting function that related the detectability of groups to group size on a population of sea otters (*Enhydra lutris*). Aerial surveys were compared to ground counts that were assumed to be true counts of the number of otters and their group sizes within an area at the time of the aerial survey. An important assumption in estimating the sighting function is that ground surveys do not miss animals.

All these improvements are sought because of inaccuracies stemming from detectability/sightability issues of individual animals or groups of animals encountered in the field. Typically visual aerial surveys conducted of animals in the field, especially in a forested habitat or partially forested habitat, are strongly skewed as a result of "visibility bias" (Samuel et al., 1987). Compounding the problem is the fact that the amount of bias is affected by a host of factors such as aircraft speed, altitude, weather conditions, spotter experience, animal group size, time of day, and ground cover, among others. It is fundamentally unsatisfying to try to remedy biased data through calculations. It would be far better to improve the detection capability by employing new technologies that directly assist in the visual acquisition of animals in the field or design new counting technologies that would provide improved detection capabilities. The concept of coloring a portion of an animal population in bright colors (marked) so that they can easily be distinguished with certainty from other individual animals (unmarked), yet still maintain the same detectability, is a non sequitur.

The practice of flushing animals during surveys and censuses should be avoided due to the stress on the animals and because the practice can induce behavioral changes that make counting them even more difficult over time. It can also easily lead to double counting, especially when using strip transects. Many of the efforts discussed above to enhance the detectability during large animal surveys deliberately included the practice of flushing animals in the survey protocol; in fact, some of them depended on this practice. Technologies are at hand and have been for many years to aid the human observer in field surveys but have been ignored because of poor performance reports that were promulgated as a result of improper use. Other factors were also indicated, such as being too expensive. We suspect that ignoring the technologies was because of these factors plus reluctance on the community of scientists to try something new. Hence, we see instead years of work by statisticians devoted to correcting visual errors in data gathering efforts.

Anderson et al. (1998) developed and evaluated sightability models for summer elk (*Cervus elaphus*) surveys in Grand Teton National Park, Wyoming. Sightability surveys were flown in helicopters with a three-person crew aboard: a pilot, a primary observer (experienced in aerial elk surveys and familiar with the survey area), and a secondary observer (with at least some aerial survey experience). An unpublished survey protocol by Unsworth et al. (1994) was followed and unfortunately there was no information provided regarding the speed or altitude of the helicopter when conducting the sightability surveys. Their summer sightability model, which excluded the influence of elk activity, was found to be similar to the unpublished work of Unsworth et al. (1994), when snow was absent. They concluded that population estimates become less precise when surveys are applied to elk in dense forest canopy (>70% vegetation cover). Accurate and precise population estimates should be obtained from their summer model if applied when elk are most observable and surveys are conducted early in the day when the elk tend to be in large groups feeding in open habitats, before they disperse into dense cover to escape the heat of the day.

A paper published by Pollock and Kendall (1987) evaluated a number of models proposed during the 20-year period prior to their work for estimating the visibility bias in aerial surveys. To gain an estimate of the visibility bias encountered during aerial surveys they examined the value offered through comparisons with complete and incomplete ground counts, subpopulations of marked animals, mapping with multiple observers, line transect sampling, and multiple counts on the same area. They point out that visibility bias results from animals being missed and is exacerbated by factors such as dense vegetation, bad weather conditions, and observer fatigue. This bias can be significant since as much as 30–50% of the animals can be missed (Caughley, 1977). They postulate that the main difficulty with aerial surveys is estimating the visibility bias so that a correction factor can be established. The amount of work and cost to establish a useful correction factor can easily be prohibitive. They felt that a complete ground count or other method that totally enumerated a subsample of the population be obtained and compared with the aerial count to get an accurate assessment of the visibility bias for the subsample; however, this method is often impractical, cost-prohibitive, or both. They surmised that the best technique for obtaining a correction factor for visibility bias is to choose the one that best fits the population, the species, the habitat, the budget, and the importance of the estimate. Keep in mind that this extra work must be done because of the poor detectability as a result of visibility bias encountered during an aerial survey.

Photographic techniques: Using photographic techniques in aerial surveys presents yet another layer of difficulty without providing the desired advantages that aerial photography would seem to provide, that is, a photographic record that could be studied in detail to improve count accuracy. When large numbers of animals (e.g., herds or roosting birds) are the target of an aerial survey the use of photography will be valuable. In general, however, aerial photography will only provide more time to count; there are still the same problems of visibility bias and vegetative cover to contend with. In fact, the same conditions are recommended for aerial photographic counting that were recommended for visual aerial surveys discussed earlier. One problem that might be ameliorated by aerial video is the issue of animals being counted that are not in the boundary of the strip transect being surveyed. If the optics and the flight altitude are adjusted to provide a field-of-view for the video camera and the field-of-view coincides with the strip width called for in the survey design, then only animals seen on the video are in the strip transect and it will be impossible to include animals outside the transect.

Thomas (1967) pointed out the difficulty of extracting accurate counts from aerial photographs when there was a lack of snow cover. Likewise, Siniff and Skoog (1964) reported that it was difficult to locate animals on aerial photographs when conditions of heavy vegetative cover prevailed. These are the same problems that are responsible for confounding aerial surveys in general. The video, however, will provide an opportunity to review the data any number of times and to have other observers carry out a count, which would allow the highest count to be extracted from the data. Aerial photographs have been used to count seabird colonies (Harris and Lloyd, 1977) with the result that the counts varied greatly among observers, undercounting being the norm. Even with repeated counts by experienced counters the photographic counts varied ±10%. The population distribution, trend, and status of Canada geese (*Branta Canadensis*) are typically determined by visual estimates made by trained observers in small aircraft during daylight flights. The trend of population distribution is toward fewer and larger concentrations of geese, primarily due to the increased availability of food in controlled goose hunting areas. Accurate estimates of the number of birds in these concentrations are becoming more difficult to make and photographic spot checks have indicated that visual estimates may be 50–75% lower than the actual number of birds in a concentration (Best et al., 1982). Garner et al. (1995) concluded that the use of standard VHS video recordings of white-tailed deer and moose were ineffective. They reported that for surveys conducted in Algonquin Provincial Park, in south-central Ontario, a total of 84 moose were recorded in the digital analysis of infrared thermography video data, and that untrained observers counted only 42 moose. Only three were observed on the standard VHS recording.

Indirect Counting Methods

The counting methods we have already mentioned involve the direct observation of animals during a survey and are intended for collecting data to be used for population estimates, but there are other methods of counting as well. The outcome of those collections of data can either be total counts or partial counts, both of which can be used to estimate animal populations. Indirect methods do not detect animals but instead detect some quantity or feature that must then be related to the density of animals that produced the detected or counted feature. For example, animal tracks can be surveyed, nests or dens can be counted, harvest counts can be made, or call counts can be made to estimate numbers of animals present in a survey plot. While these methods are not generally desirable, they sometimes are the only way to acquire the needed data to make an informed decision relating to the relative abundance of a population in some area at a particular time or over time at a particular area (i.e., it can serve as an index of relative abundance). Note that estimates of relative abundance are used when the number of animals in a population cannot be determined.

Wilson and Delahay (2001) point out that some species are notoriously difficult to survey using direct counts. They present an excellent review of field methods for estimating and monitoring the abundance of terrestrial carnivores that do not involve capture. They conclude that the use of field signs and observation techniques to estimate the abundance of carnivores that are typically nocturnal, cryptic, and that may have large home ranges requires careful consideration to minimize the confounding influences of the extrinsic (environmental) and intrinsic (methodological) factors particular to a given technique. To determine if they could use track surveys to detect changes in cougar (*Puma concolor*) populations, Beier and Cunningham (1996) used data from track transects collected in southeastern Arizona to evaluate survey designs for 8 km transects in first and second order dry washes. Cougars are nocturnal, secretive, and disperse at low densities, making it difficult to monitor changes in their populations. They point out that to index population trends over large areas at low cost, managers use hunter harvest, depredation rates, and track surveys. Track surveys may provide a better index than harvest or depredation data since trend estimates from hunter kills and depredation rates are sensitive to hunting efforts and reporting rates, respectively. They developed sampling methods that would allow managers to detect a 30–50% change in cougar abundance between two survey periods. Even though they found that a considerable sampling effort is required for track surveys to detect large population changes, they believe that track surveys are better than the alternatives since track surveys can yield statistically valid inferences about population change, whereas indices based on hunter harvest or depredation cannot. Long et al. (2008) provide an extensive treatment of noninvasive survey methods for carnivores, including chapters devoted to tracks and scat, remote cameras, hair collection, genetic and endocrine tools, and other relevant topics. Tracks have also been observed from aircraft over fresh snow where counts were made using a line intercept technique. Becker (1991) used probability sampling results to develop a new method of estimating furbearer abundance based on observing animal tracks in the snow. This method requires that good snow conditions be present during the course of the study and that all animal tracks intersected during the sampling process are observed. Using a sampling design that assumes animal tracks can be observed and followed to both the animal's location at the end of a snowstorm and to its present location, he estimated 9.69 ± 1.97 (SE) wolverines (*Gulo gulo*) for a 1871 km^2 area in south-central Alaska. Using a second sampling design that assumed that the number of different animals encountered along a set of transects could be determined and that it was possible to get movement data from a random sample of radio-collared animals resulted in an estimation of 15.09 ± 4.34 lynxes (*Felis lynx*) for a 285-km^2 area of the Kenai Peninsula, Alaska. It is often desirable to convert indices to absolute estimates that force an explicit consideration of error and uncertainty in the index, which is often overlooked when only the raw index is used. Stephens et al. (2006) describe the primary method of estimating the number of many game animals in large territories of the Russian Federation, called the winter count, which involves monitoring game species by counting sets of tracts in snow that intersect a stable network of transects. The results are largely used as relative indices of abundance but conversion into an absolute estimate is often desirable. This conversion uses an analytical relationship known as the Fromozov-Malyshev-Perelshin (FMP) formula. This method requires an estimate of the daily travel distance of the surveyed species in order to estimate the probability of an individual animal crossing a transect within the 24-h period before the survey (Kuzyakin, 1983).

It is often desirable to know the absolute densities of animal populations. Indices of abundance are seldom equivalent in different habitats or consistent when applied over large geographical areas, so when one knows with certainty that the detectability was near unity many problems are easily resolved.

Chapter 3

Remote Sensing

Chapter Outline

Humans use some form of remote sensing daily during the course of our normal activities. We see, hear, and smell things at a distance from our bodies and we make decisions based on the information we perceive through these sensors. We are continually engaged in the process of gathering information to guide us through the day. Researchers have devised ingenious ways to count, study, and monitor wildlife populations and their habitats. Remote sensing is generally thought of as imaging on a global scale for such purposes as land mapping, wetland delineation, and the like. The uses of various techniques and equipment that enhance our capability for remote sensing to count and monitor wildlife are examined in this chapter. This chapter discusses how remote sensing is an invaluable tool in the field of studying wildlife ecology.

INTRODUCTION

A number of definitions have been proposed to describe the science of remote sensing (Campbell and Wynne, 2011) and they are all tailored a bit to meet the needs of the subject to be addressed. They range from the very broad definitions of Barrett and Curtis (1992, p. 6): "Remote sensing is the science of observation from a distance" and that of Colwell (1966, p. 71): "The term remote sensing in its broadest sense merely means reconnaissance at a distance" to lengthy descriptions of what is or isn't considered remote sensing. Of particular interest is the definition and observation put forth by Lillesand et al. (2008, p. 1): "Remote sensing is the science of obtaining information about an object, area, or phenomenon through the analysis of data acquired by a device that is not in contact with the object, area, or phenomenon under investigation." They further point out that as you read these words, you are employing remote sensing. Your eyes are acting

Thermal Imaging Techniques to Survey and Monitor Animals in the Wild: A Methodology
http://dx.doi.org/10.1016/B978-0-12-803384-5.00003-8

as sensors that respond to the light reflected from this page. The "data" your eyes acquire are impulses corresponding to the amount of light reflected from the dark and light areas on the page. These data are analyzed or interpreted in your mental computer to enable you to explain the dark areas on the page as a collection of letters forming words. Beyond this, you recognize that the words form sentences and you interpret the information that the sentences convey.

Upon close inspection of texts on the subject of remote sensing one finds that they deal primarily with data extracted from aerial imaging techniques to facilitate and guide at least in part global monitoring and management. For the purposes of this book we define remote sensing as the remote acquisition of information regarding animal ecology from a distance. The information acquired is not confined to imagery but may also include information gained from nonimaging technologies as well as acoustical and radio telemetry, for example. Note that this definition includes implementations of assisted visual data acquisition so that the use of aids such as binoculars and motion detectors is included as remote sensing implementations.

ENHANCED VISUAL

In the broadest sense of enhanced visual sensing the use of eyeglasses, contact lenses, and surgical procedures to improve vision is the most basic and easiest to understand; they simply improve the "direct viewing" of the observer. Vision correction surgery, also called refractive and laser eye surgery, is any surgical procedure used to correct vision problems. Tremendous advancements have been made in this field. After refractive and laser eye surgery, many patients report seeing better than they had at any other time in their lives. Most types of vision correction surgery involve reshaping the cornea and underlying tissue, so that light traveling through it is properly focused onto the retina located at the back of the eye. Other types involve replacing the eye's natural lens. There are a number of different types of surgeries to improve vision; LASIK (laser *in situ* keratomileusis) is one of the most common. It is used to correct vision in people who are nearsighted, farsighted and/or have astigmatism. Many LASIK surgeries are done with the addition of computer imaging called wavefront technology to create a detailed image of the cornea and a guide for treatment (Marcos, 2001). The main reason for visual impairment of the human eye is that it has significant optical defects or aberrations that distort the passing optical wavefront of the incoming light on its way to the retina, resulting in blurred images and a degraded visual experience. In addition, diffraction caused by the finite size of the eye's pupil also produces blurriness (Charman, 1996). The major or first-order effects of aberration (defocus), astigmatism, and prism are the only optical defects of the eye that are commonly corrected with eyeglasses or contact lenses. If we could compensate for the minor as well as the major defects in the eye's optics, we could vastly improve the quality of the image the retina receives and would be taking the first step toward supernormal vision (Miller, 2000).

If selective changes could be surgically made to correct the aberrations caused by the defects in the optics of the eye across the dilated pupil then the vision would be the very best obtainable. Even for someone with 20/20 vision, optical defects in the refractive surfaces skew the light rays and distort the original planar wavefront, impinging on the retina. An adaptive retina camera was devised by Liang et al. (1997). It has a deformable mirror placed conjugate with the eye's pupil and a lenslet array to correct for the phase aberrations caused by imperfections in the cornea and underlying tissue. The key feature is to maintain the phase of the incident planar wavefront so that it remains planar when it reaches the retina, thereby allowing them to see the retina in enough detail to resolve single cells in the retina of the living human eye. In this camera a deformable mirror is adjusted to compensate for the phase distortions introduced by the optics of the eye (i.e., it acts as a phase conjugate mirror (Anderson et al., 1996) or retro-reflector) so that the light waves can pass through the optics of the eye distortion-free. The changes to the deformable mirror that were needed to preserve the phase provide information that could be used to determine what changes would be necessary to alter the cornea and underlying tissue to have images pass through the optics of the eye undistorted. In this situation the phase of the incoming light is preserved and would result in supernormal vision.

There are a number of "direct viewing" devices that do not utilize thermal radiation yet they can enhance the nocturnal viewing capabilities of observers in the field. The simplest of these are binoculars (Danylyshyn and Humphreys, 1982). These optical devices simply magnify the images that the observer views at the expense of much of the surrounding field-of-view. As a result the images appear brighter.

There are several specifications or optical parameters used to describe the performance capability of a particular binocular. For most fieldwork one should choose a binocular that can collect a lot of light and has an exit pupil that can easily accommodate the night-adapted iris of an observer. The number normally used to specify a binocular is determined by the magnification × objective diameter and is expressed as 8 × 50, for example. The linear magnifying power of binoculars is given by the ratio of the objective focal length to that of the eyepiece focal length. A magnification of factor 8 means that the binocular will produce an image 8 times larger than the original seen from that distance. Note that a large magnification leads to a smaller field-of-view (FOV) and makes the instrument more susceptible to shaking. The FOV is usually expressed in a linear distance corresponding to the width (yards or meters) that will be seen at 1000 yards or 1000 m or the FOV is given in degrees.

The objective diameter (diameter of the objective lens) will determine how much light can be gathered to form an image. This number directly affects performance, particularly at low-light levels. An 8 × 50 will produce a brighter and sharper image than an 8 × 40, even though both enlarge the image eight times. The larger front lenses in the 8 × 50 also produce a broader beam of light that leaves the eyepieces, making the viewing more comfortable and the binocular easier to use. The light gathered by the objective is concentrated into a beam called the exit pupil. The diameter of the exit pupil is determined by the ratio (usually expressed in mm) of the objective diameter to the magnifying power. For use in low-light-level situations this ratio should be equal to or greater than the night-adapted iris of the eye. A large exit pupil ($\sim$7 mm) makes it easier to put the eye where it can receive the light: anywhere in the large exit pupil cone of light, which means that the image can be quickly found. This is important when looking for animals that may be moving. The last important feature to look for in field binoculars has to do with the eye relief or the distance from the eyepiece to the observer's eye. Typically eye relief distances are from a few millimeters to a few centimeters. The larger eye relief distance may be necessary for observers that wear eyeglasses.

There are a number of other features incorporated into binoculars that may be necessary, depending on the intended application. For marine use they should be equipped with a floating strap that will aid in their retrieval should they go overboard. They need to be waterproof (completely o-ring sealed as well as nitrogen-purged) at a minimum. This makes the binocular waterproof, fog-proof, and serves to prevent internal corrosion. Most marine binoculars use reticle or grid rangefinders as opposed to laser rangefinders found on binoculars designed for land use. This is because the distances involved on the water are usually too far for the laser rangefinders to work. Information on the use of binoculars for marine applications can be found in (Buckland et al., 2001).

There is also a relatively broad class of "indirect viewing" devices known as image converters that have been used successfully in low-light-level conditions. Image converters are devices that operate in the near-infrared (0.8–1.2 μm). For these devices both the ambient and near-infrared radiation, which is provided via lamps, are "reflected" (not radiated) from the scene and are focused onto an infrared sensitive photocathode that in turn emits electrons that impinge on a phosphor viewing screen, forming a visible image of the scene (albeit grayish green). These are called indirect viewing devices because we are not actually viewing the animal but an image of the animal on a screen. They are called image converters because the nonvisible radiation reflected from the animal is converted into a visible image on the screen. There is also a special class of image converters called image intensifiers, which are discussed later in this chapter.

Since there is very little ambient near-infrared radiation to work with in low-light-level situations, image converters almost always require supplemental near-infrared radiation to illuminate the scene. When using these devices care must be taken to ensure that the animals under observation are not sensitive to the supplemental infrared light, which may produce behavioral modifications. Seubert (1948) conducted one of the first field experiments with an image converter using a "Snooperscope," an instrument that was developed during World War II for military use to aid in observing ground troops during low-light-level situations. Using this instrument he successfully carried out nocturnal observations of birds and mammals in their nests and natural habitats. Historically speaking, image converters were the precursor to image intensifiers (commonly referred to as night vision equipment).

IMAGE INTENSIFIERS (I^2 DEVICES)

Before we continue it is necessary to clarify the differences between infrared imaging with night vision devices (NVDs), which use amplified visible and near-infrared radiation to form images in low-light-level situations, and thermal imaging cameras, which use thermal radiation (heat) in the 3–14 μm wavelength region to form images. Image intensifiers (I^2 devices) are also indirect viewing devices but do not detect thermal radiation. As in the case with the image converter (see Section "Enhanced Visual" of this chapter) an image of the scene is formed on a phosphor screen, except this image is formed using only reflected "ambient" light (i.e., no additional near-infrared radiation is used to illuminate the scene). These devices transform images formed by visible and near visible light provided by moonlight and starlight into greatly amplified images ($\sim 5 \times 10^4$), which enables one to see on overcast nights using only the available light from cloud-filtered stars and the moon; hence the name image intensifier. A detailed description of the construction, operation, and performance of image intensifiers is given by Johnson and Owen (1995). Briefly, the objective lens (front optics) of an I^2 device collects visible and near-infrared ambient light that is reflected from a scene comprised of a spatial distribution of photons that is incident on a photocathode that converts the photon image into an electron image. The electron image is greatly amplified by passing through a microchannel plate and the electron image exiting the microchannel plate impinges on a screen coated with phosphors, providing a near perfect image corresponding to the original image formed by the spatial distribution of photons collected by the objective lens. The energy of the electrons causes the phosphors to reach an excited state and release photons that recreate the original input (albeit green) image on the screen. The green phosphor image is typically viewed through another lens, called the ocular lens, which can be used to magnify and focus the image. Alternatively, the output of the I^2 device can be connected to an electronic display such as a monitor or could be interfaced with a trip camera or video camera for data collection in the field.

Hill and Clayton (1985) give a comprehensive review of nocturnal studies and observations prior to 1985 that were made with image intensifiers for a wide variety of animals, including ghost crabs (*Ocypode quadrata*), boat-billed herons (*Cochlearius cochlearius*), and frog-eating bats (*Trachops cirrhosus*). Hill and Clayton (1985, p. 7) also point out a number of other applications: "Other applications of image intensifiers have included such diverse tasks as observing mother-infant behavior in chimpanzees and gorillas, identifying and monitoring whales, conducting censuses of nocturnal seabirds, locating and tracking schools of fish from aircraft, observing nesting behavior of alligators and sea turtles, and studying patterns of sleep in humans."

An image intensifier was used by Swanson and Sargeant (1972) to study the feeding ecology of prairie ducks on prairie wetlands in south central North Dakota. All major species of ducks breeding in the prairie wetlands were observed to feed at night between sunset and midnight, which was correlated to the emergence pattern of midges and mayflies. Six species of both adult and immature birds – blue-winged teal (*Anas discors*), green-winged teal (*Anas carolinensis*), American widgeon (*Mareca Americana*), shoveler (*Spatula clypeata*), gadwall (*Anas strepera*), and mallard (*Anas platyrhynchos*) – were observed to feed predominantly on invertebrates, which accounted for 99.8% of the foods consumed.

During a behavioral study of the activity budget of individual beavers (*Castor Canadensis*) in northeastern Minnesota, several obstacles were identified as critical to observing beavers throughout their daily activity period. Obstacles were broadly categorized as those associated with (1) maintaining visual contact, (2) minimizing alarm response to observers, and (3) observing in darkness (Buech, 1985). Night observation initially caused much difficulty. Binoculars were used when light was adequate and during darkness the observers used a military first generation (GEN I) I^2 night vision sight (MK 36-2, NVS). The resolution of the night vision sight was satisfactory under good light; however, moonlight was required to provide adequate light. Starlight by itself was adequate to classify beaver activity when they were in open water but not when they were adjacent to shore in the starlight-shadow of shoreline vegetation (where they spent most of their time). On cloudy nights, the night vision sight gave an unacceptable resolution.

There are several obvious drawbacks to using image intensifiers for animal surveys. The images are formed from low levels of reflected ambient light so that there is very little contrast between the animals of interest and their background. These low-contrast grayish-green images actually exacerbate the problem of visibility bias. However, since image intensifiers allow nocturnal observation, they provide a way to observe many species in their natural surroundings without altering their behavior. The task of detecting animals at night without an I^2 device is much more difficult than observing an animal with the device. Animal movement is a key cue when using image intensifiers to find or detect animals in the field.

The importance of the motion cue was pointed out by Rodda (1992) during observations of the foraging behavior of the brown tree snake (*Boiga irregularis*) on the island of Guam. The brown tree snake is nocturnal but preys on both diurnal and nocturnal species and can reach lengths up to 3 m. It uses both active search and ambush foraging modes to take birds, rats, and lizards but will also take the smaller members of larger vertebrate species. Both monocular and binocular image intensifiers were used to observe the snakes. In the absence of moonlight the snakes were extremely difficult to detect in natural forest or on the ground if they were not moving. The lack of detectability was particularly a problem when the vine-like snakes stopped moving for long periods of time, presumably in the ambush mode.

The nursing behavior of Mexican free-tailed bats (*Tadarida brasiliensis mexicana*) was examined during six nursing seasons in two large maternity cave roosts in Davis Cave, Blanco County, Texas, and in James River Cave, Mason County, Texas (McCracken and Gustin, 1991). Densities, movements, and roosting associations of pups in crèches were documented using a night vision I^2 device and an infrared sensitive video system (I^2 with an attached low-light-level CCTV camera) to observe female-pup reunions and nursing affiliations. The coupling of an image intensifier with a low-light-level video camera permitted the unobtrusive observation of maternity colonies.

Farnsworth and Cox (1988) mated a laser illuminator to an image intensifier that was specifically designed for nocturnal observations of pollination activity by the Tongan fruit bat (*Pteropus tonganus*) or Pacific flying fox. A laser diode with an output wavelength of 810 nm was used because the sensitivity of image intensifiers falls off rapidly near 800 nm and they wanted to be sure that they were using a source of illumination least likely to disturb pollinators yet maximize the sensitivity of the intensifier. This was necessary because for conditions approaching total darkness, such as are encountered in caves or under rain forest canopy on a dark night, an auxiliary source of illumination is necessary. Even under nocturnal starlight, auxiliary illumination can reduce image distortion by allowing the intensifier to be operated at reduced gain. The modification of the intensifier was necessary to match their study goals with the ecology and behavior (Miller and Wilson, 1997) of the Tongan fruit bat.

McMahon and Evans (1992) examined nocturnal foraging in American white pelicans (*Pelecanus erythrorhynchos*) at the Dauphin River, about 50 km from a breeding colony on Lake Winnipeg, Manitoba, Canada. From two to three times as many white pelicans foraged at night than during the daylight hours and the rates of bill dipping and the mean duration of dips were significantly greater at night than during the day. The capture rates were significantly lower during the night than during the day, presumably because of lower visual sensitivity of pelicans at night. Day observations were made using either a 16–36 ×50 spotting scope or 10 × 50 binoculars and nocturnal observations were conducted using a Javelin model #325 infrared scope with an effective range of 300 m on fog and rain-free nights.

A third generation (GEN III) night vision pocket scope (fitted with a 1280 mm f/5.6 catadioptric telephoto lens) was used with a 400,000 candlepower red filtered spotlight by King and King (1994) to observe wildlife at a commercial catfish farm in the mid-delta region of Mississippi. They observed over 20 species of animals (12 species of mammals, 5 species of birds, and 4 species of fish) and reported that with only one exception, the American white pelican, the observed animals were disturbed for only a few seconds by looking toward the red-filtered light source and then resumed their normal activity. The American white pelican bird rafts would move away from the red-filtered light but as soon as it was turned off they would calm down and reform rafts. The wildlife viewed in the night vision device appeared in various shades of green, which allowed the observer to discern light and dark color patterns, thereby making species identification possible. Prey taken by great blue herons could be discerned as channel catfish (*Ictalurus punctatus*) or other species such as shad (*Dorosoma spp.*), sunfish (*Lepomis spp.*), and Mosquitofish (*Gambusia affinis*) at distances (~300 m) across a pond. The identification of species was made under conditions ranging from a dark overcast night to bright moonlight. The maximum viewing range was approximately 2000 m under ideal conditions (clear sky, full moon, and no atmospheric dust or moisture).

Image intensifiers have been used in a number of field observations with minimal disruption to the behavioral patterns of the animals studied. Havens et al. (1995) used PVS-5 night vision goggles for the nocturnal observation of animals in both rural and urban-forested wetlands. Image intensifiers are suitable for observing and monitoring a wide variety of species in a variety of habitats since only reflected ambient light is required to observe them. Bats (*Vespertilionidae*) were the only species observed in both the rural and urban wetlands. Visitation to the urban wetland was dominated by domesticated cats and dogs and by humans. White-tailed deer (*Odocoileus virginianus*), opossum (*Didelphis virginiana*), and barred owl (*Strix varia*) were observed using the habitat provided by the rural wetland. Figure 3.1 shows typical imagery of mammals

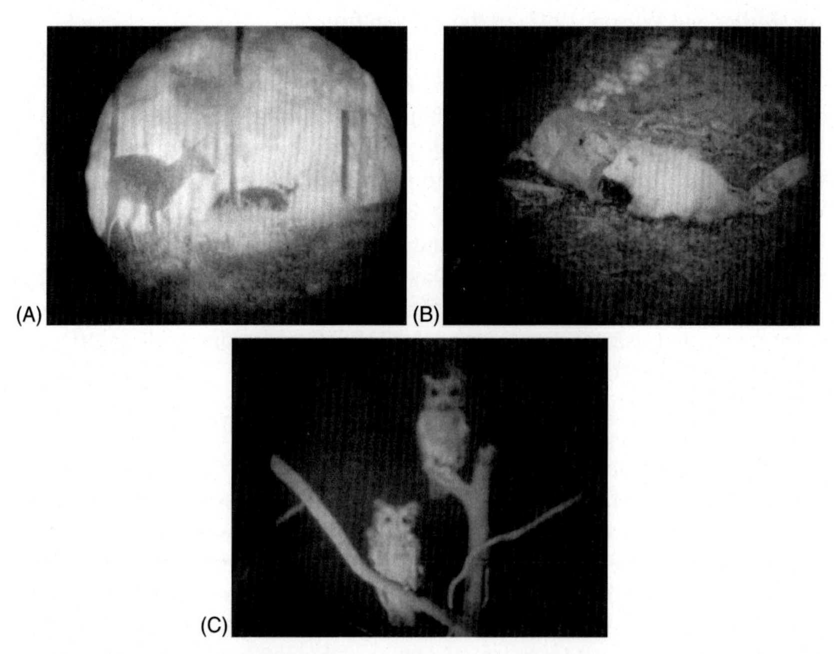

FIGURE 3.1 Photos of animals taken at night using a 35 mm camera and a 3× image intensifier lens in coastal Virginia. (A) White-tailed deer, (B) opossum, and (C) screech owls (*Megascops asio*).

(both large and small) and birds collected by equipping a 35 mm camera with a 3× image intensifier lens.

If researchers are planning to utilize the latest available models of I^2 devices then they should read the paper by Allison and Destefano (2006) that reviews equipment and techniques for nocturnal wildlife studies. They review 53 papers to examine image enhancement (i.e., night vision) and assess trends in nocturnal research techniques. A GEN III night vision scope greatly outperformed first-generation night vision binoculars during a field study on the nocturnal behavior of roosting cranes (*Grus spp.*) on five study sites in central Florida. To get an idea of the improvement of GEN III as compared to GEN I night vision I^2 devices they compare the range at which a 6-foot-tall person could be seen under a set of natural nocturnal lighting scenarios. The ranges for (GEN I and GEN III) are given in meters under the same lighting conditions as: full moon (500 and 800); half moon (375 and 750); quarter moon (0 and 700); starlight only (0 and 500); and cloudy starlight (0 and 200), indicating a vast improvement at low light levels. They report that they experienced no difficulty in distinguishing among species such as sandhill cranes (*Grus canadensis*), whooping cranes (*Grus americana*), and wood storks (*Mycteria americana*) at roost sites, and in addition they identified a number of other species at the crane roost sites including black-crowned night herons (*Nycticorax nycticorax*), raccoons (*Procyon lotor*), northern river otters (*Lontra canadensis*), alligators (*Alligator mississippiensis*), feral pigs (*Sus scrofa*), striped skunks (*Mephitis mephitis*), white-tailed deer, grey-fox (*Urocyon cinereoargenteus*), and feral dogs (*Canis familiaris*). Furthermore, observing the plumage details of whooping cranes did not pose a problem for the GEN III scope. They report that they had trouble-free usage of the GEN III for the 2-year study and suggest that night vision technologies are an exceptional, nonintrusive, functional tool for wildlife ecology studies. They caution that even the best equipment will have problems or issues with contrast, inclement weather, and large group size and density. Regardless of the specific method used and the inherent challenges they believe that GEN III, American manufactured night vision equipment can provide insight into the complete history of animals and can promote a more comprehensive approach to wildlife studies. A good review of night vision technology, especially image intensifiers, is provided by Biass and Gourley (2001) where they discuss the progress on GEN III and GEN IV devices.

Studying the activity of animals during the diurnal period may limit the quality and quantity of the data. Significant differences in animal behavior or habitat use may occur between diurnal and nocturnal periods (Beyer and Haufler, 1994). Artificial light has been shown to alter the dispersal patterns of juvenile panthers (*Felix concolor*) (Beier, 1995). This suggests that making use of night vision equipment (image intensifiers or low-light-level cameras) may be well suited for covert, nonintrusive observations of panther activity.

A remote night vision 24-h monitoring station was used for preliminary investigations to observe Louisiana black bear (*Ursus americanus luteolus*) den activity in the Atchafalaya Basin in Louisiana and to monitor game trails in the Great Dismal Swamp in Virginia (Havens and Sharp, 1995). The system used a daylight camera bore-sighted with a low-light-level camera or image intensifier camera with a motion detector/heat sensor trigger. The entire camera system was contained in an $8 \times 8 \times 12$ cm environmentally-protected container. The system included a transmitter to send the video signal to a time-lapse video recorder set up in a secure location and a light level sensor that shifted between daylight and night-time viewing. The system utilized a solar charger and battery power supply and was equipped with an infrared light source for increased illumination during nocturnal observation periods. When the trigger tripped, the camera would record a 1-min segment of video or until the activity ceased.

LOW LIGHT LEVEL CAMERAS

The cameras, films, and related components that formed the basis for traditional photographic implementations of data acquisition (now referred to as analog technologies) have been all but replaced by digital cameras and HD camcorders that are capable of high-resolution imagery (even in low-light-level conditions) that is frequently needed in the field to find, observe, and study wildlife. Collecting high quality, low-light-level photos and videos is challenging but basically requires the juggling of two parameters: image blur and image noise. Unfortunately, image blur and image noise are intertwined and if we overcompensate for the effect of one it raises problems with the other. Image blur can originate from two sources: the motion of the animal and the motion of the camera. While it is difficult and/or impractical to control the motion of the animal, you can compensate for camera motion in several ways. If it is handheld, raising the shutter speed and selecting a larger aperture to counter the shorter exposure time will help reduce blur. A larger aperture lets in more light, but it means less of the scene is in focus so there is less detail in the images. Image noise is caused by electrical interference and is the digital equivalent of film grain; it is an inevitable by-product of digital photography. Basically, a large sensor is better than a small sensor for low-light photography because larger sensors tend to generate less image noise since a larger surface area will always collect more light. The other important factor is the lens and for video, larger is better.

ISO is the level of sensitivity of digital cameras to available light and has replaced the previous standard for film speed ratings known as ASA, which was a parameter built into the film and could only be changed by changing the film. ISO is a natural consequence arising from modern digital photography since in most cameras the ISO setting can now be changed with a flip of a switch. The image sensor is the most important part of a digital camera and it is responsible for gathering light and transforming it into an image. With increased sensitivity, a camera's sensor can capture images in low-light environments without having to use a flash. Every camera has a base ISO, which is the lowest ISO number of the sensor (typically 100–200) that can produce the highest image quality with the least noise and the widest dynamic range. Higher ISO settings are created by adjusting the gain of the sensor. With more gain, the sensor becomes more sensitive, and you can capture photographs at higher shutter speeds with less light. However, this is commensurate with a decrease in signal-to-noise ratio, which is visualized as a grainier, rougher image.

This implies that it would be best to use the base ISO to get the highest image quality; however, it is not always possible to do so, especially when working in low-light conditions. If noise introduced into the collected imagery is not going to compromise the outcome of your field observations then the lower ISO setting (e.g., 200) would be best. On many newer digital single lens reflex (DSLR) cameras, there is a setting that automatically selects the ISO and it performs well in low-light environments. In these cameras the maximum ISO can be set to a certain number (say 800) so when the ISO is automatically increased based on the amount of available light, it does not rise above the set limit. This limits the amount of grain in the photography collected over a wide range of low-light-level situations encountered in the field.

Images recorded by cameras and videos vary in quality and size and as such can influence data storage capability. If the images are low resolution or in video format then storage requirements are less demanding. However, for real time observations in the field requiring quality imagery, such as web-based remote cameras or infrared cameras, image storage may become problematic. A particular camera of interest is the so-called "peep" camera (Locke et al., 2012) that consists of a black-and-white video camera with a small 2-lux, 2000 h white light source, a very short set focal length of 3.7 cm, and HFOV of 52^0. This camera, mounted on the end of a 15 m fiberglass telescoping pole, was used to facilitate the interior viewing of red-cockaded woodpecker (*Picoides borealis*) nesting cavities in real time. These cameras are important for two reasons: they minimize impacts on bird behavior and the exposure of personnel to dangerous situations (e.g., climbing ladders to observe nesting cavities; see Figure 3.2).

FIGURE 3.2 **Monitoring tree cavities to determine the presence of red-cockaded woodpeckers.**

A thermal imager can be used as an alternative method to using peep cameras to determine the presence of red-cockaded woodpeckers in nests. Results of these observations and the use of a thermal imager for this application are discussed in Chapter 11.

TRIP CAMERAS

Remote cameras (visible cameras and camcorders, image intensifiers, near-infrared and thermal infrared cameras) are a valuable resource for observing, counting, monitoring, and studying animal behavior. Remote cameras can provide long-term monitoring of animal activity and behavior that would be extremely difficult to obtain in any other way. They can be used to monitor and observe seasonal feeding habits and diet preferences, den activity, nest predation, and nocturnal activity for a wide variety of species.

The review by Locke et al. (2012) points out that the usefulness of remote cameras in wildlife ecology depends on the quality of the study design and the capabilities of the operator. They note that cameras are appropriate in research where: (1) humans would cause disturbance to wildlife behavior; (2) extended observational periods are required; (3) observation must take place in dangerous, inclement, or remote areas; (4) permanent and verifiable data are required; or (5) different capabilities from those of the human eye are required. They provide a comprehensive list of commercially-available actively and passively infrared-triggered remote cameras suitable for use in studying wildlife ecology. Included is a comparison of the advantages and disadvantages of general-use field cameras (single lens reflex and digital single lens reflex), which are discussed in Section "Low Light Level Cameras" of this chapter. Active infrared requires that an infrared beam be broken by the passage of an animal, which triggers the camera. Passive infrared requires that a sensor detect the movement of an animal that has a surface temperature different than its surroundings within the FOV of the camera. This then trips the camera and the animal is captured by photograph or video. In both passive and active infrared setups additional near-infrared lighting can be used to illuminate the animals.

It is extremely important when forming the sampling and/or observation strategy that the ecology of the target species is taken into account, including its movement and distribution in space and time. The positioning of remote trip cameras is critical. The camera should be situated so that it will record the whole animal and distinguishing features of the animal can be identified. Care must be taken when placing cameras along animal trails where animals are likely to travel but that may result in biased data because the camera placement is not random. Remote trip cameras are unobtrusive and can let researchers monitor large areas with a minimum of manpower but equipment can be costly, is subject to theft and vandalism, and can require frequent battery and data storage changes.

Karanth et al. (2004) discuss the conceptual framework needed for animal sampling of elusive mammals in tropical forests. Many tropical forest mammals that occur at low densities are cryptic and have nocturnal and secretive behaviors. They describe the methodology for using photographic trip cameras to sample because typical methods such as visual observations are impractical due to the behavior and/or low densities of the animals to be sampled. The sampling process consists of deploying a number of camera-trap units in the area to be sampled in a manner most conducive to obtaining photographs of the target species. Trap sites and camera locations are selected to provide photographs that identify the animal species. To estimate abundance of a species, the photographs need to be of a quality that allows the identification of individual animals. Cameras placed to record high quality images of both sides of an animal's pelage when the trigger is activated provide a significant benefit in identifying individual animals. Using this method of photographic capture and recapture, Karanth and Nichols (1998) were able to estimate tiger (*Panthera tigris*) densities in India and concluded that their estimates of tiger density were reasonable and generally supported predictions about the influence of prey abundance and prey community structure on the relative abundance of large felids.

Infrared-triggered cameras have been used to census white-tailed deer on a 4 ha area in Amite County, Mississippi, with the conclusion that the cameras have the potential to provide reliable estimates of white-tailed deer populations in dense woodland habitats (Jacobson et al., 1997). An innovative trip beam using an infrared source and detector was constructed to identify predators at artificial nests on the island of Guam (Savidge and Seibert, 1988). Previous attempts to observe predators were based on tracks, hair samples, or feeding characteristics, which proved to be ineffective. After identifying predators, including rats (*Rattus exulans*), (*Rattus rattus*), and (*Rattus nornegicus*), shrews (*Suncus murinus*), monitor lizards (*Varanus indicus*), and arboreal snakes (*B. irregularis*) and matching the feeding characteristics with the predator's photographed identity, nests were monitored and acts of predation were positively identified without the aid of cameras. Wilton et al. (1994) used an infrared activity monitor (Trail Master® TM 1500) in conjunction with a 35 mm camera (TM 35-1) to test whether moose (*Alces alces*) specifically use narrows when crossing large bodies of water. Strengths and weaknesses of the counter/camera system are discussed and a method to test whether or not narrows are preferred is suggested.

Cobb et al. (1996) used Trail Master TM-1500 active trail monitors with TM 35-1 camera kits utilizing modified OLYMPUS® AF-1 fully automatic 35 mm autofocus cameras with a date time imprint feature to validate the Bait Station Transect Survey, a wild turkey (*Meleagris gallopavo*) population monitoring technique used by the Florida Game and Fresh Water Fish Commission. They established research priorities for monitoring wild turkeys using cameras and infrared sensors (Cobb et al., 1997) and provided an outline, including seven primary assumptions or theoretical constraints, under which field work should be applied to ensure consistency and maximum understanding of the relationship between data collected from photographs and the movement and dynamics of the turkey population under study.

The work of Heilbrun et al. (2006) in estimating bobcat abundance using automatically-triggered cameras demonstrated the ability to accurately estimate the density of bobcats *(Lynx rufus)* in south Texas. They used active infrared-triggered TrailMaster 1500 trail monitors to trigger autofocus 35 mm cameras. The cameras photographed animals when they interrupted an infrared beam 15–30 cm above the ground that extended across the animal pathway. The cameras were positioned above the trail monitor receiver to photograph at an angle perpendicular to the expected direction of bobcat travel. They then attempted to identify the individual animals in each photograph by comparing unique spot patterns (pelage differences), facial characteristics, and body condition to those of physically captured or previously photo-captured bobcats. They concluded that the ability to accurately estimate animal abundance of wildlife populations and document certain movements without the need for physical capture has far-reaching management implications for fur-bearing animals and medium to large carnivores. Managers are likely to experience greater capture success and obtain larger sample sizes than typically found with studies involving physical capture, particularly those involving scarce, secretive, or territorial species. The use of remote photography can substantially decrease survey time and effort. Additionally, remote censusing reduces adverse effects that may be caused by more invasive methods, including complications due to capture, energy-intensive or destructive marking techniques, and behavioral changes due to the capture or marking process.

A comparison of infrared-triggered cameras and helicopter counts of white-tailed deer was made on La Copita Research Area in south Texas. Bait stations were used and a layer of complexity was added to the camera surveys by grazing cattle and other large mammals (collared peccaries and feral pigs), which confounded the results Koerth et al. (1997). As a result, no tests for accuracy could be conducted. However, both techniques resulted in a reasonable population estimate for the area, indicating that the camera technique may be a viable option for counting deer in south Texas. Martorello et al. (2001) constructed an inexpensive yet rugged and reliable bait-triggered camera station to sight black bears in capture-resight studies conducted at two sites in North Carolina: Big Pocosin (119 km^2) located on the Neuse-Pamlico peninsula in eastern North Carolina and Fontana (400 km^2) located in the Great Smoky Mountains National Park. While their cameras worked well, they conceded that monitor-triggered cameras could take multiple photographs during a single visit and would automatically advance the film, reduce field visits, and not require baiting. However, they would be more expensive.

Roberts et al. (2006) conducted a comparison of infrared-triggered camera and road survey estimates for the endangered Florida key deer (*Odocoileus virginianus clavium*) within a systematic sampling design. They point to the use of baiting camera sites in previous work to estimate population densities as a possible source of bias. They further point out that traditional methodologies such as spotlight counts, drive counts, strip counts (aerial, thermal, infrared), and mark-recapture techniques can be expensive, labor intensive, or limited to habitats with high visibility. As a result, sampling designs are often altered to obtain estimates in a nonrandom fashion, which lowers the cost and/or effort required to obtain the estimate (i.e., convenience sampling). The study was conducted on No Name Key (461 ha), Monroe County, Florida. The results of their study showed that there was a significant difference between the two methods for all seasons and years and that the infrared-triggered camera estimates were nearly twice those of the roadside estimates. Approximately 79% of all deer sightings were observed on urban roads, which comprised 63% of the survey route. This portion of the road had frequent visitors (tourists) that fed the deer so the road survey (convenience sampling) was also baited, which certainly contributed to bias. They proposed that the bias in sampling area and differences in detectability of marked deer between both methods accounts for the differences in the population estimates in their study. It could be that while No Name Key has a closed and marked population of key deer, the constant intervention by humans has greatly affected their behavior and their spatial distribution, which may well preclude the island as a viable area to do meaningful survey comparisons. Perhaps infrared cameras would be the best available method to assess key deer populations on the outer islands.

RADARS AND SONARS

Radar (radio detection and ranging) is an object-detection system that uses radio waves or microwaves to determine the range, altitude, direction, or speed of objects. For example, it can be used to detect animals (Larkin and Diehl, 2012); sea and terrain landscapes (Barrett and Curtis, 1992; Campbell and Wynne, 2011; Lillesand et al., 2008); and missiles, ships, and aircraft. The radar dish or antenna transmits pulses of radio waves that reflect off any object in their path. The object returns a small portion of the wave's energy to a dish or antenna that is usually located at the same site as the transmitter.

Sonar (sound navigation and ranging) is a detection system based on the reflection of underwater acoustic (sound) waves, just as radar is based on the reflection of radio waves in the air. An active sonar emits ultrasonic pulses using a submerged transmitter that detects reflected sound (echoes) from potential targets such as fish, underwater topographical features, sunken boats, or ships and allows the determination of the target (i.e. size, shape, location, distance, speed, etc.).

While both of these technologies rely on two fundamentally different types of wave transmission (radars-electromagnetic waves and sonar-acoustic waves) they both serve as useful tools. Radars have allowed scientists and geologists to map the topography and substrate features on the surface of the earth. Sonars have made possible the mapping of the ocean floor, leading to advances in underwater seismography, and aided in the identification of underwater energy and mineral resources. Both technologies also serve as tools in the investigative studies of wildlife ecology.

Moon-watching was one of the first methods used by ornithologists to study the migration activity of birds (Lowery and Newman, 1966; Nisbet, 1959).

Observers equipped with telescopes or binoculars viewed the silhouettes of birds crossing the face of a near full moon and then attempted to estimate migratory and flight directions from observed numbers of the birds and their directions. In these observations it was not possible to obtain range information so a uniform distribution of birds was assumed for some slice of space above ground level and was used to calculate migratory intensity. Some improvements were sought by identifying species of known size to estimate the distances and heights above ground level. The main difficulty turns out to be that data-gathering nights are subjected to a number of uncontrollable factors, including the requirement of cloudless nights occurring concomitantly with a near full moon during the migration period. The technique of using a narrow vertical pointed beam of light to illuminate passing birds while viewing with binoculars or telescopes (ceilometer technique) was demonstrated by Gauthreaux (1969) and the technique was developed further by making observations with image intensifiers (Russel et al., 1991) during a visual study of migrating barn owls (*Tyto alba*) at Cape May Point, New Jersey.

Radar, passive thermal imaging, and moon-watching studies were carried out in southern Israel by Liechti et al. (1995) to compare the three methods and to shed light on the potential and limitations of the moon-watch method. The main advantage of radar is that it can determine distance; on the downside, it cannot distinguish between birds and insects, especially at short distances (< 1 km). They determined that for distances up to 3 km all birds crossing the FOV of the infrared imager were detected. This should be expected since the warm active birds are compared against the background of a cold night sky. About one third of the birds were missed by moon-watchers for elevations below 3 km and about half were missed at distances below 1.5 km. They showed for the first time that a good thermal imager can detect even small nocturnal migrants up to at least 3 km away and many birds were detected even beyond 3 km. The maximum distance that they could observe a bird with a 40× telescope was 3.5 km, although radar showed many birds beyond this distance. They argue that the findings improve the application of the moon-watch technique, which is still the least expensive and easiest available tool to observe nocturnal bird migration. They postulate that general guidelines for the method would help to make moon-watching results comparable all over the world.

There are numerous advantages offered by the availability of reliable small radars and the tremendous amount of data from large networks of Doppler radars designed for monitoring the weather. Important applications include monitoring populations of both threatened and overabundant species, locating roosting sites and critical stopover areas for migratory birds, and following movements of flocks that depredate crops. A singular but important drawback of radar is that it does not provide information on species identification; however, this situation has been mitigated somewhat by using thermal imagers to augment data of bird migrations. Gauthreaux and Livingston (2006) devised a technique to accurately enumerate and determine the flight altitude of migrating birds. They combined a vertically-pointed stationary radar beam (marine radar with a 61 cm parabolic antenna producing a beam width of 4^0) and a vertically-pointed thermal imaging camera (100 mm lens and a $5.5^0 \times 4^0$ FOV) so that the thermal imager detected the path of a target in the *xy*-plane and the radar detected the altitude of the target in the z direction. The combined system of a vertically-pointed radar and thermal imager (VERTRAD/TI) was used to collect data at night at Pendleton, South Carolina, and Wallops Island, Virginia, and daylight data collection was carried out at McFadden National Wildlife Refuge, Texas. They point out that the orientation of the system need not be vertical and that once the beam is collimated, the radar and thermal imager could be directed at any angle between horizontal and vertical, making it suitable for many different applications. The data was analyzed using a video peak store (VPS) to display the digital videotapes, which apparently works well if a few precautions are taken. The VPS imagery allowed the identification of bird tracks that were easily distinguished from bat and insect tracks. Furthermore, the wing beat patterns of flying birds clearly produced modulations in the VPS tracks generated from the video clips of the thermal imager and produced pulsations of echoes in the radar beam, suggesting that in the future these patterns might be useful in bird identification.

Sonar detectors provide a wide range of opportunities to investigate the marine environment, including fish and marine mammal surveys. Acoustic survey in fishing is one of the research methods that can detect the abundance of target species by using acoustic detectors. For example, many pelagic fisheries are generally very scattered over a broad ocean and are difficult to detect. Survey vessels equipped with sonar emit sound waves to estimate the density of plankton and fish schools. Generally, the transducer is put underwater, which is linked to an echo sounder in the vessel that records the schools of fish as "marks" on a screen or paper trace. Then the density and number of marks are converted into biomass.

Acoustic survey has an advantage since it can be conducted over greater spatial extents in many habitats with little cost, but the drawbacks are difficulties in species identification and population size counting. Recordings of the swimming behavior and migration of pelagic fish schools were made during conventional acoustic surveys in the North Sea and the Barents Sea (Hafsteinsson and Misund, 1995). These recordings were made using multibeam sonar connected to external devices for data logging and a method for the graphical presentation of swimming tracks of schools was established for analysis of the swimming behavior of the schools. Some bias was introduced in the schools of migrating fish due to the presence of the survey vessel. The bias was more pronounced for schools of herring (*Clupea harengus harengus*) where 20% of the schools avoided the vessel and slightly less for schools of capelin (*Mallotus villosus*) where 10% avoidance of the vessel was observed. Further work by Soria et al. (1996) pointed out that the use of side scan sonar did not help much since the actual volume being sampled by the acoustic beam was not easy to evaluate. They show that vertical sounder information is also biased and suggest a methodological use of multibeam sonar in acoustic stock assessment methods.

Reed et al. (2005) give a review of historical and current research on the use of sonar in mapping, habitat characterization, and fish surveys of the deep-water

Oculina coral reef Marine Protected Area. These reefs, which stretch 167 km at depths of 60–100 m along the eastern Florida shelf of the United States, consist of numerous pinnacles and ridges, 3–35 m in height and capped with thickets of living and dead coral (*Oculina varicose*). They identified extensive areas of dead *Oculina* rubble due in part to human impacts (e.g., fish and shrimp trawling, scallop dredging, anchoring, bottom longlines, and depth charges). Trawling continues to be the primary threat to these ecosystems as evident from photographs of trawl nets and other fishing gear found on the bottom.

Finfish are not the only marine species influenced by human activity. Recently side scan sonar images of derelict crab traps have been used to verify the magnitude and impact of lost and abandoned traps on blue crab fishery in the Chesapeake Bay, Virginia. Havens et al. (2008) used side scan sonar to count and locate derelict fishing traps and assess their extent and accumulation rate. They determined that derelict (abandoned or lost) traps targeting blue crab (*Callinectes sapidus*) continued to catch blue crabs and documented that a number of other marine-oriented species were included as bycatch through the self-baiting process, which continues for years until the pots disintegrate.

Side-imaging sonar techniques and protocols developed by the Virginia Institute of Marine Science were used to locate lost pots for removal in coordination with the Virginia Marine Resources Commission and commercial fishermen (Havens et al., 2011). Crab pots are distinctive in imagery and can be differentiated from other debris based on the square shape (0.6 × 0.6 ×0.6 m). A photograph of a typical side-imaging sonar scan is presented in Figure 3.3, indicating the location of five derelict pots in 8.2 ft of water. The cursor can be

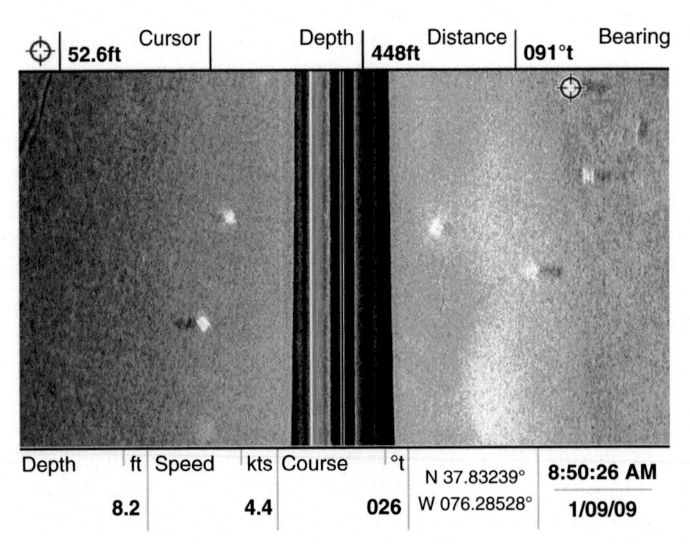

FIGURE 3.3 **A sonar image collected in the lower Chesapeake Bay, Virginia with a Hummingbird® 1197SI side-scanning imaging unit.** *(Photo credit: Center for Coastal Resources Management/VIMS)*

placed on an individual pot in the image and its location is GPS-referenced for future removal (Havens et al., 2011; Bilkovic et al., 2014). Side scan sonar is used to locate and identify other types of lost fishing gear and marine debris.

Grassroots organizations and government agencies have instituted programs to remove marine debris from recreational waters (see, e.g., the American Canoe Association "Stream to Sea" program [http://www.americancanoe.org/?page=StreamtoSea] and the National Oceanic and Atmospheric Marine Debris Program [http://marinedebris.noaa.gov/]).

THERMAL IMAGING

Thermal imaging is a very powerful remote sensing technique for a number of reasons, particularly when used to elucidate field studies relating to animal ecology. Thermal imaging data is collected at the speed of light in real time from a wide variety of platforms, including land, water, and air-based vehicles. It is superior to visible imaging technologies because thermal radiation can penetrate smokes, aerosols, dust, and mists more effectively than visible radiation so that animals can be detected over a wide range of normally troublesome atmospheric conditions. It is a completely passive technique capable of imaging under both daytime and night-time conditions. This minimizes disruptions and stressful disturbances to wildlife during data collection activities. It is capable of detecting animals which are colder, warmer, or the same as their background temperature because it does not compare temperatures but rather the emissivity of the animal against its background.

While the emphasis of this book is on the counting and observation of wildlife there are other very important applications where remote sensing via thermal imaging can be of use. For example, using thermal imagers in aerial surveys of the landscape for mapping purposes can provide some unique capabilities that cannot be gained any other way. From aircraft heights and at aircraft speeds there are no fundamental problems in achieving ground resolutions down to a fraction of a meter (Stewart, 1988). The main advantage of thermal images over visible aerial photography is that they can sense heat. For example, soil types that are absorbing differing amounts of solar radiation can be mapped as well as shading effects on north/south facing slopes on hilly or mountainous terrain. Shading can also be used to help map features of dry washes, forest edges, fence lines, agriculture fields, drainage ditches, variations in soil moisture, and evaporation and even to determine wind direction in many cases (see Figure 1.2, Chapter 1).

It is interesting to note that Quattrochi and Luvall (1999) identify a similar reluctance on the part of remote sensing scientists to adopt the powerful resources offered by thermal imaging as do we on the part of wildlife scientists engaged in studying and monitoring wildlife populations. Although numerous articles have appeared in the professional literature that have employed thermal infrared (TIR) data for the use in studying specific aspects of landscape-related processes (e.g., evapotranspiration), the direct application of TIR data for assessment of landscape processes and patterns within a landscape ecological purview is lacking. They argue that the use of TIR data from airborne and satellite sensors could be very useful for parameterizing surface moisture conditions and developing better simulations of landscape energy exchange over a variety of conditions and space and time scales. They postulate that TIR remote sensing data can significantly contribute to the observation, measurement, and analysis of energy balance characteristics (i.e., the fluxes and redistribution of thermal energy within and across the land surface) as an implicit and important aspect of landscape dynamics and landscape function.

There are three primary reasons for the lack of enthusiasm to use TIR remote sensing data for landscape ecological studies. First, TIR data are little understood from both a theoretical and applications perspective within the landscape ecological community. Second, TIR data are perceived as being difficult to obtain and work with to those researchers who are uninitiated to the characteristics and attributes of these data for applications in landscape ecological research. Finally, the spatial resolution of TIR data, primarily from satellites, is viewed as being too coarse for landscape ecological applications (e.g., Landsat Thermatic Mapper data at 120 m spatial resolution) and calibration of these data for deriving measurements of landscape thermal energy fluxes is seen as problematic. Interestingly, these reasons are very similar to those given for the limited use of thermal imagers by wildlife scientists in the preface of this book. Quattrochi and Luvall (1999) proposed ways to overcome these misconceptions regarding the use of TIR remote sensing data in landscape ecological research by providing supporting evidence from a sampling of work that has employed TIR remote sensing data for analysis of landscape characteristics.

Thermal imaging technology (see Chapter 7) developed by the military is now available from a number of commercial vendors at reasonable costs. For example, thermal imaging systems, both handheld and airborne units, are now available with sensitivities more than an order-of-magnitude better than the units used in the early experiments devoted to large mammal surveys (Croon et al., 1968; Parker and Driscoll, 1972). Thermal imaging technology provides a method for obtaining complete counts of animals with little risk of behavioral or sampling bias. Chapter 10 provides an extensive review of past thermal imaging studies conducted in the field and laboratory for a number of different applications.

RADIOTELEMETRY

Radiotelemetry systems placed on animals are typically used to track the positions of particular individuals, monitor their behavior to changes in habitat, and gather information on their physiology and current well-being. Selected data can be transmitted to remote receivers for future analysis. Samuel and Fuller (1996) caution that researchers cannot ignore sampling and design considerations in favor of sophisticated analytical methods. Furthermore, they point out that all four aspects of a telemetry study (design, equipment, field methods, and data analysis) are critical to the successful completion of a research study. Radiotelemetry has frequently been used to closely monitor animals and record instances of mortality. Study design, equipment types and attachment methods, field procedures, and analysis of telemetry data are reviewed by Samuel and Fuller (1996). Radio packages or other markers (Brodsky, 1988; Kinkel, 1989) may influence behavior and survival, and telemetry studies usually have been limited by small samples of animals and relatively short durations (Johnson, 1996).

Frequencies used in wildlife telemetry usually range from 27 MHz to 401 MHz. VHF transmitters typically give a ground-to-ground range of 5–10 km, which is increased to 15–25 km when received aerially (Rodgers et al., 1996). Lower frequencies (longer wavelengths that equate to larger transmitting and receiving antennas) propagate farther than higher frequencies since they reflect less when traveling through dense vegetation or varying terrain. The battery is a critical component since the functional life of the transmitter is determined by the battery. The battery is also one of the heavier components in the transmitter package so selecting a battery that will satisfy the experimental goals must be matched with the animal's ability to carry it without any stress.

Radio-tracking brought several new advantages to wildlife research: the ability to identify individual tagged animals and the ability to locate each animal within a group of tagged animals. These advantages have led to the wide application of radio-tracking to study the movement patterns of animals and their relationships to habitat use. Animal movement patterns are of ecological significance because of information gained regarding specific times and areas of habitat use (e.g., feeding and reproductive activity) and seasonal migratory movements. Researchers can track animals in the field through three main methods: homing in (either by ground or aerial tracking), triangulating, and global positioning systems (GPS). Cochran et al. (1965) were among the first to use passive remote tracking through automatic tracking systems.

Homing consists of following a signal toward its greatest strength. As the researcher closes in on the animal, the signal increases and the receiver gain must be reduced to further discriminate the signal's direction. The process of proceeding forward and continually decreasing the gain is repeated until the researcher sees the animal or otherwise estimates its location when sufficiently near. We tested the use of a 3–5 μm thermal imager in conjunction with telemetry to locate dens of radio-collared female Louisiana black bear (*Ursus americanus luteolus*), which were listed as threatened by the US Fish and Wildlife Service in 1992. Once common, biologists estimate the current population to be only 300–400 animals. We used a technique that allowed us to determine the precise location of denned bears. Telemetry was used to home in on bear dens in thick brush consisting mostly of blackberry (*Rubus allegheniensis*) thickets. The area was simultaneously scanned in the direction of the strongest signal with the thermal imager until a thermal image of the female bear and cubs was obtained. The bears were monitored in their dens, which were typically on the ground under a thick cover of blackberry. Monitoring was continued for a period of time to establish that the female was lethargic before a marksman would enter the den and dart the female bear. Blood samples and other vital data were collected and the cubs were then radio-tagged for identification and released back to the den. We found this method to be effective and it added a measure of safety (Havens and Sharp, 1995, unpublished data).

Triangulating involves obtaining two signal bearings from different locations (preferably at angles of about 90° to one another) that then cross at the animal. In practice, it is better to take three or four bearings because antenna directionality is imprecise. Significant error can be introduced if the bearings are not taken in a relatively short period, since the more time that passes, the greater the probability that the animal has moved. This problem can be avoided by researchers simultaneously taking bearings each from a different location. Triangulation locates an animal with minimal disturbance since the researcher can be far from the animal while obtaining a bearing. Note that the same technique could be used here as was used in the homing technique, except that there could be considerably less work and perhaps fewer people involved, so there would be less disturbance to the animals as well. Also, if the target animals are bears or panthers with young it could be much safer for the field scientists.

GPS tracking of animals is the latest major development in wildlife telemetry. It uses a GPS receiver in an animal collar to calculate and record the animal's location, time, and date at preselected programmed intervals, based on signals received from a special set of satellites. This gives researchers greater location accuracy (within a few meters) and decreases disturbances to animals when compared with VHF telemetry. Data storage and retrieval options include onboard storage for retrieval at a later time followed by downloading of data. Remote downloading options include via satellite or a portable receiver.

Nelson et al. (2004) used GPS to track white-tailed deer migration in northeastern Minnesota. GPS radio-collars were used to document dates and times of migration onsets, distances migrated, duration of migrations, daily rates of migration, 1-hr rates of travel, and patterns of travel. The deer traveled nearly a straight line course as they migrated, deviating only 1.2–4.0 km from the direct track toward their home ranges. They documented that migrating deer lingered at intermediate ranges. Individual deer were found to pause frequently for periods ranging from 1 hr to several days and while they never directly observed deer while they paused, they presumed that they were foraging and resting before traveling again. However, the large variation in the observed times between pauses and lengths of pauses had no simple explanation without some supporting direct observation. They suggested that perhaps individuals had differing energy budgets to contend with or had encounters with predators, and that weather conditions and/or fatigue may have played some role.

Compared to VHF tracking, GPS is expensive and requires more maintenance but prices are coming down and packages are getting smaller and more robust. Since GPS is the most accurate form of tracking except for visual observation it should be used if the program can support the cost. A cheaper alternative may be to use VHF and confirm with thermal imagery to maintain accuracy with limited funding available. This latter option would be preferred if a large number of animals are to be tagged since the cost of a GPS collar is about ten times that of VHF collars.

As mentioned earlier, Beier (1995) monitored radio-tagged juvenile cougars (*F. concolor*) in fragmented habitats to gain information on the spatiotemporal dispersal patterns of individuals from their natal range upon reaching the age of independence. The study area consisted of a portion of the Santa Ana Mountains in southern California comprised of three habitat corridors and several habitat peninsulas created by urban growth. The objective was to determine the factors that moderate or influence the dispersal patterns of juvenile cougars. The female initiated the dispersal by abandoning the cubs within 0–3 km from the edge of her home range. Movements were monitored and used to determine the use of natural and manmade corridors to travel to and from transient home ranges. The behavior of the juveniles in response to urban-wildland interfaces was documented. All travel in corridors and habitat peninsulas occurred at night. During overnight monitoring, the dispersers usually avoided artificial lights when in a corridor or peninsula. Dirt roads and hiking trails were especially favored as routes through dense chaparral. Adult and dispersing cougars occasionally bedded for the day 20–100 m from trails heavily used by hikers, bicyclists, and equestrians. Dispersers showed no aversion to parked vehicles, occasionally walking within 2 m of a researcher sitting in a vehicle on a dirt road with the engine off. They would also pass within 30 m of isolated homes and buildings with no outdoor lighting.

Beier (1995) noted that traffic can kill and injure dispersers and may preclude an at-grade crossing, recommending that where a heavily-used road crosses a corridor an unlighted bridged underpass would be a suitable option. Vehicle accidents within the corridor could have been prevented if the freeway had fencing to guide the animals into the underpasses. His observations showed that dispersers exploring new terrain and navigating corridors avoided city lights and oriented toward dark areas. He suggested that a corridor designed for use by cougars should be more than 100 m wide if the total distance to be spanned is <800 m and >400 m wide for distances of 1–7 km. Additionally, he suggested that dispersing cougars will use corridors that have less than 1 dwelling unit/16 ha, are located along natural travel routes that have ample woody cover, and include underpasses that lack artificial lighting and are integrated with roadside fencing at high-speed road crossings.

For decades wildlife habitats have been fragmented primarily by roads, manmade structures, and other activities. Wild animals need vast areas to accommodate their needs for food, water, reproduction, and safe cover. They also need to be able to safely pass through fragmented habitats to access these large areas of suitable habitats. The more we humans encroach on critical wildlife habitats the more widespread this fragmentation becomes and the more difficult it becomes for wildlife to safely disperse to sustainable habitats. In the short term it is expensive to build corridors around choke points or to construct/retrofit underpasses to negotiate road crossings but it may well prove to be inexpensive in the longer term. Ironically, both humans and wildlife share in the expensive, all too common, and sometimes deadly confrontations resulting from habitat fragmentation.

Using its claims data, State Farm®, the nation's leading auto insurer, estimates that 2.4 million collisions between deer and vehicles occurred in the United States during the 2-year period between July 1, 2007, and June 30, 2009 (100,000 per month). That's 18.3 % more than 5 years earlier. To put it another way, one of these unfortunate encounters occurs every 26 s (although they are much more likely during the last 3 months of the year and in the early evening). The costs associated with vehicle/animal collisions on US highways have become a major concern for insurance companies and others. Cougars are especially prone to fatality due to collisions with vehicles. While studying cougars in fragmented habitats located in urban southern Florida, Maehr et al. (1991) reported that 47% of cougar fatalities were due to vehicles. In a 1990–1992 study of juvenile cougar dispersal patterns in fragmented habitats in the Santa Ana Mountains of southern California, Beier (1995) reported that three of nine juvenile cougars were killed by vehicles, one was shot, and three died of natural causes. Only two survived to establish stable home ranges.

We must deal with the habitat damage already present in addition to improving land management and finding ways to reduce habitat fragmentation (the ideal solution). It is expensive to create underpasses for safer animal crossings, so in the short term the task becomes one of warning motorists when animals pose a danger by being on or near roadways. Passive signage is in place along most roadways to warn motorists of animal crossings but these are not adequate, as indicated by the large increases in animal/vehicle collisions. One improvement that may help to reduce these collisions could be implemented by making the signage active (flashing lights), which would be more noticeable by motorists during the evening hours and at night. This requires that the animals be detected on or near the roadway at well-used animal crossings. The detection system then automatically triggers the signage to flash a warning to oncoming motorists that an animal is in/near the roadway.

The following briefly describes a warning system that could possibly be used to alert motorists to the presence of large mammals on or near roadways. A low-cost, solar-powered thermal imager mounted on/in an artificial tree (to minimize tampering or vandalism) or other structure that provides a clear line of site of a known animal crossing can continually collect imagery of the fixed scene. When a large mammal such as a deer or cougar (Figure 3.4) enters the FOV of the imager, the semistatic thermal scene is significantly altered and a signal is relayed to a display system (signage) to warn motorists.

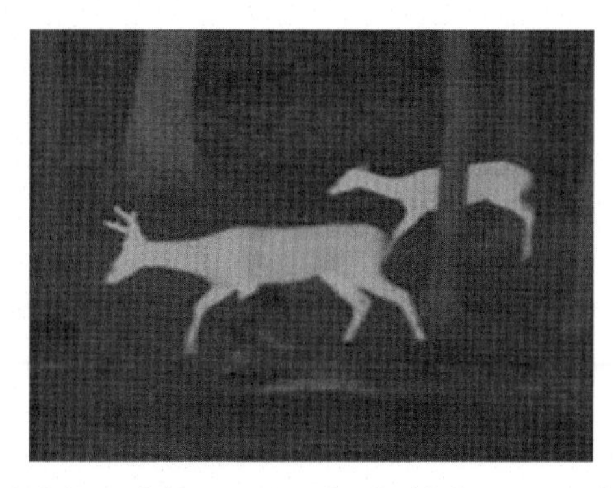

FIGURE 3.4 **A typical thermal signature associated with deer at a road crossing.**

Thermal imagery and radio telemetry were used to concomitantly track animals in the Florida Everglades. Searches were conducted in the Florida Panther National Wildlife Refuge (approximately 9500 ha) and the Big Cypress National Preserve (approximately 295,000 ha) in southwest Florida to verify if radio-collared panthers (*Puma concolor coryi*) could be detected using a handheld 3–5 μm thermal imager during an aerial survey (Havens and Sharp, 1998). The method employed the use of a Cessna fixed-wing aircraft equipped with telemetry receivers to determine the general location of individual panthers on the ground. The search helicopter (Bell Ranger), equipped with a 3–5 μm handheld thermal imager, would then survey the area indicated. On the day these surveys were conducted there were three panthers present in an area that did not require prohibitively long flight times from one animal to another. All three panthers were located almost immediately upon flyover with the thermal imager, although one thermal signature was identified only after a review of the videotape and another was recorded for a short period of time before the panther moved out of the FOV of the imager into dense undergrowth. The observation time was kept to a minimum and distances to the panthers at a maximum to avoid excessive disturbance. These images were collected in the presence of heavy tree canopies and dense ground cover such as saw palmetto. The image shown in Figure 3.5 shows a female panther (with radio collar) that is bedded down about an hour after sun-up.

The same technique was used by Havens and Sharp (1995) when locating Florida panther (*Puma concolor coryi*) dens in the Big Cypress National Preserve. Florida panthers are usually found in pinelands, hardwood hammocks, and mix swamp forests. The den of a radio-collared female panther with kittens was located by telemetry from a helicopter. She was denned in a medium-sized (~10 acres) hardwood/pine hammock covered with dense scrub. Figure 3.6 shows the simultaneous interrogation of the hammock with the radiotelemetry and thermal imaging camera, where the researchers are approximately 60–80 m from the edge of the hammock, homing in on the den site.

FIGURE 3.5 **Thermal image of a radio-collared Florida panther lying in saw palmetto, taken in a helicopter at approximately 180 m agl at 74 km/h.**

The use of transect surveys to determine densities for animals with large home ranges, such as the Florida panther, may be especially difficult because of the large areas that must be surveyed and the small numbers of such animals. However, isolated areas could be surveyed for certain species or, if signs such as tracks (Figure 3.7) or scat indicate that a target species is in a certain area, a thermal imager could be effective in quickly locating the animals. In areas where the habitat has been fragmented by anthropogenic changes to the landscape imagers could be fitted with trip beams to monitor usage by nocturnal or cryptic animals. Thermal imagers could also be used to watch for dispersing juveniles at narrow corridors that connect areas of suitable habitats.

Radio-tracking of animals is often used to test the validity of animal counts or to determine the visibility bias introduced in aerial surveys (Samuel et al., 1987).

FIGURE 3.6 **Simultaneous searching for a Florida panther with radio telemetry and a thermal imager.**

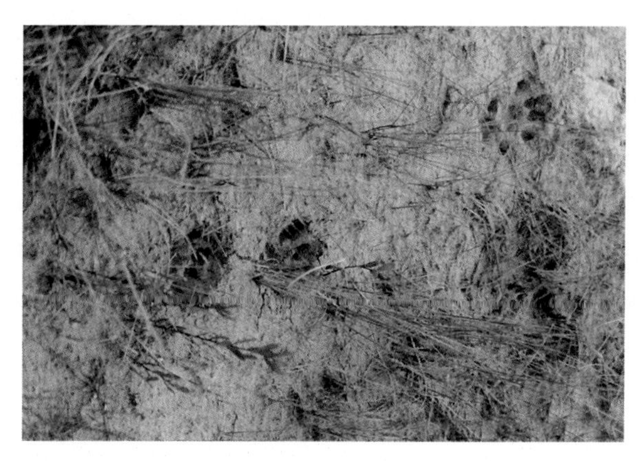

FIGURE 3.7 **A photograph of Florida panther tracks near a hammock in the Big Cypress National Preserve, Florida.**

Radio- collared elk (*Cervus elaphus*) were used in north central Idaho to assess the importance of visibility bias during aerial surveys. The radio collars were used to monitor elk groups during winter helicopter surveys to determine if elk were observed or missed during counts. They determined that visibility was significantly influenced by vegetation cover and group size but not so much by snow cover, search rate, animal behavior, or different observers. They used the data obtained by radio-tracking to develop a detectability model that would predict the probability of observing elk groups during winter aerial counts.

IMAGE INTENSIFIERS OR THERMAL IMAGERS?

Several comparisons of the performance of image intensifiers and thermal imagers have been conducted. Belant and Seamans (2000) addressed the problem of deer-aircraft collisions on airport runways by comparing three techniques for the detection of deer at night. They compared the effectiveness of thermal imaging, spotlighting, and I^2 night vision goggles to monitor the abundance of white-tailed deer along a 10 km route in Ohio during 12 nights in the winter and summer of 1997. They concluded that the thermal imager provided the best overall detectability of deer (825 in winter and 570 in summer) as compared to spotlights (716 and 445) and night vision goggles (243 and 152). Also, the thermal imager was less affected by inclement weather than spotlights and was not obtrusive. They suggest that biologists working in suburban areas or on airports should use thermal imaging technology to detect deer in areas where a spotlight would be inappropriate. Under the conditions tested, they did not recommend using the GEN II goggles to "detect" white-tailed deer at night.

Note that this work was to determine the detectability under conditions that would be encountered on airport runways and their recommendations took into

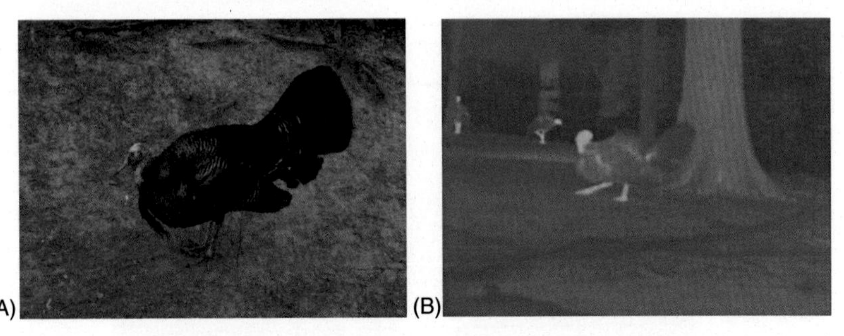

FIGURE 3.8 (A) A photograph of a male turkey and (B) a thermal image of a male and two hens.

account the additional factors of minimizing disturbances to pilots and the effects of weather. The point is that detection was the key objective of the study. In many applications biologists may need to be able to distinguish the pelage or plumage of groups of animals to aid in species identification. Thermal imagers detect heat and if an animal is radiating uniformly (see Figure 3.4) with uniform intensity it may be difficult to determine features in its pelage or plumage. In situations where the surface coloration differences need to be observed, an intensifier may be the instrument of choice (e.g., Allison and Destefano, 2006). Figure 3.8 compares a daytime photo and a night-time thermal image of male turkey (*M. gallopavo*) that shows the limitation of a thermal image in capturing the coloration and features of the turkey's plumage at night. However, there is no problem detecting the birds nor in determining that they are indeed turkeys since they exhibit a unique thermal signature associated with the thermal emission from their head and legs. Keep in mind that when viewing the video of the thermal imagery there is an additional level of recognition cues available for determining the behavioral patterns of target species.

Thermal imagery can be used to capture some features of the pelage under certain conditions (McCafferty, 2007); however, this is not usually the case and images similar to those shown in Figure 3.2 are used to identify animals. Chapter 9 deals with the topic of thermal image interpretation and discusses the factors that influence their utility. An excellent comparison of thermal imaging and image intensifier performance for improving the efficiency and accuracy of nocturnal bird surveys is given by Lazarevic (2009).

While we are primarily concerned in this book with the substantial benefits thermal imagers can provide to biologists, we point out that image intensifiers provide a unique and inexpensive tool for the nocturnal observation and study of animals that do not radiate thermal energy as a consequence of their metabolic activity such as reptiles, insects, and amphibians. If the methodology for surveying poikilotherms can be structured so that animals are viewed against a uniform nonreflecting background (insects) against the sky, or amphibians or reptiles against a uniformly reflecting surface of water or mud flat, then accurate surveys might be easily conducted with image intensifiers.

Chapter 4

Heat Transfer Mechanisms

Chapter Outline

The transfer of heat is actually the transfer of energy from one object to another. Heat energy originates from other kinds of energy according to the first law of thermodynamics (conservation of energy). The second law of thermodynamics governs the transmission of heat energy. The three modes of heat exchange – conduction, convection, and radiation – always transfer heat energy from a body at a higher temperature to a body at a lower temperature. Phase changes associated with water, such as evaporation, can augment these processes. We need to be keenly aware of these thermal processes as we develop and design a field project that would use a thermal imager for data collection and we need to ask ourselves the following question: How can we use these processes to our advantage when setting up data collection in the field? To answer this question it is necessary to examine the processes that influence the radiative component of heat emanating from the animal and from the background to which it is compared.

BACKGROUND

When we refer to the temperature of the background we are not referring to the ambient or air temperature. We are referring to the apparent temperature of the backdrop with which we are comparing the apparent temperature of an animal of interest. That backdrop can be the surface of a landscape, a body of water, a landmass (bare/vegetated or a combination thereof), the sky (day or night), or any combination of these. The nuances of heat exchange within the background are complex and nearly always in a state of flux with heat transfer due to conduction, convection, radiation, and phase changes that continually modify the true temperature of the background. The true temperature of the background is not as important as the apparent temperature difference between the background

Thermal Imaging Techniques to Survey and Monitor Animals in the Wild: A Methodology
http://dx.doi.org/10.1016/B978-0-12-803384-5.00004-X

and an animal. The apparent temperature is based solely on the apparent temperature differential ΔT_A between the radiative component of the background and the radiative component of an object or an animal.

Conduction

Conduction is the process by which heat is transferred in a solid. If the end of a solid rod is heated, thermal energy is imparted to the atoms or molecules locally at the point of heating. This causes those atoms to become excited (increases their internal vibrations or oscillations). These excited atoms impart some of their newly acquired energy to neighboring atoms that in turn impart energy to other neighboring atoms and so on until the unheated end of the rod begins to warm. In this way the energy of thermal motion (phonons) is passed along from one atom to the next, always from the hotter to the cooler end of the rod. The actual mechanism for the energy exchange depends on the material of the rod. For example, in a good electrical conductor such as metal "free electrons" participate in the exchange of thermal energy as well as the excited atoms. In fact, the basic law of heat conduction is analogous to the law of electrical conduction. Conduction will occur whenever there is a temperature gradient across a solid or whenever objects at different temperatures are in contact with one another. The heat flow is always from the point with the highest temperature to the point with the lowest temperature or from the warmer object to the cooler object. In its most basic form the rate of conduction ($\mathrm{d}Q/\mathrm{d}t$) is determined by the differential form of the Fourier equation

$$\mathrm{d}Q = -kA(\mathrm{d}T/\mathrm{d}x)\mathrm{d}t \tag{4.1}$$

Here $\mathrm{d}Q$ is the heat conducted in the direction x during the time interval dt, k is the thermal conductivity, and A is the cross-sectional area of the conducting surface. The rate of conduction is important in that we will be relying on it to bring thermal uniformity to the background. The thermal conductivity, k (W/m-K), is a material property, and if k is large (copper, 385 W/m-K) or small (air, 0.023 W/m-K) the rate of conduction will be high or low if all else is the same. Real objects in the field will have values of k somewhere in between: rock ~ 4 W/m-K, water ~ 0.6 W/m-K, and wood ~ 0.2 W/m-K.

Convection

Convection is the process whereby heat is exchanged and transported from one place to another through the actual motion of material (liquid, air). When a solid at temperature T_S and a fluid at temperature T_F are in contact there are a plethora of factors involved in determining the rate of heat transfer. At the solid/fluid interface the fluid molecules are heated by conduction from the surface of the solid and in turn these molecules give up heat to adjacent molecules during the mixing process. The mixing can be very turbulent, as in boiling water.

The mixing is caused by density changes in the fluid as a result of the induced temperature differentials within the fluid (natural convection). The rate of heat transfer can be increased significantly by forced flow of the fluid (forced convection). Since many forced flow schemes are possible fluid dynamics will play an important roll in the heat transfer process as well as properties of the fluid and solid, including geometrical and material properties. As a result the modeling and computational methods to account for convective heat transfer are very complex (Jakob, 1949).

For our purposes we can get a feel for convection by combining this complexity into a simple proportionality constant, h. Equation 4.2 gives a simplified expression for the rate of convective heat transfer ($\mathrm{d}Q/\mathrm{d}t$). It is simply proportional to the temperature difference at the solid/fluid interface

$$\mathrm{d}Q/\mathrm{d}t = hA(T_S - T_F) \tag{4.2}$$

In the field T_F would represent all possible fluids (e.g., water, air, mud) that have flow properties. The value of h increases dramatically as forced convection takes over from wind or flowing water from heavy rainfall, for example. Rain will bring the background to a uniform temperature very quickly and forced convection as a result of wind can quickly alter the temperature of objects in the field. Conduction, convection, and radiation are the processes that bring the background into thermal equilibrium such that all material comprising the background is at the same temperature. While we have little control over these processes in the outdoors we must always use them to our advantage. This equates to just picking the correct time for the field study. Practices that will aid the field scientists in choosing this opportune time are taken up in the section discussing background clutter in Chapter 11.

Radiation

Heat radiation is identical to light radiation and is only mentioned here in context with the other modes of heat transfer. The heat transfer mechanisms of conduction and convection are expressed in equations (4.1) and (4.2) and are shown to be dependant on thermal gradients dQ/dT as a required force for heat transfer to occur. Heat transfer by radiation differs from that of conduction and convection in three fundamental ways: (1) it does not require a temperature gradient, (2) it does not require a medium for transport, and (3) it occurs at the speed of light. Whereas conduction and convection require the presence of matter in the form of solid objects, gases, or fluids to accomplish heat transfer, radiation does not. Heat transfer by radiation takes place across a vacuum at the speed of light. All objects radiate heat if their temperature is above absolute zero (0 K). This is the point at which all molecular motion ceases. Since radiation is the only source of energy that is used by a thermal imager to form an image it is treated in detail in Chapter 5. For now, keep in mind that in the field, all objects are radiating heat and this radiation is being reflected, absorbed, and reradiated by all other objects around it. This process will continue indefinitely unless there is some way to isolate the system from all outside sources (e.g., solar radiation). If the system, comprised of n objects at temperatures T_1, T_2, ...,T_n, could be thermally isolated it would come to thermal equilibrium at a temperature T_e, yet the system would still continue to radiate because T_e is not absolute zero.

Thermal imagers can detect objects only when their emitted power P_O is different from the emitted power from the background P_B and this difference is large enough to be within the sensitivity range of the detector. It is important that we have a thorough understanding of the mechanisms affecting the emitted power of objects and their backgrounds since the emitted power depends upon the relative temperatures of the object and the background. In general, these temperatures are constantly changing and choosing the time when they exhibit the greatest relative difference will be a crucial part of formulating a plan adequate for field studies. There are a number of things that can affect the magnitude of the background radiation. In addition to the prevailing meteorological conditions, the background radiation detected by the thermal imager is also strongly influenced by the time of day, location, altitude, and the direction of observation relative to the earth's surface. If an object is at the same temperature as its background then there is no temperature difference. However, a thermal imager may see an apparent temperature difference because it is measuring radiation and since the emissivity of the object is in general not the same as the emissivity of the background there appears to be a temperature difference. These topics are covered in detail in Chapter 5.

Phase Changes

The thermal signatures of objects of interest in a scene are created by an apparent temperature difference between the objects and their backgrounds. These objects are thermally connected to their background and their environment and cannot be studied in isolation since they are continually exchanging heat with their environment via conduction, convection, and radiation. Phase changes (changes of state) refer to the change in a material that carries it from one form to another, such as from a solid to a liquid (melting), liquid to a solid (freezing), liquid to a gas (evaporation), gas to a liquid (condensation), or a solid to a gas (sublimation). Changes of state can be responsible for either adding or subtracting heat within a system. For surveys and observations conducted in the natural environment the phase changes associated with water (primarily evaporation and condensation) will be the most significant in altering the thermal differences between an object and its background. We have all felt the dramatic effects of evaporation at one time or another during our daily activities. The evaporation of sweat from any activity that produces even the smallest amount of perspiration lets us feel the cooling effect from the removal of latent heat.

Evaporation/Condensation

The amount of water evaporating on a global scale is approximately the same as the amount of water delivered to the earth as precipitation. A moist atmosphere is largely transparent to visible light and is an excellent absorber of infrared radiation. The water in the atmosphere, distributed almost entirely (99.5%) in the form of vapor, comprises only 0.25% of the total mass of the atmosphere (Stevens and Bony, 2013). At the top of the atmosphere the solar and terrestrial irradiances are balanced. However, while 74% of the incoming solar irradiance reaches the surface of the earth, the net terrestrial irradiance at the surface is only 26% of its value at the top of the atmosphere. The radiative deficit (48%) is balanced by surface turbulent fluxes of enthalpy, arising mostly from evaporation, which transport warm water vapor from the surface to the troposphere, where it cools and condenses. Evapotranspiration is defined as the combined moisture lost to the atmosphere from soil and open water (evaporation) and from plant life (transpiration). While evapotranspiration is an important variable in understanding the local operation of the hydrologic cycle it is difficult to measure accurately.

Evaporation and condensation of water can play a major role in determining the background temperature and, perhaps more importantly, the uniformity of the background temperature. Here the background temperature is not typically the ambient air temperature measured by a thermometer but rather the temperature of a radiating surface such as bare soil, vegetation, standing water, rocks, deadfall forest litter, or a complex mixture of these that might be a background for a thermal image of interest. As an example of how conduction, convection, evaporation, and radiation come together to establish a relatively uniform background with regard to thermal reradiation, consider the following example. Figure 4.1 is a sketch of a small cross-sectional slice of a porous layer of soil containing air (case I) or as shown with water (case II) between the particles of soil.

Assume that this layer of dry soil (case I) has been heated during the day and its temperature is being monitored after sunset. We find that the true temperature

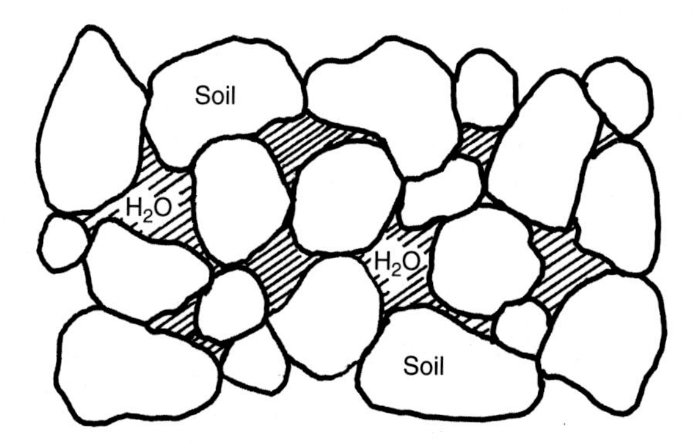

FIGURE 4.1 **Illustration of a cross-sectional segment of porous soil near the surface that can be dry (i.e., no water) or wet (i.e., moist with water filling the void between soil particles).**

of the dry soil that was maximized by solar heating during the day is now being modified by thermal conduction and radiation through air pockets to other soil particles. The thermal conductivity of different soil compositions can vary significantly and influence the magnitude of the overall heating or cooling process. We should expect that the thermal inertia of dry soils will be influenced significantly by soil porosity and that all soils go through a diurnal thermal inertia cycle that is necessarily related to its porosity. On a macroscopic scale there is also a component of heat transfer via conduction through the small pockets of air separating the soil particles, but it is not very effective since air is a poor conductor of heat. We see that reradiation, conduction, and convection at the surface of the soil layer will serve to cool the surface and tend to bring it into thermal equilibrium with the rest of the environment. The heat transferred to the surface of the soil layer by these processes is transported away from the immediate surface via radiation and convection as air passes over the surface.

On the other hand, when we examine the thermal modifications to a section of moist soil (case II), as shown in Figure 4.1, we find that the dissipation of heat is much more effective. In the segment of moist porous soil near the surface the voids between the soil particles are now filled with water and the moist surface is coming to thermal equilibrium. Radiation, convection, conduction, and evaporation all contribute to modify the true temperature of the moist soil as it begins to cool. The heat transfer process for the damp soil is much more efficient since the water in the voids between the soil particles conducts heat on the order of 25 times better than air, so the conductive component of the transfer process is much improved.

Keep in mind that evaporation is a dynamic process that occurs when liquid molecules escape into the vapor phase. Evaporation occurs because all of the water molecules are constantly in motion and colliding with each other and when these collisions take place some molecules gain a bit of energy from other molecules. Molecules near the surface that pick up energy from other molecules begin to travel faster and eventually they may escape from the surface of the water. Since the molecules that escape the wet surface have above average energies, those left behind have below average energies and the damp surface becomes cooler. These molecules are swept away by air passing over the damp surface and the cooling of the surface continues until they come to equilibrium or all of the water evaporates (Feynman et al., 1963). The energy needed to change the unit mass of water from its liquid state to vapor without a temperature change is the latent heat of vaporization, which is large compared to the specific heat of air. In fact, the heat released by condensing 1 gram of water vapor is sufficient to produce a 2.5°C temperature increase in a kilogram of air. The process is so efficient that a variety of heat exchangers have been designed to take advantage of the process of evaporation, which forms the basis for a wide range of efficient coolers and refrigerators used in drier climates. When a wet or damp surface is in thermal equilibrium with the air above it the relative humidity at the air/water interface is unity and at this point net evaporation ceases. In this situation the net evaporation ceases only because it is balanced by returning molecules through condensation.

These same arguments can be made for organic litter comprising the forest floor where bits and pieces of organic matter would replace the porous soil in Figure 4.1 above. In general, the pore space between soil particles is smaller than those between pieces of organic matter and the thermal conductivity of soil particles is approximately an order of magnitude larger than that of organic material. Keep in mind that these arguments are over-simplified yet will still serve to meet our needs for obtaining a more thermally uniform background with which to compare thermal signatures of interest. For a more detailed analysis of heat flow in soils see Monteith and Unsworth (2008). A detailed treatment of thermal conductivity of porous and loose insulating materials is found in Jakob (1949, p. 83).

A good example of the effects of evaporative cooling on a much larger scale can be seen with a thermal imager. Under certain conditions it can actually be used to determine the prevailing wind direction so that one can visually see the effects of evaporative cooling on the background. Figure 1.2, Chapter 1, provides an excellent thermal image showing an aerial view of an agricultural landscape in estuarine lowland that is comprised of a number of fields bordered by hedgerows. The soils of the area are moist and the thermal imagery is particularly sensitive to moisture content near the surface. The photograph of the imagery presented clearly shows that the surface temperatures in the lee of the hedgerows (sheltered from the wind) are considerably higher (lighter in the imagery) than the exposed areas of the fields. The open areas are being preferentially cooled by the prevailing wind (forced convection) and subsequent evaporative cooling and these areas appear much darker in the imagery.

In addition to the inert materials forming the background required for thermal imaging observations, plants such as grasses, shrubs, and trees are also present in the background imagery when a thermal imager is used in the field. In many cases the vegetative components in the background are the dominant features. When we collect data the background must be as uniform as possible with regard to reradiation so that we can obtain good thermal imagery. The same processes of conduction, convection, radiation, and evaporation along with transpiration serve to modify the temperature of these plants. Plants posses a degree of control over the amount of heat gained or lost from their leaf surfaces through changes in the transpiration rate by stomatal control and through the adjustment of solar energy absorbed by leaf orientation. Transpiration rates vary significantly depending on the soil type and drainage conditions (saturation deficit) plus a host of atmospheric conditions. The temperature, amount of sunlight, humidity, wind, and precipitation can all modify the transpiration rate at any given time. The plant cells controlling the stoma cause them to open at higher temperatures so water is released to the atmosphere. As the relative humidity increases the transpiration rate decreases. If air movement (wind) around the leaves is sufficient to blow the humid air away so that it is replaced by drier air the transpiration rate will increase again. The complex interdependencies of the transpiration rate, solar radiation, temperature, wind speed, and saturation deficit are treated in some detail by Monteith and Unsworth (2008). Vegetative surfaces will typically have lower temperatures than adjacent bare soil areas during periods of strong solar irradiation (solar loading) because the plants can lose heat through transpiration. Soil temperatures can easily be 20°C higher than the air temperature while vegetative surfaces will be near or below the air temperature. Tibbals et al. (1964) found that the leaves of coniferous species are cooler than the leaves of broadleaf species when subjected to the same conditions. Precipitation can quickly modify the temperature and radiative properties of the background whether it is bare soil, vegetated, or a spatial distribution of the two.

ANIMALS

The temperature that appears to emanate from an animal in the field depends on its emissivity (covered in Chapter 6). The true temperature of animals in the field can be influenced by conduction, convection, radiation, and changes of state commensurate with environmental factors. Most animals are poikilothermic and their body temperature is regulated by heat gained from or lost to the environment, rather than by heat produced by the animal's own metabolism. Mammals and birds are homothermic and their body temperatures (typically 36–38°C for mammals and 40–43°C for birds) are controlled primarily by metabolic activity, which is modified by heat exchange with the environment. Many invertebrates such as wasps and bees survive very cold climates by resorting to exercising (shivering) to generate heat in times of need to protect the colonies from freezing. Temperature control (thermoregulation) is covered in Chapters 10 and 11.

Conduction

Conductive heat losses are minimal for a standing or foraging animal, such as a deer, but can increase if the animal is lying down (bedded). Consider the example of the thermal image of a sitting cat, shown in Figure 4.2a. In this figure we see evidence of the heat loss by the cat, Sniper, due to conduction by direct

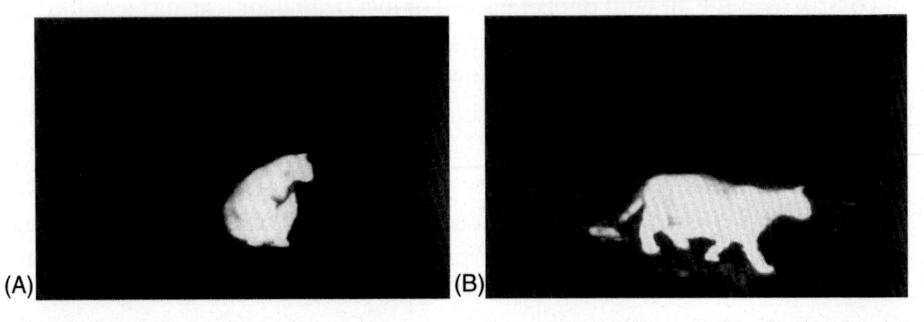

FIGURE 4.2 Thermal images of a house cat: (A) in a seated position and (B) after just leaving the seated position.

contact while sitting on the ground yet the loss is not evident in the thermal images of the cat. Sniper's previous seated location is clearly indicated by the warm spot radiating from the soil behind him in Figure 4.2b.

Convection

Convection can become a problem if heat is removed from an animal such that the apparent temperature difference T_A gets smaller. Note that T_A only depends on the radiated component of heat. If the animal seeks a thermal comfort level by convective cooling, then the metabolic heat loss to achieve the same effect will be diminished. Heat loss due entirely to free convection from an animal such as a deer will be minimal because of the insulation properties of the deer's hair. However, wind passing over an animal can reduce its apparent temperature by removing heat through convective cooling, (Moen, 1968; McCullough et al., 1969; Marble, 1967) and if the animal is in a high state of metabolic activity heat loss can occur through evaporation of sweat as well.

Radiation

In general, radiation from homothermic animals is continually occurring. However, in some animals the insulation properties are so good the radiation is not easily detected by thermal cameras. This type of thermoregulation allows some birds and mammals to maintain nearly constant body temperatures during the year even when there are dramatic changes in temperature from season to season. The winter insulation of the artic fox (*Alopex lagopus*) is so effective that the thermal neutral zone can extend downward to −40°C with relatively little metabolic drain (Vaughan, 1978). In the example below the change to summer pelage has reduced insulating qualities in the animals. Even so, the thermal signature of the two artic foxes shown in the photo (Figure 4.3) clearly indicates that the radiative losses commensurate with their thermoregulatory surfaces yet

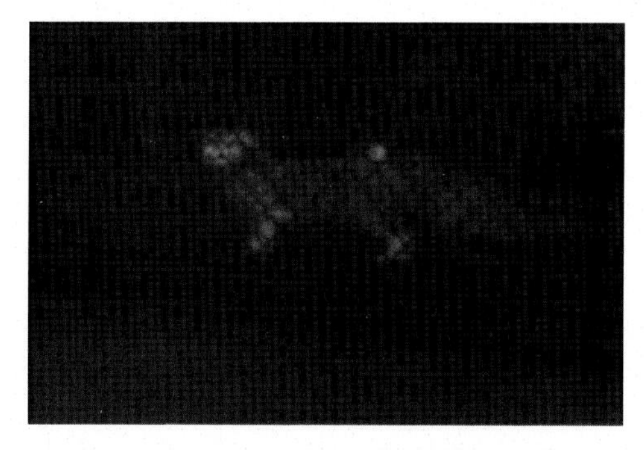

FIGURE 4.3 **Thermal image of two artic foxes (*Alopex lagopus*) taken with a 3–5 micron thermal imager during summer at a small zoo in coastal Virginia.**

these animals are still difficult to observe with thermal imagery. The view of one fox is partially obscured by the fox in the foreground. Here the brighter areas in the photograph are the areas identified by Klir and Heath (1992) as thermoregulatory surfaces and these areas are radiating enough heat to be imaged by the thermal camera. The heat signature for these foxes is probably maximized for a situation where they are not active and in a very warm climate.

A polar bear's (*Ursus maritimus*) body temperature and metabolic rate stay at the normal level even when the temperature drops to −37°C (−34°F). Stirling (1988) reported that polar bears out on the ice are so well-insulated that infrared images did not show pictures of a bear but only a spot just ahead of the bear's head, which was caused by the bear's breath. After a bear warms up walking at 7 kM/hr his body temperature rises from ~98°F to ~100°F and excess heat is liberated at "hot spots" on the nose, muzzle, ears, footpads, and insides of the thighs. Brooks (1970) attempted to acquire thermal images of polar bears by using both 3–5 micron and 8–14 micron imagers but was unsuccessful because of the superior body "insulation" of the bear's pelage. However, he was able to detect the presence of bears with the 8–14 micron imager. If the bear was moving over new snow (1.5 in.) covering the ice pack it would leave a "warm trail" that the imager would record. This distinctive trail would persist for a few minutes after the bear's passing and lead to the position of the bear producing the track. It is likely that disruption of the thin layer of snow caused by the bear's passing altered the emissivity of the surface to be sufficiently different from the uniform snow cover such that it was recorded by the imager.

Phase Changes

While most work done in the past did not consider the importance of phase change contributions to the temperature of animals being surveyed, they can be very important with regard to the apparent temperature difference that includes the background. In general, the contribution to an animal's true temperature from phase changes (evaporation) will be inconsequential unless the animal is in a state of high activity and is capable of perspiring. Best et al. (1982) considered the contribution of heat loss by Canada geese (*Branta canadensis*) due to evaporation to be minimal. The physiology of different animals varies greatly and each survey undertaken must include the temperature-regulating processes that the species possesses. In many cases the thermoregulatory behavior (Chapters 10 and 11) of an animal is responsible for the uniqueness of a thermal signature (spatial distribution of the signature) associated with an animal (see the eyes and medial regions of the legs in the photo of the artic fox) and can be used for not only detecting an animal but also for identification of the animal.

Mammals such as cattle, horses, sheep, and man sweat. The evaporation of sweat cools the surface and ultimately the true temperature of the animal is reduced. However, not all animals sweat. For example, dogs lose moisture from evaporation from the tongue and upper respiratory tract through panting. When a dog is under severe thermal stress it can increase its normal respiratory rate 10-fold (from 30 to 300 breaths per minute) to keep its body temperature in a safe range. A third mechanism for cooling, while not as effective as panting or sweating, has been observed in kangaroos, and to some extent in rabbits, rats, and cats (Schmidt-Nielsen, 1970). These animals wet down the fur of the limbs and belly by licking them and they are subsequently cooled by evaporation of the moisture.

Birds lose heat primarily through conduction and forced convection through their head, legs, and feet and from the areas under their wings. Birds have no sweat glands, and if exposed to elevated temperatures, they increase the evaporation from the respiratory tract similar to the panting dog (Schmidt-Nielsen, 1970).

The temperature of poikilotherms tends to follow the temperature of their surroundings but the animals will seek out environments that are commensurate with their activity and in this way they contribute to their own true body temperature at any given time. In poikilothermous animals, radiation, evaporation, and conduction normally will dissipate the metabolic heat as rapidly as it is produced.

If animals are exposed directly to rainfall or snow then phase changes associated with melting ice and evaporating water can play a significant role in the apparent temperatures of the animals, at least in the short term. In these cases, not only is the true temperature of the animal modified by the thermal energy exchanged due to the phase changes but perhaps more importantly the emissivity is also affected by the presence of water and/or snow on the surface of the animal. The high thermal conductivity of water and evaporation both serve to reduce the apparent temperature difference in the scene. In fact after heavy rainfall all the heat transfer processes are active for both the animal and its background. Conduction, convection, radiation, and phase changes all contribute to changing the apparent temperature difference of the background and the apparent temperature of the animal and the apparent difference between the two.

Here it is important to point out that during the discussion in this chapter we omitted contributions to both the background and the animal due to radiative effects. These contributions are important and will be taken up in Chapter 5. It is obvious that the radiation emitted from the background and from the animal are important since we use the difference between these two to form our thermal images. Not so obvious, however, are the contributions to this process from absorption, reflection, and scattering of thermal radiation from the atmosphere and from objects in the local environment that are not necessarily in the background; this will be covered in the Chapter 5.

Chapter 5

Optical Radiation

Chapter Outline

Thermal imagers can form images of animals that have cooler temperatures than their backgrounds, although in most cases the radiant heat loss exhibited by animals exceeds the radiant heat loss of their backgrounds. The primary factors contributing to heat loss of an animal in the field are conduction, convection, radiation, and in some cases the evaporation of surface moisture. Of these loss mechanisms, only radiation losses are able to directly contribute to the thermal signature detected by the imager. As with any radiation, this energy transfer occurs at the speed of light. The energy emitted from the surface of an animal depends not only on the temperature of its surface but also on its collective surface characteristics (smooth vs. rough, hair, feathers, skin, hide), which moderate the emissivity of the surface. We will see in this chapter that the emissivity plays a major role in the formation of thermal images. The important factors that determine the values of the emissivity typically encountered in the field are discussed in Chapter 6.

Thermal imagers can be considered to be remote sensing devices that can perform noncontact thermal mapping of spatially extensive objects. All materials emit radiation in a broad, continuous spectrum of wavelengths. The amount of radiation available to carry out the thermal mapping depends on the amount of radiation emitted by the objects in the scene and the amount that actually arrives at the detector. Before we launch into an analysis and evaluation of the performance parameters of thermal imaging cameras we must first gain an understanding of what we are attempting to measure. Once we measure it, what do we actually have?

Thermal Imaging Techniques to Survey and Monitor Animals in the Wild: A Methodology
http://dx.doi.org/10.1016/B978-0-12-803384-5.00005-1

KIRCHHOFF'S LAW

When radiation is incident on a surface, the properties of the surface and the wavelength of the radiation determine the amount absorbed, reflected, and transmitted by that surface. Energy conservation requires that the total incident flux (watts) be constant regardless of the surface condition or the wavelength of the radiation. That is, the sum of the absorbed flux, reflected flux, and transmitted flux must be equal to the incident flux.

$$\Phi_A + \Phi_R + \Phi_T = \Phi_I \tag{5.1}$$

By normalizing this relationship we can relate the parameters of absorptance (α), reflectance (ρ), and transmittance (τ) as

$$\alpha + \rho + \tau = 1 \tag{5.2}$$

since an object (such as a rock, tree, or animal) is opaque the transmittance is zero ($\tau = 0$). Light reflected from the surface of this object can be specular or diffuse, depending on the quality of the surface. If the surface of the object does not reflect any of the incident radiation then $\rho = 0$ and the absorptance equals unity ($\alpha = 1$). For the reflectance to be zero the surface would have to be perfectly black (a blackbody); however, such surfaces do not exist in nature. Nonetheless, if the opaque object is in thermal equilibrium with its surroundings the amount of energy absorbed must equal the amount radiated or emitted (emittance) from the surface of the object. If this were not the case then the object would either be heating up or cooling down. Therefore, when objects are in equilibrium with their surroundings the absorptance is equal to the emittance ($\alpha = \varepsilon$) or a good absorber is a good emitter. To get an idea of what the emittance actually is we need to examine the Stefan–Boltzmann Law.

STEFAN–BOLTZMANN LAW

In 1879 Stefan discovered the relationship between the total emission of radiant energy of a blackbody and its absolute temperature. Boltzmann proved this relationship theoretically from thermodynamic principles in 1884 for a perfect blackbody.

$$\Phi_b = A_1 \sigma T^4 \tag{5.3}$$

Here Φ_b is the radiant flux for a perfectly black surface of area A_1 in unit time. The Stefan–Boltzmann law is a basic law of nature and the factor σ is a true natural constant. This expression provides a simple way to compute the power per unit area radiated (radiant emittance) into a hemisphere by a perfect blackbody or a "full" radiator. Note that this says nothing about the spectral distribution of that radiation. However, since there are no perfect black surfaces in nature this relationship can be expressed in a more useful form that relates the total power emitted by a non-blackbody to its true (absolute) temperature measured in Kelvin. The true temperature must be determined with a thermometer, thermocouple, or optical pyrometer.

$$P_t = \varepsilon \sigma T^4 \tag{5.4}$$

Where P_t = total power emitted by the nonblackbody (W/m^2), ε = emittance of the surface (ε = 1 for a blackbody), σ = Stefan–Boltzmann constant (5.67×10^{-8} Wm^{-2} K^{-4}), and T = absolute temperature in K.

When the Stefan–Boltzmann law is expressed in this way it can be used to compute the radiant emittance of many common nonblackbody or "gray" surfaces since the emissivity, ε, is independent of wavelength. The emissivity of a surface is defined as the ratio of the energy absorbed by the surface to the energy absorbed by a blackbody. As a result, all surfaces have an emissivity between 0 and 1. The term "emissivity" is intended for use when the surface of an object is near perfect and "emittance" should be used when the surface is less than perfect (i.e., real materials, objects, animals). However, since the term emissivity (which is actually defined for a unique case) is more commonly used than emittance in the infrared radiation literature (Seyrafi, 1973; Holst, 2000; Kruse et al., 1962; Monteith and Unsworth, 2008) we also adopt its use here. From equation (5.4) we see that the emissivity, ε, turns out to be a very important parameter for thermal imaging since the total power emitted, P_t, depends on the temperature T and a constant σ leaving the emissivity to determine the radiated power from an object that may be at the same temperature as its background. For this reason a detailed discussion of emissivity and the physical parameters that can affect its value is presented in Chapter 6.

PLANCK RADIATION LAW

Historically, the development of the radiation laws of Kirchhoff and Stefan–Boltzmann were arrived at through classical thermodynamics and were confirmed by experiments. However, an explanation of how at a particular temperature the intensity of radiation was distributed to wavelengths in the blackbody spectrum could not be answered by thermodynamics. The temperature and surface condition of the objects determine the total amount and the spectral distribution of the emitted radiation. Perfect emitters are known as ideal blackbodies and are described by Plank's blackbody radiation law. Objects above absolute zero emit infrared radiation as a result of the vibrational and rotational oscillations of the atoms comprising the emitting material. In 1900 Max Planck found that he could account for the spectral distribution of electromagnetic energy (light) radiated by a thermal source by postulating that the energy of a harmonic oscillator is quantized. This remarkable observation marked the beginning of the quantum theory of light (Loudon, 1983).

Planck's radiation law gives the spectral radiant emittance, P_{bb} (W/m^2 μm), of a perfect blackbody as a function of the wavelength and temperature.

$$P_{bb}(\lambda,T) = (c_1 / \lambda^5)[e^{(c_2/\lambda T)} - 1]^{-1} \quad (5.5)$$

Here, λ (μm) is the wavelength and T (K) is the absolute temperature of the blackbody. The constants, $c_1 = 3.741 \times 10^8$ W-μm^4/m^2 and $c_2 = 1.438 \times 10^4$ μm-K, are known as the first and second constants of Planck's law. From this equation we are able to plot the intensity and spectral distribution emitted by a blackbody as a function of absolute temperatures (Figure 5.1). We notice several important features about these curves. The value of $P_{bb}(\lambda,T)$ increases as the temperature increases and we see that for very large or very small values of the wavelength that $P_{bb}(\lambda,T)$ goes to zero.

We further note that the integration of $P_{bb}(\lambda,T)$ over all possible λ gives the Stefan–Boltzmann equation (5.6) for blackbody radiation.

$$\int P_{bb}(\lambda,T)\,d\lambda = \varepsilon\sigma T^4 \quad (5.6)$$

The area under each of the curves shown in Figure 5.1 calculated from the Planck radiation law (5.5) for a particular temperature is equal to the emission given by the Stefan–Boltzmann law at that temperature. An important relationship depicted by the plots (Figure 5.1) of the blackbody radiation curves show that as the temperature increases/decreases the peak of the radiated power

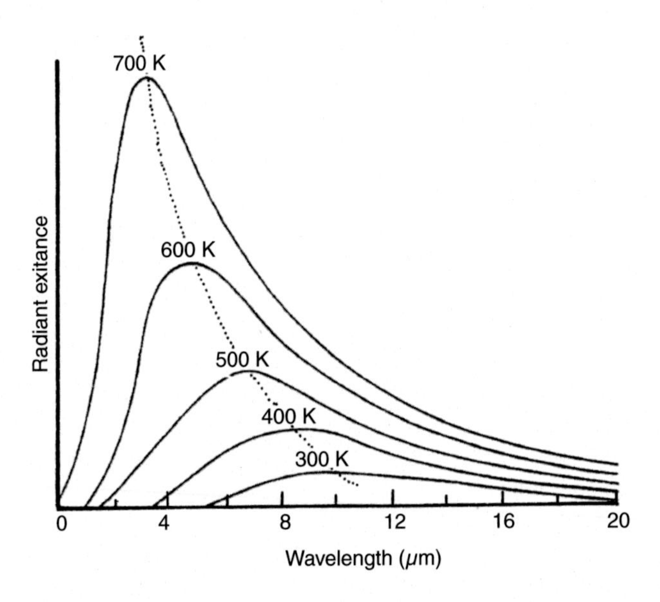

FIGURE 5.1 **The spectral distribution of power radiated from a blackbody source at different temperatures, which also shows the Wien displacement (dotted line).**

curves shift to shorter/longer wavelengths. The wavelength λ_{max} for each peak value of $P_{bb}(\lambda,T)$ at each temperature is given by the Wien displacement law, which is obtained by finding the values of λ_{max} that force the derivative of $P_{bb}(\lambda,T)$ with respect to λ at any given temperature to be zero. That is, when $dP_{bb}(\lambda,T)/d\lambda = 0$ evaluated at λ_{max} we find $\lambda_{max} = 2893/T$ μm. Additionally there is an overall decrease in the emitted radiation when the temperature decreases or there is less radiation emitted and correspondingly fewer photons to collect at lower temperatures. This is an important physical consideration when collecting field data in cold environments and in the past posed problems for researchers using the early line-scanning imagers for animal population surveys in the late 1960s and early 1970s. Since that time the sensitivities of thermal imagers have been vastly improved with advances in focal plane array development and therefore their use at lower temperatures is not nearly as problematic as it was in the past.

For most field applications the temperature range of animals and their backgrounds will be close to ambient temperatures or somewhere between −22°C and 48°C (250 K and 320 K). In this temperature range almost all the radiation emitted in the field by natural objects (vegetation, water, rocks, soil, animals, and even snow) occurs in the broad infrared spectral band of 3–100 μm and therefore natural objects behave like full radiators or blackbodies in this spectral range. Figure 5.2 shows the spectral radiant emittance for several temperatures in this range plotted on a linear scale as opposed to the logarithmic scale presented in Figure 5.1.

Thermal imagers are designed to operate in two specific wavelength bands in the infrared. The first thermal imaging band covers the 3–5 μm or the mid-wave

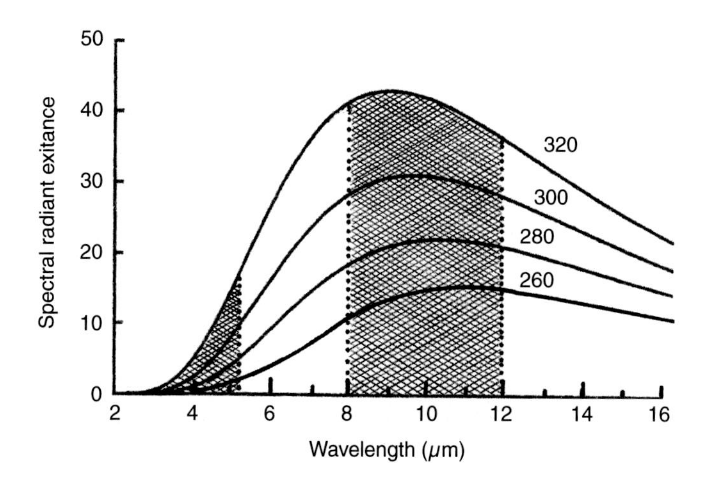

FIGURE 5.2 **MWIR and LWIR bands relative to the atmospheric windows.**

IR band (MWIR) and the second thermal imaging band covers the 8–12 μm or the long wavelength IR band (LWIR). In general, a system is a LWIR system if it is responsive somewhere in the LWIR region and likewise a system is an MWIR system if it responds somewhere in the MWIR region. When imagery is collected with either system it only uses a small portion of the total emitted radiation, as represented by the blackbody curves in Figure 5.2. The MWIR and the LWIR spectral bands are shown as cross-hatched in this figure and note that detectors will respond to infrared radiation in those regions of the spectrum rather than the full spectrum obtained from integrating equation (5.5) over all possible wavelengths.

Thermal imaging systems were designed to operate in these two spectral bands because they are the two primary transmission windows in the atmosphere suitable for thermal imaging applications (Figure 5.3). The effects of selective atmospheric attenuation are evident in the transmittance curves for the two path lengths shown in this figure. The strong attenuation in the MWIR band centered near 4.2 μm is due to absorption by CO_2 and the strong attenuation between 6 and 7 μm is due to H_20 absorption. There are a number of factors that must be considered when choosing a thermal imager for use in either of these two wavelength bands. For example, both MWIR and LWIR thermal imagers can be used to provide a two-dimensional temperature distribution in a scene to provide better visibility through smoke, mists, dust, and aerosols, and since the scene provides its own radiation through emission it does not need to be illuminated with visible or infrared radiation for the image to be recorded. These

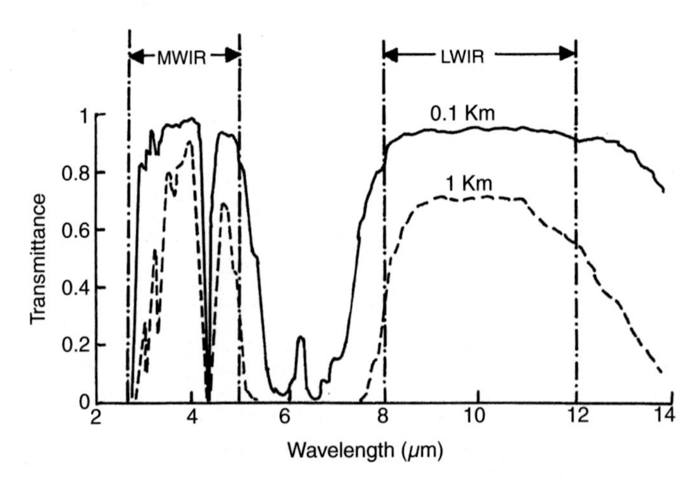

FIGURE 5.3 A sketch of the atmospheric transmittance over a 100 m and 1 km path length. The actual transmittance is continually changing depending upon the presence of airborne particulates (dust, fog, and haze) and gases and the path length from the object to the thermal imaging system. This is a representative curve since in reality the conditions, which give rise to the atmospheric transmittance, can change within minutes.

topics will be addressed in Chapter 8 along with a discussion of the performance of imagers designed for each of these wavelength bands.

BACKGROUND TEMPERATURE

If we consider an isolated macroscopic segment of the environment that contains an inanimate object (e.g., a rock or deadfall) that is warmer than its surroundings from past solar loading we know that the object will radiate heat into the cooler background that surrounds it. As the object continues to radiate it will eventually reach thermal equilibrium with its environment and at this point heat transfer ceases. A similar argument can be made for a cooler object that finds itself in a warmer background. The cooler object will absorb heat from its background until they reach equilibrium. In either case when this point is reached both the object and the background are radiating with the same power density in accordance with equation (5.4). Intuitively we would expect that the thermal imager would not be able to distinguish the object from its background. In general this is not the case because in the real world we are not dealing with blackbodies. While the background may be in thermal equilibrium in that the elements comprising the background (forest litter, soil, rocks, and vegetation) are all at the same temperature, these same elements have different emissivities and it is quite easy to get a good thermal image of the background. In field situations where the radiative power emanating from the surface of biotic objects (animals) is to be compared with the radiative power emanating from their backgrounds the situation is somewhat different for two reasons: animals and their backgrounds generally will have significantly different values for their emissivity and the animals also have a source of internal heat through metabolic activity. Nonetheless, objects can only be detected if there is a difference in the emitted power of the object P_O and the emitted power of its background P_B. Figure 5.4 depicts a scene of a deer and its surrounding background that are in a constant state of change with regard to the distribution of heat. The instant the image is captured by the IR camera the difference in the radiative component of the background and that of the deer is compared. An image is formed of the scene from the radiation emitted by the deer P_O (which has a component due to metabolic activity) and the background radiation P_B, which is reradiating previously absorbed solar radiation.

If we assume blackbody behavior (or identical emissivities) and an object at temperature, T_O, which is embedded in a cooler background at temperature, T_B, then from (5.4) we see that the difference in emitted power is

$$\Delta P = P_O - P_B = \varepsilon\sigma(T_O^4 - T_B^4). \tag{5.7}$$

An imager cannot detect the absolute temperature of the object or the absolute temperature of the background but only the difference between the emitted powers, $\Delta P = (P_O - P_B)$, which could be very small to the point of taxing the

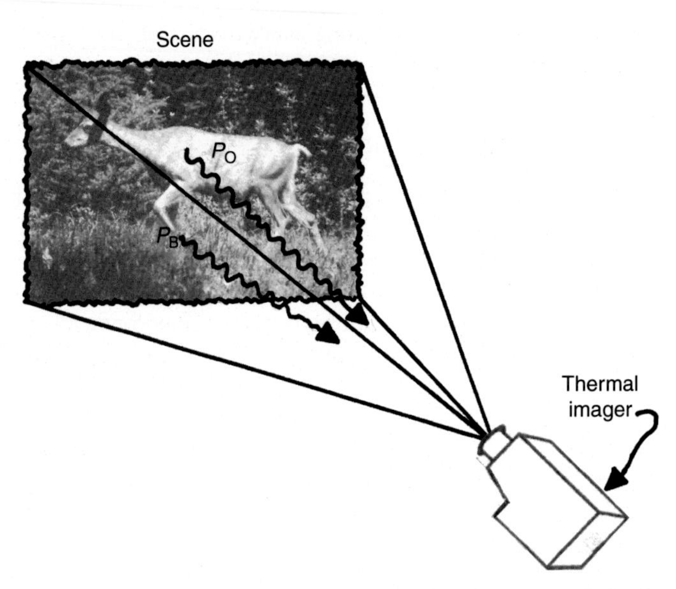

FIGURE 5.4 The two radiative components of interest in this scene are the background, P_B, and the deer as an object of interest, P_O.

sensitivity of the thermal camera. In the natural environment the radiation emitted by an animal or by the background is proportional to T^4 yet the energy exchange between the animal and its environment at temperatures T_O and T_B can be very small and the power differential ΔP becomes proportional to $\Delta T = (T_O - T_B)$. When we substitute for T_O in (5.7) and expand $(\Delta T + T_B)^4$ we get

$$\Delta P = \varepsilon\sigma(\Delta T^4 + 4\Delta T T_B^3 + 6\Delta T^2 T_B^2 + 4\Delta T^3 T_B + T_B^4 - T_B^4).$$

Keeping only first order terms in ΔT (i.e., ΔT is small) we find the emitted power differential to be:

$$\Delta P \sim 4\varepsilon\sigma T_B^3 \Delta T \tag{5.8}$$

from this relationship we see that the emitted power differential, ΔP (the signal producing a response in the imager), is proportional to the cube of the background temperature, T_B^3. This is a very important result that says, "For a given ΔT the signal available to the imager increases as the cube of the background temperature," or warmer background temperatures are desirable. Simply put, a warm forest floor would provide a better background for surveying large mammals than complete snow cover, provided the material comprising the forest floor is in thermal equilibrium (i.e., the background is not only warm but it is also uniform). We are not talking about the ambient or air temperature here but rather the radiant temperature of the background that the object temperature is compared with to get the

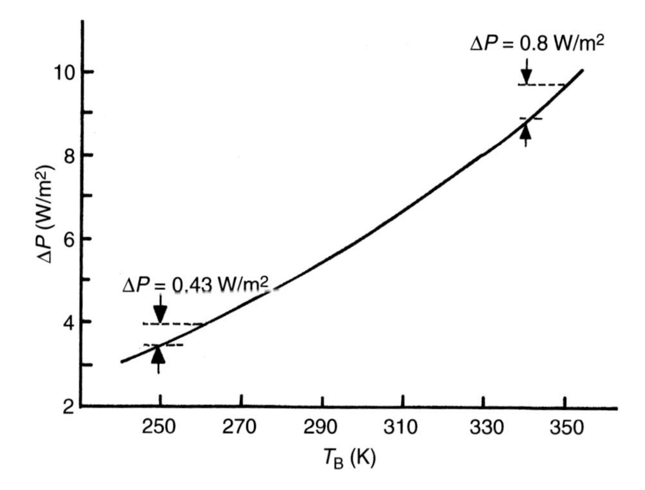

FIGURE 5.5 The T^3 dependence of the background temperature T_B. Note that the radiant power differences are larger for warmer temperatures.

object-background temperature difference $\Delta T = (T_O - T_B)$. The emitted power differential ΔP from equation (5.8) is plotted as a function of the background temperature (T_B) in Figure 5.5 where we used $\varepsilon = 1$, $\sigma = 5.67 \times 10^{-8}$ Wm^{-2} K^{-4}, and $\Delta T = 1$ K to compute the values of ΔP (W/m^2).

Note that the radiant power differences are larger for warmer temperatures and that the power differential at 250 K is approximately half of the power differential at 350 K for the same ΔT. Table 5.1 gives the values of ΔP for

TABLE 5.1 The Emitted Power Differential for $\Delta T = 1$ K for a Range of Typical Background Temperatures

ΔP (W/m^2)	T_B (K)
3.54	250
3.97	260
4.45	270
4.96	280
5.52	290
6.12	300
6.67	310
7.42	320

The values of the emitted power differential ΔP calculated from equation (5.8) for a range of background temperatures T_B that would be typical for fieldwork. $\Delta T = 1$ K, $\varepsilon = 1$, and $\sigma = 5.67 \times 10^{-8}$ W/(m^2/K^4).

sample background temperatures over a range likely to be encountered in the field (−20°C to 45°C).

The reduction in the radiant power difference at lower temperatures is due primarily to the lack of photon production by colder objects and their environments. In order for thermal cameras to function well the detector must be capable of producing a response from the radiant temperature increment in the object scene that is sufficient to overcome the noise of the system. This is a very important performance parameter called the Noise Equivalent Temperature Difference (NETD), which is defined as the temperature difference in the object scene that produces a change in output signal that is equal to the system root mean square noise. The NETD is a function of the background temperature and is usually expressed as a temperature difference at ambient temperature. This and other performance parameters of thermal imaging systems will be discussed in Chapter 7.

APPARENT TEMPERATURE

In field studies we will be dealing with real objects (not blackbodies). A thermal imager is not capable of determining the true temperature of an object, nor can it determine whether the radiation emanating from the surface of the object is emitted, scattered, or reflected radiation. It will respond to all of them collectively. If the scene is composed of objects and a background that are near thermal equilibrium their true temperatures will be similar and the reflected radiation from the background will cause the contrast of the images to be reduced. This being the case, we can now consider the concept of "apparent" temperature and "apparent" temperature difference.

Objects are thermally connected to their environments in a number of ways and as such cannot be studied as separate entities. They are continually exchanging heat with their environments via conduction, convection, and radiation. For the field applications we will be addressing with thermal imagers it is not important that we know the true temperature of the animals or the true temperature of the backgrounds. In fact, we will show in equation (5.17) that the animal and its background can have the same true temperature and the thermal imager will be able to detect them if their emissivities are different. For opaque objects recall that Kirchhoff's Law tells us that good absorbers are good emitters and that poor absorbers are good reflectors. The real situation for fieldwork will be for objects (animals) somewhere in between these two extremes, that is, they will be moderate absorbers so they will in turn be moderate emitters and moderate reflectors. For opaque objects we saw that $\alpha + \rho = 1$, and when these objects are in equilibrium with their surroundings then $\alpha = \varepsilon$ and $\varepsilon + \rho = 1$. This relationship says that if we had a perfect blackbody ($\varepsilon = 1$) then the reflectance would be zero ($\rho = 0$). Since we are working with objects that have emissivities less than unity the reflectance must have a value between 0 and 1. In other words, the radiation that appears to emanate from the surface of a nonblackbody (i.e., all real objects) must include the power that is reflected from its surroundings. For most daytime field studies the chief component of the reflected power will come from reflected solar radiation. Since this component can be quite large (generally much larger than the emitted component) special care must be taken to minimize its effect. This is an important consideration when collecting thermal imagery during the daylight hours and will be covered in detail in Chapter 11.

Apparent Temperature of the Object

We can now express the total radiative power emanating from the surface of an animal by including the emitted $\varepsilon_0 \sigma T_O^4$ and the reflected component of radiation emanating from various elements in the environment surrounding the animal. The reflected components are continually changing as an animal moves through its environment so it is impossible to account for them individually. However, we can lump them together in a general term as $\rho_0 \sigma T_E^4$ where ρ_o is the reflectivity of the animal and T_E is the collective temperature of the environment (surroundings) of the animal. Figure 5.6 depicts the components of the radiative power P_0 that appear to be coming from an animal situated in some natural environment that is being detected by a thermal imaging camera. As we mentioned earlier, the total radiation detected also includes the radiative component from the environment that is reflected from the animal. The key point is

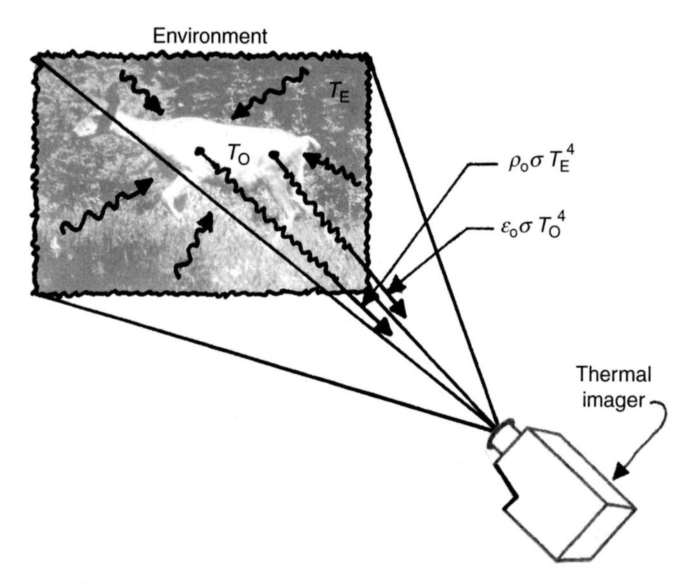

FIGURE 5.6 **The apparent temperature of the deer formed from two radiative components: one due to the radiation being emitted from the deer and one comprised by the radiation emitted from the environment that is being reflected from the deer.**

that the reflected component $\rho_o \sigma T_E^4$ originates from emissions from the animal's environment, which can be significant at times. The total radiated power from the animal can now be written as

$$P_O = \varepsilon_o \sigma T_O^4 + \rho_o \sigma T_E^4 \quad (5.9)$$

the second term $\rho_o \sigma T_E^4$ is just the radiation emitted by the environment σT_E^4 at temperature T_E that is reflected with reflectivity ρ_o from the object. Using equation (5.4) we can express the total power emanating from the object in terms of an "apparent temperature," T_{OA}, and replace ρ_o with $(1 - \varepsilon_o)$.

$$P_O = \sigma T_{OA}^4 = \varepsilon_o \sigma T_O^4 + (1 - \varepsilon_o) \sigma T_E^4 \quad (5.10)$$

We can now write an expression for the apparent temperature of the object T_{OA} in terms of the object emissivity ε_o given by

$$T_{OA} = \left\{ \varepsilon_o T_O^4 + (1 - \varepsilon_o) T_E^4 \right\}^{1/4} \quad (5.11)$$

it is interesting to note that the apparent temperature depends on the emissivity of the object surface ε_o and on both the true temperature of the object T_O and the true temperature of the environment T_E surrounding the object. Even though we do not know the values of the emissivity or the true temperatures, the apparent temperature is still a very useful parameter. If the object were a perfect blackbody ($\varepsilon_o = 1$) then ($T_A = T_O$) or the apparent temperature is the true temperature of the object; it would be a perfect absorber and would not reflect any radiation originating from its environment. Likewise, if the emissivity of the object is zero, then the surface is a perfect reflector and ($T_A = T_E$) or the object appears as the same temperature as the environment. Values of the emissivity between these two limiting cases will determine the apparent temperature in the field.

Apparent Temperature of the Background

In a similar fashion we can write the power emanating from the background P_B as the sum of the radiation emitted by the background and that emitted by the environment that is reflected from the background.

$$P_B = \varepsilon_B \sigma T_B^4 + \rho_B \sigma T_E^4 \quad (5.12)$$

The second term $\rho_B \sigma T_E^4$ is just the radiation emitted by the environment σT_E^4 at temperature T_E that is reflected with reflectivity ρ_B from the background. We can now express the total power emanating from the background in terms of an apparent background temperature, T_{BA}, and replace ρ_B with $(1 - \varepsilon_B)$.

$$P_B = \sigma T_{BA}^4 = \varepsilon_B \sigma T_B^4 + (1 - \varepsilon_B) \sigma T_E^4 \quad (5.13)$$

We now have an expression for the apparent temperature of the background T_{BA} in terms of the background emissivity ε_B given by

$$T_{BA} = \left\{\varepsilon_B T_B^4 + (1-\varepsilon_B) T_E^4\right\}^{1/4} \quad (5.14)$$

Apparent Temperature Difference, ΔT_A

Since we are actually measuring temperature differences with thermal imagers, it is necessary to find the apparent temperature difference ΔT_A. The total emitted power difference ΔP between an object with an apparent temperature T_{OA} and its background with apparent temperature T_{BA} can be written as $\Delta P = P_O - P_B$. Here P_O is given by equation (5.9) and P_B is given by equation (5.12).

$$\Delta P = \sigma(\Delta T_A)^4 = P_O - P_B \quad (5.15)$$

Combining equations (5.9) and (5.12) we get an expression for the apparent temperature difference, ΔT_A.

$$\Delta T_A = \left\{(\varepsilon_o T_O^4 + \rho_o T_E^4) - (\varepsilon_B T_B^4 + \rho_B T_E^4)\right\}^{1/4} \quad (5.16)$$

We can replace ρ_o with $(1 - \varepsilon_o)$ and ρ_B with $(1 - \varepsilon_B)$ and we get an expression for the apparent temperature difference in terms of the emissivity of the object and the background as

$$\Delta T_A = \left\{\varepsilon_o (T_O^4 - T_E^4) - \varepsilon_B (T_B^4 - T_E^4)\right\}^{1/4} \quad (5.17)$$

We now see that ΔT_A will have a value even when the object temperature, T_O, is the same as the background temperature, T_B, if the emissivities of the object and background are different. However, the scene contrast will be reduced. Another consequence of equation (5.17) is that the contribution from the environment T_e^4 is subtracted from both T_O^4 and T_B^4 so that its role is noticed in ΔT_A only through the difference in the emissivities of the animal and its background. The emissivity and the apparent temperature of the animal determine if sufficient thermal contrast exists between the animal and the background so that the thermal imager can detect it. Equation (5.17) does not include any changes introduced by atmospheric effects. The signal-to-noise ratio (S/N) of a system operating during average atmospheric conditions can be written as $S/N = \Delta T\ (\beta^R)/\text{NETD}$ (see Chapter 7). Here the signal is just the thermal contrast, $\Delta T = (T_O - T_B)$ between an object and its background that is attenuated by the average atmospheric transmission at some range R. The noise is given by the noise equivalent temperature difference, which is a function of background temperature. We will discuss the effect of the signal-to-noise ratio on the performance of an imager in Chapter 7.

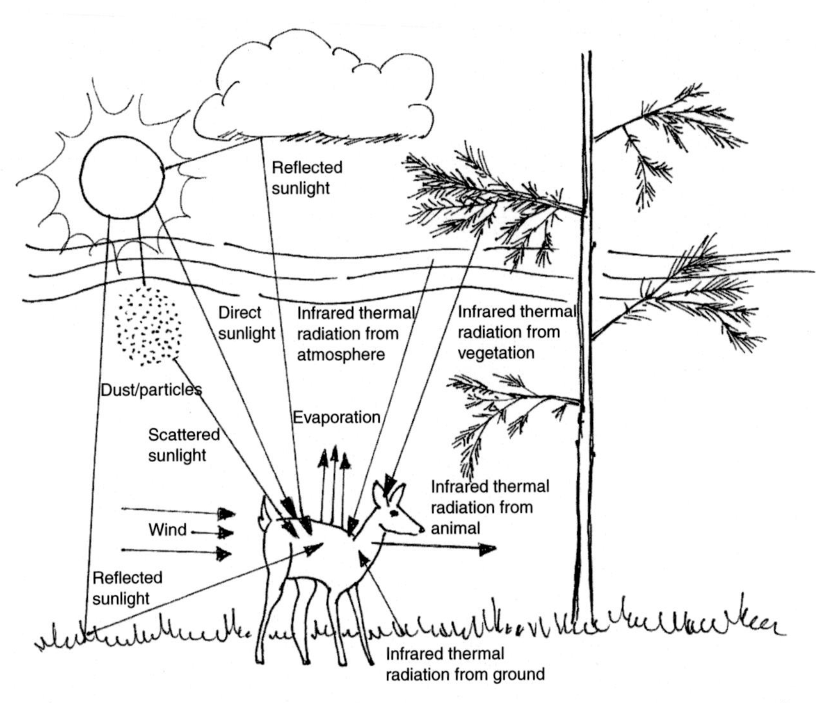

FIGURE 5.7 Some sources represented by the terms of equation (5.16) that can contribute to a thermal image in the field.

Assuming that there are no significant manmade sources of heat contributing to the object and/or background radiation, the four components comprising the expression for the apparent temperature difference are illustrated in Figure 5.7. In this figure the source of the environmental effects is attributed entirely to solar heating for the sake of simplicity, but in reality, there could be a number radiating objects near the object of interest and a host of atmospheric effects that could contribute to the modification of ΔT. For our purposes these sources of radiation are all considered to be emanating from the environment and they are lumped together and shown as a reflected term $\rho_0 T_E^4$ from the animal and as a reflected term from the background $\rho_E T_E^4$.

Other factors that deserve consideration when making a determination of the sources contributing to the thermal signature of an animal in the field would be those that can reduce or even eliminate portions or the entire signature. These blocking and shading effects, including clouds, dust, mist, rain, snow, and any opaque objects, such as trees or shrubs, have been discussed by Parker (1972), Marble (1967), and McCullough et al. (1969). It is important to keep these shading effects in mind when collecting data as they tend to confound the interpretation of field data. We discuss ways to eliminate or minimize the effects caused by shading and shielding by opaque objects that may tend to confound thermal images in Chapter 11.

Chapter 6

Emissivity

Chapter Outline

Emissivity measurements of materials conducted in the laboratory are quite elaborate and nontrivial (see, e.g., Palmer, 1995). As a result the emissivity is typically estimated when thermal imagers are used for measuring temperatures for industrial and manufacturing applications. This uncertainty in the emissivity leads to a concomitant uncertainty in the measured temperature (Holst, 2000, p. 190). That being said, it follows that it is virtually impossible to determine the emissivity of animals in the field because the emissivity depends on a collection of continually changing parameters, including the temperature, viewing angle, and a host of surface characteristics (defects, irregularities, hair, feathers, dirt, condensation, or other features). It is also very difficult to measure the true temperature of objects in the field as it is also continually changing. Trying to measure the emissivity and true temperatures with a thermal imager has been proven to be very difficult (Leckie, 1982). Fortunately, we do not need to know the values of the emissivity or the true temperature of objects T_O and their environment T_E to conduct field studies or carry out surveys of animals because in Chapter 5 we were able to define apparent temperatures, T_{OA} and T_{BA}, to account for the total radiation emanating from the surface of the animal and from its background. We also saw that the emissivity plays an important role in this accounting.

The emissivity ε, while simply defined as the ratio of the energy absorbed by the surface to the energy absorbed by a blackbody at the same temperature, is a very complex quantity. Since it is unlikely that the pelage of animals will behave as selective reflectors or emitters we can take $\varepsilon(\lambda,T)$ to be a constant and assume gray body behavior where ($0 \leq \varepsilon \leq 1$). The Stefan–Boltzmann law was introduced in Chapter 5 and this basic law of nature defines the relationship between the total emission of radiant energy of a blackbody and its absolute temperature. Since this relationship is wavelength independent and since there are no true blackbodies (perfect absorbers) in nature, we saw that the Stefan–Boltzmann

Thermal Imaging Techniques to Survey and Monitor Animals in the Wild: A Methodology
http://dx.doi.org/10.1016/B978-0-12-803384-5.00006-3

law could be written in a very useful form as $P_t = \varepsilon\sigma T^4$ equation (5.4) where ε is called the emissivity and is equal to unity. We also noted that for opaque objects that are in equilibrium with their surroundings the emittance was equal to the absorbance and ($\varepsilon = \alpha$). For all cases in the field the emissivity of animals and their backgrounds will have values between 0 and 1. It can also be shown (Kruse et al., 1962) that $\varepsilon(\lambda) = \alpha(\lambda)$ for each spectral component of the radiation as well. The Stefan–Boltzmann law can now be used to compute the radiant emittance of many common non-blackbody or "gray" surfaces since the emissivity is independent of wavelength. For our purposes the emissivity is now a property of an animal that can depend on a number of factors, including the quality of the surface, the shape of the animal, the temperature of the animal, and the viewing angle.

Collectively, the surface characteristics along with the viewing angle control the emissivity of the animal at any given temperature. Since thermal imagers are noncontact devices they cannot just look at the animal and ignore its surroundings. Typically, animals are approximately in thermal equilibrium with their surroundings such that the amount of energy absorbed equals the amount radiated. This relationship and other local environmental effects and their importance in conducting accurate field studies will be discussed in later sections. In general, animals emit only a fraction of the radiation described by Plank's law. For example, human skin ($\varepsilon = 0.98$) is a very good emitter whereas polished metal ($\varepsilon = 0.1$) is not. Hammel (1956) pointed out the fact that bare human skin that has near blackbody emissivity may range in color from dark to light in the visible part of the spectrum but in the infrared spectral region all human skin is black. He reported that the emissivities of tanned winter pelts of artic animals, which included an array of differing pelage colors from white to dark brown, all showed emissivities $\varepsilon \approx 1$. For example, the specimens included the pelage of a snowshoe hare (*Lepus americanus*) in its white winter coat, a gray wolf (*Canis lupus*) with light gray pelage, and a beaver (*Castor canadensia*) with dark brown pelage. Also included was a willow ptarmigan (*Lagopus lagopus*) in its white winter plumage. While the coloration of an animal's exterior surface may not influence its emissivity the physical condition or quality of its exterior surface does affect its emissivity and in some cases does so substantially. We must keep in mind that even though the emissivity of animal coats is near unity there will be times when there is very little loss of heat radiated to the environment. The temperature of the animal's pelage can be nearly the same as the temperature of the air yet its body temperature will be unchanged. The basis for this insulating mechanism has been argued to be due to trapped air in the animal's pelage, which causes a reduction of conductive heat losses to the surface of the animal because of the low heat conductivity of trapped air in their coats. Simonis et al. (2014) suggested a different mechanism for dealing with the dilemma of not being able to detect animals with high emissivity pelages with a thermal imager such as has been the case with polar bears (*Ursus maritimus*). They showed that when two bodies at different temperatures, say ($Ta > Tb$), are separated by still air, the heat transfer from a to b takes place by two mechanisms (conduction and radiation) and the heat transfer rate by radiation is much faster (an order of magnitude) than that of conduction. They postulate that since the thick polar bear pelage contains several different sizes of hair with a high density of interfaces that the infrared radiation emitted from the animal's skin into the pelage will undergo sufficient scattering, absorption, and reflection to retrodiffuse the radiant heat, thereby minimizing the radiative losses to the environment. The key point here is that the heat losses being prevented by the animal's pelage are radiative losses and this makes the animal difficult to observe with thermal imaging cameras.

A similar observation can be advanced for humans wearing clothing, which is our primary method for regulating our temperature in extreme weather. By examining thermal images of humans wearing several layers of clothing we can see that less heat is emitted through our improvised pelage. As our true temperature drops we add more clothing to reduce heat loss so that we can remain in our comfort range. Note that this appears as a reduction in the apparent temperature so to some extent our radiative heat losses are being curtailed. That is, there are probably both mechanisms at work to reduce heat loss: conductive blocking due to trapped air (lack of convection) and retrodiffusion of radiant heat. The greatest distinction between the two effects is the rate of heat transfer.

Figure 6.1 shows a thermal image of a man in appropriate clothing for a warm and sunny fall day in coastal Virginia. The clothing is doing two things for him: shielding his body from direct sunlight (a good outcome) and limiting radiative heat losses to his surroundings (not necessarily a good thing on a warm fall day). Remember that darker regions of the image indicate that the camera is recording less emitted or reflected radiation in that region of the thermal image. Imagine now if he were to don a heavy wool hat and a hooded parka. He would essentially stop radiating heat to his environment and quickly become overheated as his metabolism drives up his body temperature.

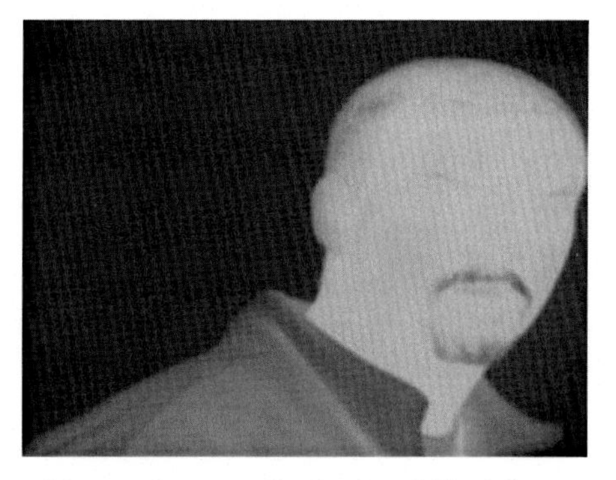

FIGURE 6.1 **A thermal image of a man taken in coastal Virginia on a warm sunny day.**

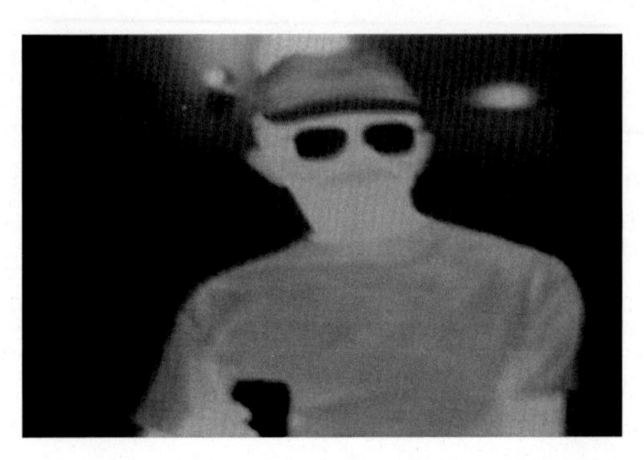

FIGURE 6.2 **A thermal image of a man wearing eyeglasses and holding a cold beverage.**

Note that this type of an effect is not "thermal shielding," which we will discuss in Chapter 11. An example of thermal shielding is shown in Figure 6.2, where the true temperature of the animal (human in this case) is unaltered but spatially there are regions of the image that are devoid of emitted radiation in comparison to that emitted by the skin ($\varepsilon = 0.98$). Here the eyeglasses are almost totally reflecting ($\varepsilon \approx 0$) and remain at near air temperatures because there is very little energy exchange from the man to his glasses through conductive transfer. The emissivity of the cold drink is near zero as well so they show up very dark as compared to the emitted radiation from the exposed skin of the man. However, the man's temperature is unaffected. There is a different mechanism involved with the cold drink in his hand. Here the true temperature of the drink is much less than that of the man and his surroundings so it will absorb heat from the background and from the man's hand as it comes to thermal equilibrium.

QUALITY OF THE SURFACE

Any defects or surface imperfections affect the value of ε and in the case of animals the myriad of surface conditions include those associated with various skins, shells, furs, feathers, hair, or scales covering part or the entire animal. Furthermore, these surface conditions can be significantly modified by naturally occurring materials such as water, dust, mud, or various organic materials normally found in the habitat. Anything that would change the reflective or absorptive properties of the animal's surface will affect the emissivity, according to equation (5.2).

For achieving detectability in thermal imaging it is only necessary to see the animal of interest; it is not necessary to know the emissivity nor to calculate or know the true temperature of the animal. The following example is provided to illustrate the dramatic effect on the emissivity of an animal by simply wetting its pelage. Consider the thermal images of river otters (*Lutra canadensis*), which were collected within minutes of one another. In Figure 6.3 there are two

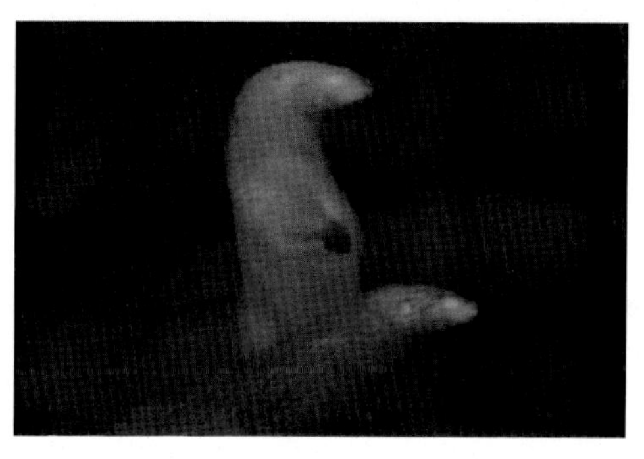

FIGURE 6.3 **A thermal image taken with a 3–5 mm MWIR camera of two river otters (*L. Canadensis*) on the bank that emerged from the water only minutes before the image was taken.**

otters that were photographed together. One of them (standing on hind legs) had emerged from the water a few minutes earlier than the otter in the foreground and its pelage is nearly dry. The emissivity of the standing otter is notably higher than the otter that left the water later. The otter in the foreground is partially dry (head moreso than the rest of its body) and its emissivity is correspondingly reduced as compared to the standing otter. In Figure 6.4 an otter had just emerged from the water and its pelage was wet to the point that the emissivity of the animal was very low at the time the thermal image was recorded. In fact, with the exception of the otter's eyes, ears, and head the background emissivity is greater than that of the otter's body.

Finally, in Figure 6.5 we see a thermal image of the same otter as in Figure 6.4, taken a few seconds later with its mouth open. The open mouth creates a pathway for radiant energy to escape from the animal's interior since it is no longer being shielded by the animal's wet skin and pelage. This is an example of drastic changes in the emissivity, which can be in a state of change or completely changed during an observation time period. These changes can occur in a matter of minutes as the animal's surface dries or becomes wetted and

FIGURE 6.4 **A thermal image (3–5 mm) of a river otter that has just emerged from the water.**

FIGURE 6.5 **A thermal image of the same otter as Figure 6.4 except that its mouth is now open, showing the effect of wet pelage on emissivity.**

when the emissivity of the animal is modified to the point that it is comparable to the background, detection with a thermal imager could become problematic. As an aside, it is important to recognize the fact that during the time the otter's emissivity was less than that of its immediate background we were still able to record very good thermal images of the animal. The imager records pictures based on the apparent temperature difference between the animal and its background and it doesn't matter which has the higher apparent temperature. We will see later that this is the fundamental reason why it is possible to record poikilotherms using a thermal imager.

VIEWING ANGLE

The geometrical factors that influence the emissivity can be somewhat complex when trying to determine the emissive or absorptive characteristics of the surfaces of various materials (Kruse et al., 1962). Kirchhoff's law $\alpha + \rho + \tau = 1$ for a rough opaque surface or animal (considered gray bodies) reduces to $\varepsilon + \rho = 1$. When the viewing angle is normal to the surface of an animal the emissivity is high and there is very little reflected radiation from the surface of the animal. If the viewing angle is now changed to say 45° then both the emissivity and the reflectivity have modest contributions. As the angle is increased even further such that we are viewing the animal at a grazing angle (near 90° with respect to the surface normal) the reflectivity is large and the emissivity is minimal. A blackbody radiates isotropically into a hemisphere as a diffuse emitter and the radiance is independent of the direction in which it is emitted; however, this is not the case for gray bodies. Gray bodies exhibit a strong angular dependence as the viewing angles become large with respect to the surface normal, eventually tending to zero at 90°. The two-dimensional polar plot shown in Figure 6.6 illustrates the differences in the directionality of the radiance for an isotropically emitting blackbody ($\varepsilon = 1$) and for a gray body, showing the angular dependence of the emissivity.

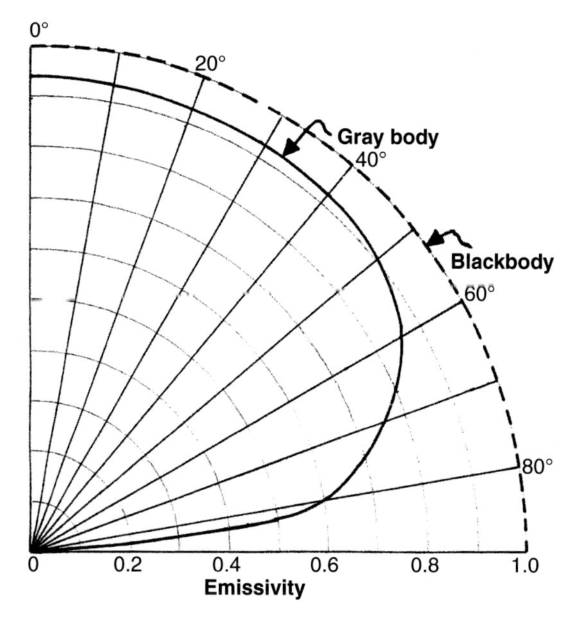

FIGURE 6.6 **A polar plot for a cross-section of a quadrant of the hemispherical radiance distribution of a gray body.** It shows the directionality associated with the radiance distribution as a function of the gray body emissivity. Note the fall off in the value of emissivity as the angle approaches 90°.

The dependence of the emissivity on the viewing angle has been noticed primarily in thermoregulation studies where thermal cameras are employed. Since most of these studies are conducted under laboratory conditions the effects of changes in the emissivity across a thermal signature are much more noticeable. If the imagery is to be used for clinical diagnostics a loss of signature uniformity (due to angular dependence of the emissivity) becomes very important and must be taken into account if a proper conclusion is to be reached.

SHAPE OF THE OBJECT

We also need to be aware of the changes produced in the spatial distribution of infrared images by the emissivity as a result of the animal's shape. Recall that only the radiation within the field-of-view (FOV) of the thermal imager will contribute to the image. Imagine an object with a uniform surface that is flat with a viewing angle perpendicular to the surface of the object. An image of uniform intensity with sharply-defined edges will be recorded by the camera if the object is within the FOV of the imager. If, however, the object is not flat but has a three-dimensional aspect, say spherical for simplicity, then the image will appear brighter in the center and fading toward the outer edges as a result of the angular dependence of the emissivity. At the ranges encountered in the field during surveys this effect is not so important from a mission point of view (i.e., the animals can still be detected and counted) but for thermoregulation and thermographic applications (Chapter 11) it would be very important.

When animals are imaged in a closed environment at relatively short distances during thermoregulation studies the angular effects on the emissivity are much more pronounced. McCafferty et al. (1998) used a liquid nitrogen cooled (3–5 μm) thermal imager to perform infrared thermography to measure the radiative temperature and heat loss of barn owls (*Tyto alba*). They reported that an examination of recorded thermal images showed that there was a decrease in apparent radiative temperature toward the outer edges of their images. This is commensurate with a reduction in emissivity where the surface of the animal becomes curved relative to the viewing angle of the imager. The thermal imager sees all radiation emanating from an object that falls within the FOV of the imaging camera. This includes the radiation reflected from its surface from all other sources radiating toward the object. In a laboratory setting strong contributions from reflected radiation can be minimized and the changes in the emissivity caused by the curved surfaces of an object become evident. The implications for clinical thermography were considered by Watmough et al. (1970) where they calculated that the apparent temperature differences in thermographs of a curved isothermal anatomical surface will be small and can generally be neglected for viewing angles less than 45°. As a result they suggested that in clinical examinations the skin should be viewed normal to its surface to minimize deviations in the apparent temperature difference to 0.5°C. They further noted that for regions that were viewed obliquely the emissivity reduction as a function of viewing angle could easily mask a hot spot of 4°C or more and would not appear in the thermographs. The contribution of the reflected radiation from the environment was included in the analysis of possible errors of a thermographic temperature measurement by Clark (1976). It supported the conclusions reported by Watmough et al. (1970) with regard to the fact that errors in thermographic observations of skin temperature distributions are likely to be small for viewing angles less than 45° from the normal but also pointed out that reflected radiation from adjacent smooth anatomical surfaces will have a major influence on the error observed for viewing angles greater than 45°.

APPARENT TEMPERATURE VERSUS VIEWING ANGLE

We saw in Section "Apparent Temperature of the Object" of Chapter 5 that a thermal imager cannot distinguish what portion of the total radiation emanating from the object is reflected and what portion is due to self-emission. For all practical purposes we can minimize the effects of the change in apparent temperature across the FOV by keeping the maximum viewing angle below 45°. That is, we should try to keep $\theta_F \leq 45°$ in Figure 6.7. For angles greater than 45° the reflectivity begins to increase markedly while a concomitant decease occurs in the emissivity. It should be noted that at angles approaching the grazing angle ($\rho \sim 1$ and $\varepsilon \sim 0$) even rough black surfaces tend to become good reflectors.

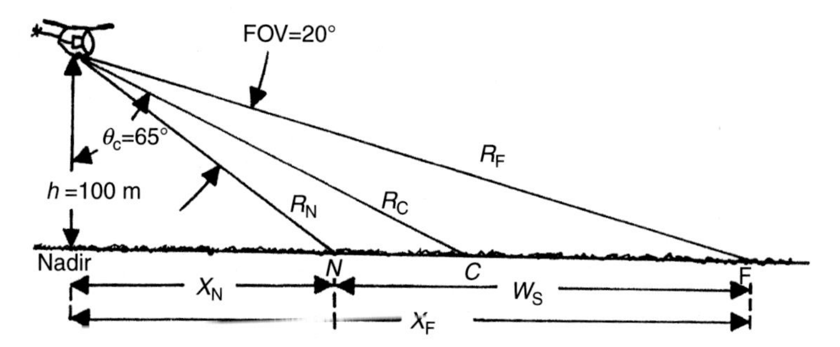

FIGURE 6.7 A sketch showing an aerial observation platform used to survey the landscape below. Note the difference in the contributions to the swath width from the right and left sides of the viewing angle θ_C.

Fortunately, we won't have to be greatly concerned for most fieldwork with regard to the angular dependence of the emissivity, particularly at the ranges used to conduct most surveys. Good absorbers are objects with surfaces that have emissivities greater than 0.9, while good reflectors have emissivities less than 0.2. Most mammals have emissivities greater than 0.9, making them good absorbers and poor reflectors. For example, human skin has an emissivity of 0.98. Hammel (1956) showed that the emissivities of many arctic fauna are greater than 0.98. Best and Fowler (1981) found that the mean emissivity of Canada geese (*Branta canadensis*) was 0.96 and was not significantly different from other species of geese. The emissivities were obtained for a number of small mammals by Birkebak (1966), including Cottontail rabbit (*Sylvilagus floridanus*), 0.97; Woodchuck (*Marmota monax*), 0.98; and gray squirrel (*Sciurus carolinensis*), 0.99. The fact that homoeothermic animals exhibit high emissivity is certainly helpful in achieving a large apparent temperature differential ΔT_A. We still, however, must concern ourselves with the background, which can and does reflect significant amounts of radiation, both from other objects in the background and from the environment (i.e., sun, clouds, sky, and the earth). Solar radiation can contribute significantly to the thermal signature of animals in the MWIR due to reflected sunlight. Gates (1980) gives the mean solar reflectance and absorptances for the dorsal and ventral aspects of a wide range of animals, including birds, mammals, amphibians, and reptiles. Solar radiation at ground level is ~15 times greater in the MWIR than in the LWIR so care will be needed when collecting data with MWIR imagers during daylight hours on clear days.

When doing aerial surveys or looking at animals from an elevated platform this strong angular dependence can give rise to imagery that clearly exhibits this angular dependence. Several features in the imagery will show these effects. We rewrite the apparent temperature of an animal given by equation (5.11) to examine the effect of T_{OA} on the viewing angle θ_C.

$$T_{OA} = \left\{\varepsilon_o T_O^4 + (1-\varepsilon_o) T_E^4\right\}^{1/4} = \left\{\varepsilon_O (T_O^4 - T_E^4) + T_E^4\right\}^{1/4}$$

As the value of the emissivity ε_O is varied across the footprint on the landscape defined by the FOV of the camera the apparent temperature of the animal T_{OA} will change. As the viewing angle is increased with respect to the nadir the value of ε_O will become smaller and at the grazing angle $\varepsilon_O \approx 0$. This simply says that as $\varepsilon_O \rightarrow 0$ the apparent temperature of the animal $T_{OA} \rightarrow T_E$ or the contrast in the imagery tends to zero. In general the emissivity of an object is highest when viewed at normal incidence and for most surfaces the emissivity approaches zero as viewing angles approach the grazing angle. From the plots of Figure 6.6 a good rule of thumb would be to keep θ_C less than 45°. Additionally, as θ_C approaches the grazing angle, the scattering of radiation along the line of sight close to the surface will also serve to reduce the image contrast (scattering effects are taken up in Section "Atmospheric Effects" of Chapter 11). Figure 6.7 is a sketch of an aerial platform consisting of a helicopter or drone fitted with a thermal imager to survey the landscape below. Here the camera is positioned to illustrate the problems associated with the viewing angle and what would not be recommended for field use.

In Figure 6.7 the distance from the nadir to the nearest point in the swath width (W_S) is

$$X_N = h[\tan(\theta_C - \theta_{VFOV}/2)].$$

The distance from the nadir to the farthest point in the swath width is given by

$$X_F = h[\tan(\theta_C + \theta_{VFOV}/2)].$$

The swath width from $X_F - X_N$ as can now be found as

$$W_S = h[\tan(\theta_C + \theta_{VFOV}/2) - \tan(\theta_C - \theta_{VFOV}/2)].$$

At an altitude of $h = 100$ m the camera line of sight is directed at a highly oblique angle of 65° with respect to the nadir. If the camera has a FOV of say 20° then the point N is at $X_N = 142$ m from the flight line or nadir and the point F is at $X_F = 373$ m. They are separated by $W_S = 231$ m, which would be the swath width for this viewing configuration. These turn out to be fairly extreme values for an altitude of 100 m. If the viewing angle is increased just 5° the changes to the swath width nearly double. Note that this problem can be corrected by reducing the viewing angle θ_C with respect to the nadir. Maintaining the same altitude and reducing the viewing angle θ_C will dramatically reduce the swath width, which may not be acceptable for the application at hand. If for example we pick the viewing angle to be 40° and maintain the same 100 m above ground level (agl) altitude then $X_N = 58$ m and $X_F = 119$ m, which means the swath width has been reduced to $W_S = 61$ m. This may or may not fit in with your planned survey but the problems produced because of the directionality associated with the emissivity for a gray body will be minimized. That is to say that T_{OA} over the swath width will maintain its uniformity and the effects of scattering from the atmosphere will be reduced as well. The swath width can be returned to the

~400 m range by increasing the altitude to $h \approx 650$ m while keeping the camera position so that the viewing angle is fixed at 40°. Here we only considered the effects of the viewing angle on the emissivity and its subsequent effects on the imagery. Other effects of the viewing angle on the imagery will be covered in detailed discussions in Section "Angular Dependencies and Effects" of Chapter 11.

From an applications point of view there has been some very remarkable research carried out to achieve tunable emissivity in the far infrared. For example, electrochromic materials such as tungsten oxide have been used to modulate the emissivity of thermal radiators by applying a voltage to induce a change in the optical and infrared properties. Thermochromic materials whose optical properties are temperature-dependent can also be used to modulate the emissivity by exploiting its frequency dependence. Kats et al. (2013) demonstrated that the insulator-metal transition (IMT) in thin films of vanadium dioxide (VO2) that was deposited on single crystal c-plane sapphire can be thermally induced near room temperature. They demonstrated that the optical interaction between the film and the substrate associated with the IMT displayed perfect blackbody emissivity over a narrow wavelength range, which surpassed the emissivity of the standard black-soot emissivity reference used for comparison. This peak remains significant even when the emittance spectrum is integrated over the 8–14 μm atmospheric window and as a result the structure also features a broad temperature (>10°C) region. The structure displays a large negative differential emissivity such that the sample emits significantly less thermal radiation even as it is heated up. These anomalous emittance properties can find uses in infrared camouflage, thermal regulation, and infrared tagging for battlefield identification friend or foe.

A comprehensive development of the relationships presented in this section from basic principles can be found in several good texts on infrared technology, heat transfer, or optics (i.e., Kruse et al., 1962; Jakob, 1949; Vollmer and Mollmann, 2010; Zalewski, 1995). For our purposes we only need to know that thermal imagers detect thermal "radiation" only. As we have seen, this form of heat has many sources and it is important that each source is recognized as a contributor for each field application we wish to undertake. The animal of interest can reflect, radiate, and reradiate quantities of heat that it has self-generated through metabolic activity or has acquired from other sources through absorptive, conductive, or convective transfer processes. We must maintain an awareness of these sources as we formulate a working methodology for use in the field.

Chapter 7

Thermal Imagers and System Considerations

Chapter Outline

There are many different types of infrared detectors but those used in thermal imagers need fast response time, ruggedness if they will be used outdoors, and the ability to be fabricated into arrays so that the requirements associated with thermal sensitivity can be met. These detectors fall into two classes: thermal detectors and photon or quantum detectors. When conducting animal surveys in the field it is necessary to be able to detect animals at relatively long ranges based primarily on the animal's size and its temperature difference with respect to the background. This is basically a surveillance application that depends on the performance specifications of the imager being used coupled with several special conditions to deal with the local atmospheric conditions. For surveillance applications the primary goal is to optimize the ability to see spatial distributions based on the temperature difference in the scene being viewed. Other applications might require the monitoring of thermal distributions and making temperature measurements to seek information about the well-being of living organisms or to provide methods of monitoring products for quality assurance in manufacturing. For these types of applications the most important performance parameters for a thermal imaging camera can be reduced to the spatial resolution and the noise equivalent temperature difference (NETD) although using an imager for making absolute temperature measurements must respond linearly with temperature and be equipped with an absolute temperature reference.

BRIEF HISTORY (BASIC CONCEPTS)

Thermal imaging systems are not new. In fact, they are used regularly in a number of application areas. Originally developed by the military for detecting, recognizing, and identifying enemy personnel and equipment, they are now

Thermal Imaging Techniques to Survey and Monitor Animals in the Wild: A Methodology
http://dx.doi.org/10.1016/B978-0-12-803384-5.00007-5

routinely used in police work, firefighting activities, border patrol crossings, perimeter surveillance, and environmental work (monitoring for pollution control and energy conservation) as well as a variety of medical, industrial, and construction applications. Thermal imagers can be used to image objects both large and small such as tanks deployed on the battlefield (landscape scale) to components that comprise a printed circuit board. They are infrared radiation detectors that may be used to perform noncontact thermal mapping of any device, system, object, or animal that emits infrared radiation (heat).

Infrared detector focal plane arrays (FPAs) can be either photon detectors or thermal detectors that are capable of converting infrared energy into electrical signals, which after multiplexing can be read out in a video format. Thermal detectors (bolometers) (Wolfe and Kruse, 1995; Vollmer and Mollmann, 2010) can be either resistive or capacitive in nature and sense the incident radiation by way of absorption and ancillary temperature change in the detector that produces an electrical charge or signal that mimics the incident infrared energy distribution. The capacitive devices are known as pyroelectric vidicons. Pyroelectric detectors require the incoming scene to be modulated, usually by a mechanical chopper, so a change in the capacitance of the device will produce a change in the stored electrical charge pattern on the detector (Garn and Sharp, 1974). Bolometric systems are now commercially available with uncooled detector arrays that offer moderate performance at a moderate price. A number of these systems have been used in the field for ground-based surveys and behavioral studies (Boonstra et al., 1994, 1995; Focardi et al., 2001, 2013; Franzetti et al., 2012; McCafferty et al., 1998; McCafferty, 2013). Recently these imagers have been used to detect signs of rabies infection in raccoons (Dunbar and MacCarthy, 2006) and were used in airports to scan and detect passengers infected with the swine flu (Shenkenberg, 2009). While imagers equipped with thermal detectors are suitable for a wide range of applications, the sensitivity and performance necessary to conduct surveys under a wide variety of field conditions (particularly airborne surveys) requires that we look to infrared (IR) imagers using photon detectors as opposed to thermal detectors.

The first thermal imaging systems employing photon detectors were in use nearly 35 years ago. The precursors of today's thermal imagers had moderate resolutions and were large, heavy, complex, cryogenically cooled, and very expensive. Since that time new technologies have been developed that have resulted in a new breed of infrared imaging systems featuring high resolution; the lightweight units are no larger than modern camcorders. For the most part, these advances were made due to the development of high-resolution FPAs and concomitantly the development of Stirling cycle micro-coolers that allow the detectors to be cooled to ~75 K (−189°C), which eliminated the need for cryogenic cooling (Kozlowski and Kosonocky, 1995). These new IR imagers can produce hundreds of images per second. The units are now commercially available from several vendors and while the cost of the units has come down considerably they are still somewhat expensive. Before we continue with describing what features and performance characteristics a camera must have to produce good thermal images, we need to look at a thermal image and think about what would actually make it suitable for a particular application.

Thermal Image

All objects above absolute zero ($T > 0$ K) radiate heat and the amount of heat radiated is determined by the temperature and surface condition of the object. Modern thermal imaging cameras are IR cameras that are capable of measuring the heat radiating from objects. Since heat transfer by radiation occurs at the speed of light images of the objects can be formed. One can record thermal images captured by the IR camera on video, view the camera display on a TV monitor, or simply view the objects of interest through a viewfinder as one could with a conventional camcorder. The only difference here is that the IR camera senses and displays a spatial distribution of thermal (heat) energy instead of visible light. This allows one to see in total darkness, through smoke, and other low visibility, low-contrast situations. These cameras can also be used during daylight hours to see heat-generated images when visual observation is inadequate to distinguish a heat-emitting object from its background. The imager sees all objects in a particular scene that radiate infrared energy. A protocol must be established to determine if the objects imaged are objects of interest for any particular survey.

Generally speaking, the camera provides a display of the spatial distribution of the thermal energy emanating from a scene (thermal image) and the individual objects of interest within the scene are referred to as thermal signatures. Each signature is a composition of the spectral distribution (wavelength band of infrared-radiation), spatial distribution (size and shape), and intensity distribution (temperature) that allows us to distinguish it from other signatures and the background of the scene. There is also a temporal component associated with these distributions, which can be used to aid in the interpretation of collected imagery. The objects of interest may be moving, for example. There are a number of factors that can influence these distributions that will be considered in later sections. It is instructive to examine a typical thermal image to get a feel for some of these factors. Since thermal images are formed by the radiated intensity difference between objects and their backgrounds it is important to recognize that the temperature and the surface of the object play a major role in the recorded energy distribution that the camera captures. In other words, it is more than just a temperature difference but an "apparent" temperature difference that produces the image since the surface moderates the emissivity and the resulting energy distribution emanating from the object. We will also see that reflected and scattered ambient radiation also contributes to the energy distribution responsible for the imagery.

A thermal image is shown in Figure 7.1. The image contains a thermal signature of a house cat (Sniper), which is depicted as a white silhouette on a

FIGURE 7.1 **A thermal image of a house cat in a forested setting.** The thermal signature of Sniper was captured with an InfraCAM® model A 3–5 μm infrared camera. The photo shown here was taken with a 35 mm camera of a single frame from 8 mm video of the thermal imagery displayed on a TV monitor.

somewhat darker (cooler) background, yet the image was collected during daylight hours. Other objects (trees, deadfall, and ground litter) in this scene are also radiating thermal energy and those images are also recorded. The gain of the 3–5 micron thermal imager was reduced in this image to provide a suitable apparent temperature difference for this infrared image. Note that the large log in the photograph is radiating a significant amount of heat and is comparable to that being radiated by Sniper yet the thermal signature of Sniper exhibits nice contrast when compared to the shady cool under-story provided by the forest canopy. The log was heated during the day due to the absorption of considerable amounts of thermal radiation from the sun (solar loading). The key difference here is that Sniper is his own source of radiant energy while the other objects are merely reradiating energy that was previously absorbed from sunlight. It should be noted that the polarity of the thermal camera could be reversed so that the collected imagery could also be presented as black signatures on a nearly white background.

There are a number of books with good discussions of infrared imaging systems and their uses (Maldague, 1992; Kaplan, 1999; Burney et al., 1988; Holst, 2000; Williams, 2009; Vollmer and Mollmann, 2010). However, they contain very little information on the use of infrared cameras for observing animals in the wild.

There is a plethora of factors that can influence the quality of thermal images collected in the field. An infrared imager is a complex electro-optical system (Seyrafi, 1973) that derives its performance from the intrinsic properties of detector materials (quantum efficiency), signal processing techniques, and human observation/interpretation of signals near the noise level of the system, all of which is extracted from a simple optical input. These cameras are rugged devices that meet the specifications for physical and mechanical robustness demanded by the military. Even so, a few precautions taken during the data-gathering phase will help to secure quality imagery. If the imagery is being recorded on and from a moving observation platform (vehicle, aircraft, watercraft) then the recording equipment should be soft-mounted to eliminate noise and vibrations associated with the moving platform. In the field an observer is typically viewing a TV monitor whose display must be of sufficient brightness and contrast and be close enough to the observer so that visual acuity is preserved. In this phase of data collection, as with visual surveys, observer experience will play a role that may be influenced by training, workload, and fatigue. The significant difference here is that the data is also being recorded to video for comprehensive analysis at a later time. Fortunately, the issues involving image observation/interpretation can be significantly reduced or eliminated by using the proper methodology that will in turn negate the need for extensive training or experience on the part of the observer.

A review of past work in which IR thermal imagers were used for animal surveys reveal two significant facts: (1) the performance of the IR imagers used prior to 1995 was only marginal in actual field studies and surveys, and (2) for the most part the methods used to collect the imagery in these efforts did not fully exploit the strengths that the earlier imagers actually offered. There was, however, significant progress made in these early efforts toward identifying the limitations of the technology. Advances made in the performance capability of the cameras and changes in the proposed methodology for using these cameras have minimized these limitations significantly.

PERFORMANCE PARAMETERS

The performance of thermal imaging cameras can be characterized by a handful of parameters. However, the extraction of useful imagery from a state-of-the-art infrared imager is a complex process. It involves much more than just conducting surveys with the best imager available. It is not enough to simply view an image to determine if it is optimized. In addition to the performance capability built into the cameras we must also consider a number of other miscellaneous factors that ultimately affect the image quality. We need to keep in mind that there is a clear distinction between seeing a spatial distribution of thermal energy (detection) and seeing the detail in that thermal energy distribution (resolution). We will discuss later how the atmospheric transmittance, the scene content, observer experience, subordinate equipment, aircraft type, ambient illumination, ambient temperature, and other factors affect image quality. These factors are not necessarily mutually independent of one another, which contributes to the complexity of interpreting image quality. For example, the atmospheric transmittance (an optical phenomenon) and the ambient temperature (a thermal phenomenon) both affect the image quality captured by an infrared imager and both of these are affected by rainfall. If we just consider the performance parameters associated with the imaging camera we intuitively want the IR camera to be capable of forming thermal images of small objects that have small temperature differences with respect to their backgrounds.

Additionally, for animal censusing, we would want to be able to resolve images from a considerable distance and over a considerable area (i.e., from an aircraft flying transects). As a first step in choosing an IR imager for use in animal studies and population surveys in the wild we would want to examine the performance parameters provided by both mid-wave IR band (MWIR) and long wavelength IR band (LWIR) imagers, which feature high thermal sensitivity, spatial resolution, and thermal resolution. Another parameter of importance is the signal-to-noise ratio that can be used to give physical insight into the significance of these parameters and how they relate to features we see in images, such as contrast. In addition to these obvious parameters there is the minimum detectable temperature difference (MDTD), which is important for animal counting applications when the species is known. Perhaps the most important parameters are the minimum resolvable temperature difference (MRTD), which depends on the spatial resolution of the imager, the noise equivalent temperature difference (NETD), and human observation/interpretation. The MRTD allows discrimination of animals based on the range and atmospheric conditions in the field.

Spectral Response

Thermal imagers are designed to operate in two specific wavelength bands in the infrared. The first thermal imaging band covers the (3–5 μm) or MWIR and the second thermal imaging band covers the (8–12 μm) or LWIR. In general, a system is an LWIR system if it is responsive somewhere in the LWIR region and likewise a system is an MWIR system if it responds somewhere in the MWIR region. When imagery is collected with either system it is only using a small portion of the total emitted radiation represented by the blackbody curves presented in Figure 5.1. Thermal imaging systems were designed to operate in these two spectral bands because they are the two primary transmission windows in the atmosphere suitable for thermal imaging applications.

The atmospheric transmittance depends on the amount of water vapor, gases, and aerosols that are present in the atmosphere. They interact through absorption and scattering with the radiation emanating from an object as it travels from the object to the imaging system. This loss in radiation is the atmospheric transmittance. For example, the strong drop in transmittance at 4.2 μm is due to absorption by carbon dioxide (CO_2). Both the upper and lower spectral limits of the MWIR and LWIR regions are determined by absorption due to water vapor. Since the makeup of the atmosphere is continually changing it is impractical to assign a value to the transmittance unless the measurement is made in a laboratory environment and as a result the curves in Figure 7.2 are only representative of typical atmospheric conditions.

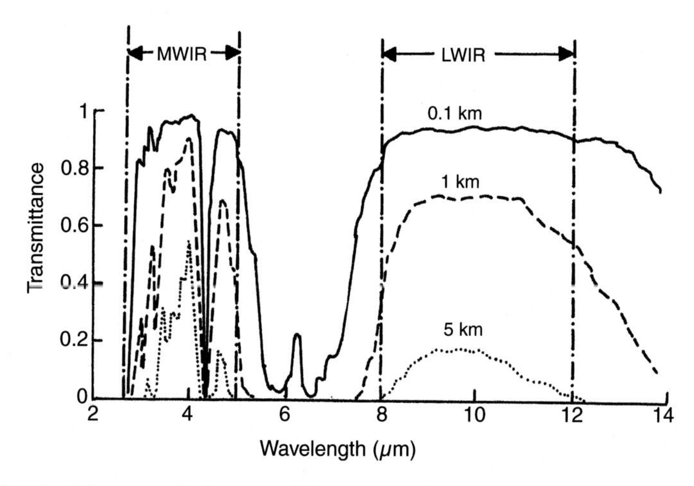

FIGURE 7.2 The atmospheric transmittance as a function of wavelength over 100 m and 1-km path lengths. The actual transmittance is continually changing, depending upon the presence of airborne particulates (dust, fog, and haze) and gases and the path length from the object to the thermal imaging system. This is a representative curve since the conditions in the field, which give rise to the atmospheric transmittance, can change within minutes.

The atmospheric transmission is high when there are only small amounts of water vapor, gases, and aerosols present in the atmosphere. At these times the sky will appear cold to the thermal imager since deep space is very cold. Under these conditions (viewing the zenith) the self-emission of the atmosphere is low and the average temperature of the background viewed by the imager is low. As the imager is lowered from the zenith to smaller elevation angles more aerosols and gases at ambient temperature are encountered along the line of sight, resulting in an average temperature close to that of the earth's surface. For most of the detection and survey work encountered in the field, small elevation angles are the norm. Down-looking angles from aircraft will experience the same high average temperatures associated with atmospheric layers close to the surface of the earth. Under these circumstances, atmospheric constituents degrade the thermal resolution. Both scattering and absorption in the line of sight reduce the magnitude of the thermal signature of the object to be detected, located, recognized, or identified. On the other hand, scattering of radiation into the line of sight and self-emission from the atmosphere in the line of sight both serve to reduce the thermal contrast between the object and the background. Since nearly all fieldwork involving censusing will utilize relatively long path lengths, the self-emission and scattering by the atmosphere in the path could be significant and must be taken into consideration. For short path lengths (<20 m) the transmittance for both the LWIR and MWIR windows can be considered as unity. The directional dependence and the amount of background radiation (both emitted and reflected) that the imager will encounter is another important factor that must be incorporated into the overall methodology.

Signal-to-Noise Ratio

The signal-to-noise ratio (S/N) is more important in detection systems than it is in imaging systems (Seyrafi, 1973, p. 269). For even low signal-to-noise-ratios a human observer can overlook a great deal of random noise flickering in an image displayed on an observation monitor. The human eye easily pulls very weak well-defined signatures from noisy images because it provides a powerful temporal and spatial integration capability. Observers will be able to identify relatively persistent brightness contours and shapes that form easily recognized geometric shapes (e.g., circles, triangles, rectangles, stars, or alphanumeric characters).

We can put the sensitivity improvements achieved with staring FPAs into perspective by examining the signal-to-noise ratio of a system when operating under the same average atmospheric conditions. The Beer-Lambert law determines the attenuation of the atmospheric transmittance, τ, by absorption and scattering. Figure 7.2 shows the results of the strong attenuation associated with absorption and scattering as a function of wavelength. We can write the atmospheric transmittance, τ, as

$$I = I_0\, e^{(-kR)} \tag{7.1}$$

where k (1/km) is the extinction coefficient and R (km) is the path length or range. The extinction coefficient is wavelength dependent and accounts for both absorption and scattering losses in the atmospheric transmittance for any given path length. $k(\lambda)$ can be separated into its absorptive $k_A(\lambda)$ and scattering $k_S(\lambda)$ components as $k(\lambda)> = k_A(\lambda) + k_S(\lambda)$. The transmittance curve in Figure 7.2 shows the wavelength dependence of the attenuation caused by absorption (primarily CO_2 and H_2O vapor) and scattering from the atmosphere. The most prominent attenuation mechanisms caused by the atmosphere are linear absorption, elastic scattering (Rayleigh scattering), and scattering from particulates (Mie scattering). Equation (7.1) can be rewritten by defining an average attenuation coefficient, $\beta = e^{-k}$. As a result we can write the transmittance as

$$I = \beta^R = e^{(-kR)} \tag{7.2}$$

Both β and k have the units of 1/km and in the limit of a completely transparent atmosphere ($k = 0$) then $\beta = 1$ and $I = 1$. From Figure 7.2 we see that generally for $R \approx 1$ km the values relating to $k_A(\lambda)$ and $k_S(\lambda)$ are small in the 3–5 μm and 8–12 μm range. However, the average attenuation coefficient β has values ranging from 0.70/km to 0.95/km, depending on whether the conditions are poor or excellent. Topics relating to the importance of atmospheric effects on collecting thermal images and their quality will be discussed in more detail in Chapter 11.

The signal-to-noise ratio (S/N) of a system operating during average atmospheric conditions can be written as

$$S/N = \Delta T(\beta^R)/NETD. \tag{7.3}$$

Here the signal is just the thermal contrast $\Delta T = (T_O - T_B)$ between an object and its background that is attenuated by the average atmospheric transmission (β) at some range R. The noise is given by the noise equivalent temperature difference that is a function of background temperature. As the temperature of the environment drops NETD increases and system noise increases, which reduces the signal-to-noise ratio. Both MWIR and LWIR detector performances are degraded due to the lack of photon production at the lower temperatures (the MWIR more so than the LWIR).

We can get an idea of how the local atmospheric conditions can influence the range of an imager by using equation (7.3) and the specifications for a typical thermal imaging system to compare values of the range for different average attenuation coefficients. Holst (2000) provides average attenuation values β for weather conditions ranging from poor to excellent for a hypothetical local environment, which can be used to compare range estimates. The weather quality and average attenuation β are given as: poor (0.70/km), fair (0.80/km), average (0.85/km), good (0.90/km), and excellent (0.95/km). For the 3–5 μm thermal imager (InfraCAM®) the following specifications apply: NETD = 0.7°C @ 23°C; IFOV = 0.6 mrad; FOV = 8 × 8 degrees; and we let S/N = 5, $T_B = 23$°C, and $\Delta T = 4$°C. We can rewrite equation (7.3) as

$$R = \{\log(\text{S/N}) - \log(\Delta T) + \log(\text{NETD})\} / \log(\beta) \tag{7.4}$$

Table 7.1 gives the calculated values of the range R obtained from equation (7.4) for the attenuation coefficients representing the weather conditions from poor to excellent and the parameters given for NETD, ΔT, and S/N.

From these rough calculations one can easily see the dramatic effect that the atmospheric attenuation can have on the range. Clearly water vapor is one of the most problematic obstacles that confront researchers using thermal imagers in the infrared.

TABLE 7.1 Estimates of the Range Calculated from Equation (7.4) for Various Atmospheric Attenuation Conditions

Range (m)	β/km	Conditions
370	0.70	Poor
590	0.80	Fair
820	0.85	Average
1270	0.90	Good
2600	0.95	Excellent

The values of the range R calculated from (7.4) for hypothetical atmospheric attenuation coefficients associated with various local weather conditions. S/N = 5, $\Delta T = 4$°C, NETD = 0.7°C

Thermal Sensitivity

The output of an IR detector is a manifestation of a complex relationship involving many interdependent quantities. This makes it very difficult to compare imagers of different designs and different detector materials. In an infrared imager the detector is the most important component in the unit. For the most part, the significant improvements gained in the thermal sensitivity of modern IR imagers come about from advances in detector technology, specifically FPA. Imagers equipped with FPA technology do not require mechanical scanning mechanisms or choppers as do conventional IR imagers. Typically, a mosaic can be constructed consisting of 65,536 discrete detectors arranged in a 256 × 256 pattern to form an FPA. Each element (pixel) in the array appears as a point or resolution element in each image. Conventional IR imagers used in the early experiments for animal surveys had a single detector or a small array of detectors that were rapidly scanned over the entire scene so that the detector only received radiation during a very short period of time. By comparison each detector in an FPA stares continuously at the scene so it is "on" continuously. This allows it to receive a larger number of infrared photons from the scene, thereby providing a very high thermal sensitivity.

Thermal sensitivity refers to the smallest temperature differential that can be detected by the imager. A laboratory measurement of the thermal sensitivity is the NETD, defined as the temperature difference in the object scene that produces a change in output signal that is equal to the system root mean square noise. Basically, the radiant temperature increment in the object scene must be capable of producing a response from the detector that is sufficient to overcome the noise of the system. For most comparisons it is sufficient to use the temperature difference that produces a signal-to-noise ratio of unity at the sensor output.

In Chapter 5 we noted that the emitted power differential ΔP from an object and its background can be very small and that the sensitivity of the camera can be marginally effective. As a result, the camera must have sufficient gain so that the difference in the detector output voltage between the object and the background is a measurable quantity. This is a very important parameter and determines the fundamental limit of the camera. It is therefore a fundamental goal to reduce detector noise, which is usually defined as the noise equivalent power (NEP) and is expressed as NEP = V_{RMS}/R_d where V_{RMS} is the root mean square noise and R_d is the detector responsivity. The basic output signal from the detector is a voltage V_d proportional to the power incident on the detector according to $V_d = \kappa R_d P$ where κ is a constant particular to the camera design and R_d is the responsivity of the detector. For small ΔT the differential camera output is

$$(\Delta V_d/\Delta T) = kR_d(\Delta P/\Delta T). \tag{7.5}$$

Recall from above that NEP = V_{RMS}/R_d and that the NEP is the signal level that produces a S/N of unity. Ignoring the atmospheric conditions ($\beta = 1$) we can use equation (7.3) as

$$\mathrm{S/N} = \Delta\mathrm{T}(\beta^{\mathrm{R}})/\mathrm{NETD}$$

to get NETD = ΔT when S/N = unity. Since the NETD is the input ΔT that produces the S/N of unity then $(\Delta V_d) = V_{RMS}$ where V_{RMS} is the root mean square value of the noise. We can rewrite equation (8.1) as

$$V_{RMS} = kR_d(\Delta P/\Delta T)\text{NEDT}. \quad (7.6)$$

where ΔT is the noise equivalent temperature difference NETD and the detector signal level ΔV_d is the root mean square value of the noise V_{RMS}. The sensitivity NETD is a function of the background temperature T_B since $(\Delta P/\Delta T)$ is evaluated at a specific background temperature.

$$\text{NETD} = V_{RMS}\{kR_d(\Delta P/\Delta T)\}^{-1}. \quad (7.7)$$

Note that the NETD decreases as the background temperature increases and therefore the minimum detectable temperature difference decreases or the system becomes more sensitive at warmer temperatures. Conversely, as the power differential available to the detector gets smaller due to the decrease in background temperature, the NETD increases or the system gets noisier and the temperature resolution decreases. Remember that small NETD is preferable and that generally the NETD becomes excessively large in the MWIR region as the background temperature drops below 0°C.

An FPA allows the detectors to collect photons for a longer period of time than a scanning system. This allows materials with modest quantum efficiency to easily produce signals above the noise of the system. The NETD is a function of the background temperature and is usually expressed as a temperature difference at ambient temperature. The NETD is also a function of the radiation collecting optics or the lens f-number, so that changing the lenses of the system can change the NETD. A typical value of the NETD for a 3–5 μm spectral band staring focal plane array fabricated either from indium antimonide (InSb) or from platinum silicide (PtSi) is 0.025 or 0.07°C @ 23°C. This can be compared to 0.2°C @ 23°C typical of the 8–12 μm scanning IR imagers that have been used in most of the animal survey work to date. This is approximately an order of magnitude improvement in thermal sensitivity over the scanning units used in earlier animal surveys conducted from the early 1960s to the middle of the 1990s.

Spatial Resolution

The spatial resolution of an IR imager does not uniquely determine the image quality. However, high spatial resolution is needed to acquire quality images since it furnishes important information with respect to the finest detail in an image that can be discerned. Since the spatial resolution of the system (detector array and optics) is unrelated to the system noise it is therefore independent of the thermal sensitivity discussed earlier. The spatial resolution of an IR imager is usually specified by the instantaneous field-of-view (IFOV) of the system or by the diameter of the bright central disk of the diffraction pattern (Airy disk) formed by an ideal optical system. While either may be used it is important to note that both contribute to the resolution of the system, which can be detector limited (IFOV) or optics limited (Airy disk).

Detector-Limited Resolution

The size of the FPA ($D_{HORIZONTAL} \times D_{VERTICAL}$), the individual detector size, the number of pixels, and the center-to-center spacing (detector pitch) of the pixels characterize detector arrays. The IFOV specifies the angular resolution of the camera and is expressed as the angle in milliradians (mrad) over which the detector senses radiation. The parameters needed to determine the IFOV are the focal length, f, of the optical system and the width of the detector's active area, w. From this, the angle θ_D subtended by the detector can be estimated in the small angle approximation as $\theta_D = \theta_{IFOV\text{-}H}$ (horizontal instantaneous FOV).

$$\theta_D = w/f \tag{7.8}$$

If the detector is not square then the angular resolution would be an average θ_a of the horizontal and vertical resolutions or $\theta_a = (\theta_{IFOV\text{-}H} \times \theta_{IFOV\text{-}V})^{1/2}$ where w and h are the horizontal and vertical dimensions of the detector (see Figure 7.3). In this case θ_a would be the instantaneous field-of-view of a square detector with the same area as a rectangular detector having area (wh). Figure 7.3 is an illustration of section of a focal plane array delineated into an array of individual pixels with detector elements. The "fill-factor" or the ratio of the area of detector to the area of a pixel is shown for the array and the difference in the subtended angle by each is compared. The comparison of the detector angular subtense θ_D and the pixel angular subtense θ_P is shown for only one dimension (width).

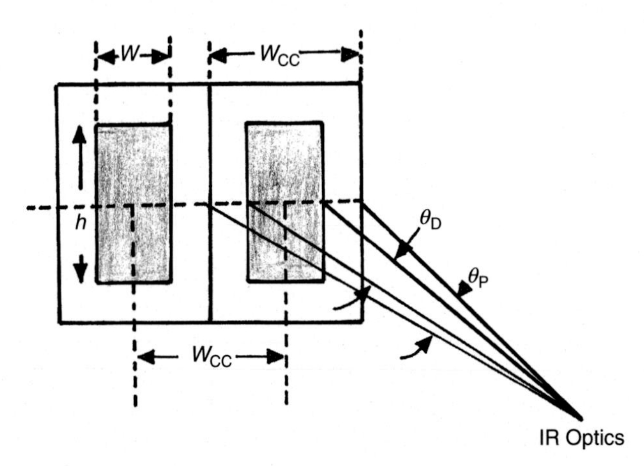

FIGURE 7.3 An illustration showing two pixels in a staring focal plane array showing a fill-factor (less than 100%) and the resulting differences in the subtended angles associated with a pixel and a detector.

For a 100% fill factor $\theta_D = \theta_P$. However, in general the detector angular subtense θ_D is smaller than the pixel angular subtense θ_P and this difference is proportional to the fill-factor. For example, a 60% fill-factor means that $\theta_D = (0.6)\theta_P$ and the relationship between the IFOV for a pixel as given above for a staring FPA is different than the IFOV for a detector that does not fill the area defined as a pixel. State of the art photon detector focal plane arrays have fill factors of over 90% and microbolometers focal plane arrays greater than 80%, so for modern imagers $\theta_D \approx \theta_P$.

In order to determine the IFOV of an IR imaging system using a staring FPA we need to know some specifics about the FPA. If the number of pixels and the array size is known the pixel size (center-to-center spacing) can be calculated. For a staring focal plane array the IFOV is determined by the angle subtended by a pixel θ_P so the IFOV is given by

$$\theta_{IFOV} = 2\tan^{-1}(w_{cc}/2f) = \theta_P = W_{CC}/f \tag{7.9}$$

where w_{CC} is the detector pitch or center-to-center spacing of the pixels. This is a very useful parameter since it allows us to estimate the size of the smallest perceptible thermal signature we can detect with the camera. Conversely, if the size or spatial extent of the object is known we can determine the maximum distance at which the object can be resolved. In this context the spatial resolution of the system is given by the following relationship

$$X_{min} = \theta_{IFOV}(R) \tag{7.10}$$

Where R is the distance from the thermal object and X_{min} is the minimum size object that can be detected for a given distance R. If the focal length f of the lens system is changed then the value of X_{min} changes according to

$$X_{min} = (w_{CC}/f) \times (R). \tag{7.11}$$

We can now make an estimate of the smallest object X_{min} (Figure 7.4) that will be detected at a given range R for a 3–5 μm thermal imager with a staring focal plane array having an IFOV = 0.6 mrad by using equation (7.11).

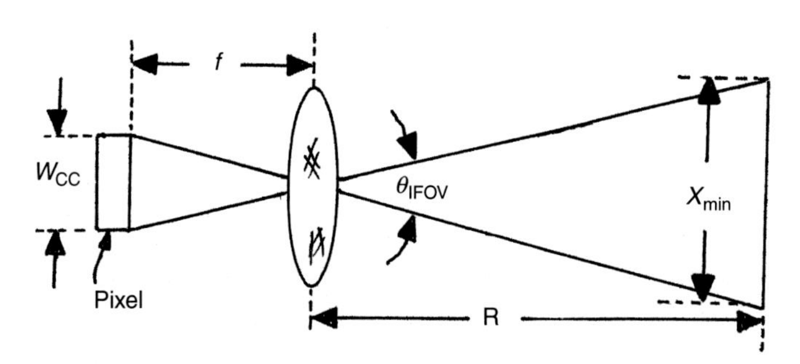

FIGURE 7.4 The instantaneous field-of-view and spatial resolution as determined from the detector angular subtense.

TABLE 7.2 The Calculated Estimates from Equation (7.11) of the Minimum Detectable Object Size for a Series of Ranges

Range, R (m)	Minimum Object Size, X_{min} (m)
50	0.03
100	0.06
200	0.12
500	0.30
1000	0.60
2000	1.20
5000	3.00

The influence of atmospheric attenuation was not taken into account for these estimates. Minimum detectable object size X_{min} for a thermal imager with a focal plane array having detector pitch w_{cc} = 30 μm, focal length f = 50 mm, and IFOV = 0.6 mrad for different ranges R.

Table 7.2 gives the minimum detectable object sizes X_{min} for a set of ranges from 50 m to 5 km for a thermal imager with these specifications. Note that these are optimal values since we have not included any effects due to atmospheric conditions (e.g., we have assumed that the atmospheric transmittance $\tau = 1$), which usually is never the case (see the range estimations given in Table 7.1 when the atmospheric attenuation is included in the range calculations). In Chapter 11 we discuss ways to achieve the best ranges possible for a variety of conditions.

Optics-Limited Resolution

The diameter of the bright central disk (Airy disk) of the diffraction pattern formed by a good optical system can be determined from diffraction theory (Hecht and Zajac, 1979, pp. 350–354). For MWIR imagers with an average wavelength of $\lambda_a \sim 4$ μm and a system lens diameter D, the Airy disk diameter θ_{Airy} (expressed as an angle in mrad) is

$$\theta_{Airy} = 2.44(\lambda_a / D) \tag{7.12}$$

which gives a value of X_{min} as

$$X_{min} = \theta_{Airy}(R) = 2.44(\lambda_a / D)(R) \tag{7.13}$$

From this we can calculate a value for X_{min} at a specified range and that it only depends on the optics and wavelength of the radiation and has no reference to the detectors at all. Furthermore, by increasing the size (D = diameter of the optics) we can improve the resolution by reducing the lens f-number (F#). $F\# = f/D$, where f is the focal length of the lens. Most infrared detectors employ optical systems with F#s between 1 and 4. For example, a typical imager such as the 3–5 μm Inframetrics InfraCAM® is equipped with a (F# 1.3) 50 mm lens and the FLIR Systems MilCAM-Recon® 3–5 μm camera uses a 50 mm F# 4 lens system as standard. As the ratio of the F# and the width of the detector changes the factors determining the limiting resolution also change. In actuality the system resolution is a combination of the optics and the detector and the transition from optic-limited to detector-limited resolution is gradual. The detector output should always be as large as possible so that objects with a small ΔT can be detected and reducing the F# will increase the detector output. However, since most IR optical systems are equipped with F#s larger than 1 and given that large diameter infrared optics are impractical from an economic standpoint we can, for all practical purposes, use the IFOV (either θ_D or θ_P) rather than θ_{Airy} to specify the spatial resolution of a staring array.

The spatial resolution and thermal sensitivity are independent of one another since the resolution does not include the effects of system noise nor thermal contrast but the thermal sensitivity does. However, the image resolution achieved with cameras using staring FPA technology is considerably better than that achieved with scanning systems because the thermal sensitivity is better. That is, the system resolution is greatly improved by allowing smaller IFOVs to be used with FPAs than with scanning systems. FPAs are comprised of large numbers of individual detector elements, each with differing signal responsivities (gain and offsets) and these differences must be accounted for to obtain quality imagery. The spread in the gain and offset values causes a spread in the detectivity for different pixels receiving the same incident radiant power and if this nonuniformity is large then the imagery obtained is generally not very useful. A correction process referred to as the nonuniformity correction (NUC) is carried out during the camera calibration process where the gain and offset of each pixel is adjusted and the correction parameters are stored in the camera. For example, the FLIR SC8000 HS® Series MWIR camera comes equipped with a 1024×1024 InSb FPA with an 18 μm pitch and has built-in NUC for on camera nonuniformity corrections.

Typically, the scanning systems used in early animal surveys had IFOVs between 1.0 mrad and 3.0 mrad. With the introduction of 3–5 μm and 8–12 μm staring FPAs, longer focal length optics were incorporated into the systems and the spatial resolution was improved by more than an order of magnitude to values ranging between 0.05 mrad and 0.6 mrad. Modern 8–12 μm imagers using a 4-element HgCdTe (mercury cadmium telluride) detector are equipped with both wide field-of-view (WFOV) and narrow field-of-view (NFOV) optics, providing IFOVs of 1.2 mrad and 0.3 mrad. These small IFOVs allow surveys of much smaller objects to be conducted from higher altitudes over greater areas. Since modern infrared imagers are equipped with superb optics we find that using the IFOV to estimate ranges or the smallest discernable object size commensurate with a given field application is appropriate in most cases.

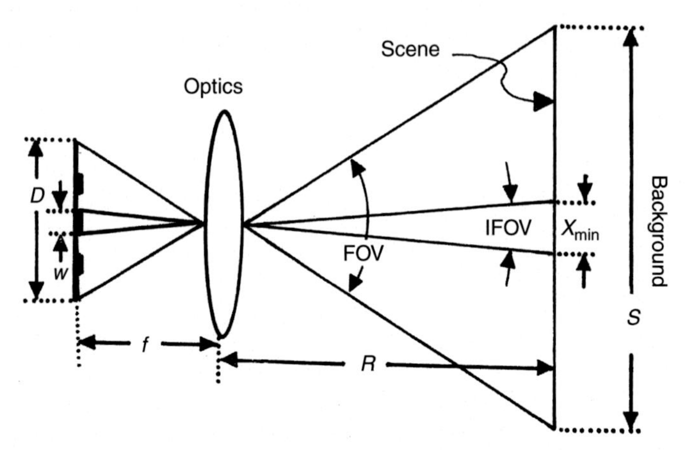

FIGURE 7.5 **A two-dimensional view of the geometrical relationships of the FOV, IFOV, and the focal plane array in a thermal imager.**

The FOV of the system (Figure 7.5) is the total viewing area of the system measured in radians. The FOV is determined by the size of the focal plane array *D* and the focal length of the optics *f* as

$$\mathrm{FOV} = 2\tan^{-1}(D/2f) \tag{7.14}$$

The FOV determines the swath width S when censusing from aircraft and is sometimes given as a length at a certain altitude above ground level (agl). The swath or scene dimension is determined by the FOV of the imager and range R as $S = 2R\tan(\mathrm{FOV}/2)$. If the focal plane array is not square (see Section "Detector-Limited Resolution" of this chapter) then the FOV of an imager is given for both the horizontal (HFOV) and vertical (VFOV) views in degrees and the swath will take on horizontal and vertical dimensions.

Most systems are equipped with a WFOV and an NFOV capability. This is normally accomplished by changing the lens on the front of the camera. When the lens is changed, however, the IFOV of the system is also changed since IFOV depends on the focal length of the lens system, according to equation (7.6). The smaller the IFOV for a given FOV, the more detailed the images. We can therefore select the IFOV and choose either a narrow mode (narrow ground swath but good resolution) or a wide mode (wide ground swath but less resolution). If there were an object in the background of a scene that was in the FOV of the imager it would be detected if its size exceeded X_{min}.

Range

What does it take to see through adverse atmospheric conditions and to what extent or how far can we see with thermal imagers? What is the range under conditions of fog, haze, aerosols (particulates of all sizes), rain, and dust? For fieldwork, these are very important questions that can greatly influence the outcome. For the MWIR CO_2 and H_2O vapor is the most attenuating species. The actual range of an IR camera is affected by a number of things, including the camera and the atmospheric conditions at the time of usage. Range predictions can be estimated based on the signal-to-noise ratio (Table 7.1) based on attenuation and we can also include the effective contribution from radiation scattered forward into the imaging camera from the atmosphere (path radiance), which is essentially part of the background radiation and will diminish contrast in the imagery.

In the above expressions X_{min} gives a value for the minimum object size for 50% probability of detection. We shall address this situation in Chapter 9 when we discuss the issues concerning differences associated with detection, recognition, and identification of objects when using thermal imagers under a specific set of representative conditions (Johnson, 1958). Within this context, detection does not mean finding so called "hot spots" in a scene but rather finding "thermal signatures" of interest that warrant a closer look according to some pre-established criterion. For example, we can compare the values of X_{min} computed for imagers used in early experiments with those used more recently. For R = 1000 m and θ_{IFOV} = 2.5 mrad (typical for early thermal imaging fieldwork) we can detect with 50% probability an object having a spatial extent of 2.5 m. Using an imager with a $\theta_{IFOV} \approx 0.5$ mrad we could easily detect objects of 0.5 m spatial extent with the same probability. New cameras have IFOVs an order of magnitude less than we used in this work. The actual threshold for the spatial resolution achievable with cameras using new FPAs is determined by the pointing stability of the camera. Ways of dealing with this threshold point will be addressed in Chapter 11.

Thermal Resolution

The minimum resolvable temperature difference (MRTD) depends on the thermal sensitivity, NETD, the background temperature, and the spatial resolution. The MRTD depends on the background temperature in the same way as the NETD: as the background temperature increases the MRTD decreases. As such it is the most meaningful parameter for comparing image quality from one imager to another. The MRTD is probably the most important measure of an infrared system's ability to detect and identify a thermal anomaly in a particular scene. The MRTD is not an absolute value since it takes into account the visual acuity of the observer and quality and settings of the display device (typically a TV monitor). As a result, the MRTD should be considered as a perceivable temperature differential relative to a given background and not something that is mandated by optics and electronics. Williams (2009) presents several concepts that have been formalized to replace the subjective measurement with ways to objectively measure parameters suitable to model the performance of a human observer. These models provide a relatively good match to the subjectively measured MRTD and may produce far more repeatable and reliable results resolving targets than the subjective measurements. Additionally, the minimum detectable temperature difference (MDTD) can also be predicted from a model similar to that of the MRTD. The parameters are treated with a slightly different algorithm because of the difference between being able to recognize (resolve) a target and being able to detect the presence of a target in a scene.

The MRTD of a system is determined in a laboratory by human observers who make a determination of the minimum temperature that they can recognize by viewing a standard 4-bar pattern (Johnson, 1958). The spatial frequency (cycles/mrad) or spacing between the bars of the pattern and the temperature of the bar pattern relative to the background temperature are varied. Human observers determine the temperature difference at which the bar pattern is indistinguishable from the background by the imaging system. The temperature differential is determined as a function of the spatial frequency in both the horizontal and vertical directions and can then be compared to other systems at a particular spatial frequency. When comparing MRTDs a low temperature difference at a high spatial frequency is desirable. Typical values for today's handheld 3–5 μm imagers using InSb or PtSi focal plane arrays are 0.05°C @ 0.5 cy/mrad or 0.025°C @ 0.5 cy/mrad when 50 mm optics are used. By comparison a modern commercially available 8–12 μm imager using a 4-element HgCdTe (mercury cadmium telluride) detector has a MRTD of 0.22°C @ 0.375 cy/mrad.

Note that the MRTD may be considered as being inversely proportional to the spatial resolution since as the separation of the bars in the pattern becomes smaller (higher spatial resolution) the MRTD becomes larger. The degradation in the MRTD by increasing the spatial resolution results due to the incident radiation required to produce a response by the detector above the system noise, which depends on the thermal sensitivity or NETD. As the IFOV decreases there are fewer IR photons impinging upon the detector and at some point the system noise overshadows the thermal sensitivity. For these very small IFOV cases there are simply not enough photons reaching the detector to produce a signal regardless of the thermal contrast between the animals and their background. This particular problem was especially restrictive of the performance for scanning imagers since they have their detectors exposed to the scene radiation for only a short period of time (scan time) as compared to staring FPAs that continually stare at the scene and collect photons.

There is a big distinction between thermal imagers that utilize quantum detectors as the sensitive element as opposed to detectors that rely on thermal effects to generate an image. The differences are enormous for censusing and surveying, particularly on a landscape scale. For the most part, for applications requiring higher sensitivities and/or time resolutions, such as in R&D, cameras with photon detector FPAs cooled to liquid nitrogen temperature ($T = 77$ K) are used for the MWIR and LWIR spectral region. These photon detector arrays exhibit a much larger responsivity/detectivity and a spectral response varying with wavelength. Cameras using photon detectors have smaller NETD values and are also capable of higher frame rates than cameras fitted with microbolometers arrays. The practical issues concerning the use of both photon detectors and bolometric cameras will be contrasted in Chapter 8. Physical features (weight and size) as well as cost issues associated with the desired performance will be discussed. There are a number of interrelated issues to be aware of when deciding what imager would be best suited for any particular application.

Chapter 8

Imager Selection

Chapter Outline

INTRODUCTION

Thermal imaging is simply the process of converting infrared (IR) radiation (heat) into visible images that depict the spatial distribution of temperature differences in a scene viewed by a thermal camera. The imaging camera is fitted with an infrared detector, usually in a focal plane array, of micron-size detecting elements or "pixels." The detector array may be cooled or uncooled, depending on the materials comprising the array and the camera's intended use. A lens system focuses scene radiation onto the detector array and appropriate processing electronics display the imagery. Infrared radiation is attenuated by the atmosphere and the degree of attenuation depends greatly on the local atmospheric conditions at the time the imagery is collected. It is important to match the detector response with either of the two atmospheric windows: midwave IR band (MWIR) or long wavelength IR band (LWIR), as shown in Figure 7.2. Remember that for most survey work and field observations we are not concerned with the measurement of temperatures related to objects in the scene but only the apparent temperature differences between objects in the scene. Note that this is a big step back from the level of difficulty that most thermographers face when measuring temperatures in the field, but nonetheless it comes with its own set of demanding requirements.

It is important to create the best images possible to extract meaningful data regarding the detection, recognition, and identification of animals of interest in the field. This is exactly the purpose of surveillance applications, which the military has been laboring over for years. The first thermal imaging cameras

Thermal Imaging Techniques to Survey and Monitor Animals in the Wild: A Methodology
http://dx.doi.org/10.1016/B978-0-12-803384-5.00008-7

were developed in the 1950s by the military; they were large, heavy, and very expensive. The camera technology at that time required that they be cooled with liquid nitrogen. Improvements in camera development continued over time and new detector materials, array fabrication techniques, coolers, optics, electronics, software, and packaging have resulted in reliable high-performance thermal imaging cameras. By the mid-1990s focal plane arrays were mostly of the cooled variety. Uncooled arrays were beginning to make their appearance and were incorporated into new imagers. The technology today encompasses both cooled and uncooled focal plane arrays. Most books devoted to thermal imaging technology are focused on the development of thermal imagers and detector arrays for uses in commercial applications rather than those of concern to the military such as long range surveillance and target identification/acquisition. Civilian uses of thermal imagers are usually devoted to temperature measurements of components used in industrial applications, building inspections, surveillance for security, police, and fire applications, and robotic vision. Wildlife ecologists planning to use thermal imagers in the field will be more interested in long range acquisition and identification of different animal species. Other researchers may be interested in using infrared imagers to study different aspects of animal physiology such as thermoregulation. By locating and monitoring thermal abnormalities in different parts of an animal's anatomy one can infer underlying circulation that may be related to physiology, behavior, or disease. McCafferty (2007) has reviewed 71 empirical studies that used infrared thermography for research on mammals. In either case the advancements in detector array technology and the improvements gained in the infrared imagery for both cooled and uncooled imagers will find use in both applications.

THERMAL DETECTORS VERSUS PHOTON DETECTORS

When considering the performance parameters associated with the imaging camera, we intuitively want the IR camera to be capable of forming high-quality thermal images of small objects that have small temperature differences with respect to their backgrounds. In Chapter 7, the key performance parameters were introduced that are required for obtaining quality thermal images. However, there is much more involved in determining the actual quality of an image obtained in the field. Purchasing an imager with the "best" performance parameters is no guarantee that one will obtain high-quality images but it is certainly a good place to start. There are many closely interrelated factors that influence the quality of the imagery besides the characteristics of the system. For example, as a first step in choosing an IR imager for use in animal studies and population surveys in the wild, we would certainly want to examine the performance parameters provided by both MWIR and LWIR imagers. Recall that the minimum resolvable temperature difference (MRTD) or thermal resolution depends on the spatial resolution of the imager, the noise equivalent temperature difference (NETD) or thermal sensitivity, and human observation and image interpretation. The MRTD allows discrimination of animals based on the range and atmospheric conditions in the field. The performance parameters of a system are usually given for a set of optimized conditions, which is sometimes very difficult to obtain in the field. Nonetheless, high-quality imagery is certainly obtainable by verifying that all factors that influence the imagery are optimized. In some survey applications where the species is known and is the only species present in the survey area, the minimum detectable temperature difference (MDTD) is important for animal counting applications.

Infrared detectors that meet the requirements for thermal imaging must have a fast response time, be robust, and be able to be fabricated into arrays of detectors to obtain adequate sensitivity. There are two broad classes of detectors that meet this criterion: thermal detectors and photon or quantum detectors.

Thermal Detectors

As the term implies, thermal detectors utilize a material with some temperature-dependent property such that when the incident radiation flux is absorbed, the temperature of the material increases. This temperature increase produces a measurable physical change, generally an electrical output signal, resulting from the temperature dependence of the resistance, thus serving as a thermal transducer. Importantly, these detectors respond to the absorbed radiant flux irrespective of its spectral distribution. A number of different temperature-dependant physical properties have been utilized to construct thermal detectors. For example, some liquid crystals have very sensitive temperature-dependant properties that have been used to produce thermal images. Thermal imagers based on pyroelectric detectors that convert spatial temperature distributions into spatial distributions of spontaneous electrical polarization have been demonstrated for a broad range of materials such as ferroelectric crystals (e.g., triglycine sulfate), polymers (e.g., polyvinylidene fluoride) PVF_2, and ceramics (e.g., lanthanum doped lead zirconate titanate, PLZT) (Garn and Sharp, 1974).

Microbolometers

Microbolometers are the thermal detector of choice for many applications as a result of a much lower cost and large detector arrays. A good treatment of microbolometer technology is given by Vollmer and Mollmann (2010) and by Budzier and Gerlach (2011). A microbolometer consists of an array of pixels, each pixel being made up of several layers. The focal plane arrays can be made from metal or semiconductors and operate as nonquantum devices since incident radiant energy causes a change in state in the bulk material rather than a change that produces charge carriers. The performance of microbolometers is determined by the readout noise, the thermal coefficient of resistance (TCR), which is a measure of the change in resistance as a function of temperature, and the degree of thermal isolation between pixels and between pixels and their surroundings. The reduction of convective energy exchange between pixels is accomplished by a vacuum seal of the detector package. There is a wide variety of materials that are used for the detector element in microbolometers and material properties influence the value of the detector responsivity (a measure of how well the detector is able to convert the incident IR radiation into an electrical signal). The sensitivity is partly limited by the thermal conductance. The speed of response is limited by the thermal heat capacity divided by the thermal conductance. Reducing the heat capacity increases the speed but also increases thermal temperature fluctuations or noise. The interplay of these parameters are all considered and managed in the design of these arrays. Most recently, the requirement for temperature stabilization with a thermoelectric (TE) device has been relaxed for microbolometer FPAs made from both VO_x (vanadium oxide) and α-Si (amorphous silicon) so that they are now being delivered without TE devices.

The two most commonly used IR radiation-detecting materials in microbolometers are α-Si and VO_x. Many phases of VO_x exist although it seems that $x \approx 1.8$ has become the most popular for microbolometer applications. The quality of images created from microbolometers has continued to improve. The microbolometer array is commonly found in two sizes, 320 × 240 pixels or less expensive 160 × 120 pixels. Current technology has led to the production of devices with 640 × 480 or 1024 × 768 pixels (Coffey, 2011). There has also been a decrease in the individual pixel dimensions. The pixel size was typically 45 μm in older devices and has been decreased to 17 μm in current devices. As the pixel size is decreased and the number of pixels per unit area is increased proportionally, an image with higher resolution is created.

Some advantages of thermal cameras include: (1) robustness with long mean times between repairs, (2) the 0.8–14 μm spectral band, which allows penetration through obscurants such as smoke, dust, water vapor, and smog, (3) small and lightweight, (4) lower power consumption than cooled detectors, (5) real video output immediately after power on, and (6) less expensive than cameras with cooled detectors. Additionally, progress is being made to modify microbolometers so that they can show wavelength tunable responses at selectable wavelengths. Maier and Bruckl (2009, 2010) demonstrated that the spectral response and dynamic behavior of microbolometers can be tuned by using integrated resonant absorber elements (metamaterial absorbers) and that the presence of a continuous metallic shielding layer affects heat conduction, which leads to faster thermal response times. The downside is that they are less sensitive than cameras with cooled detectors.

Photon Detectors

As mentioned in Chapter 7, the first thermal imaging systems employing photon detectors were in use in the 1960s. They had moderate resolutions and were large, heavy, complex, and cryogenically cooled as well as very expensive. Technologies have been developed that have resulted in new infrared imaging

TABLE 8.1 Detector Type, Material, Wavelength Range of Operation, and Cooling Requirements for Commonly Used Thermal Imaging Cameras

Common Detectors for use in Thermal Imaging Cameras			
Detector	**Material**	**Wavelength**	**Operating Temp.**
Photon	InSb	MWIR (3–5 μm)	77 K
Photon	PtSi	MWIR (3–5 μm)	77 K
Photon	HgCdTe	LWIR (8–14 μm)	77 K
QWIP	GaAs	LWIR (8–14 mm)	70 K
Microbolometer	VO_x or α-Si	Broadband	Uncooled
Pyroelectric	TGS	Broadband	Uncooled

systems. These systems feature high resolution. They are lightweight units no larger than modern handheld video cameras due to the development of high-resolution FPAs and the concomitant development of suitable solid-state coolers (Peltier devices) for MWIR detectors operating at temperatures near 200 K and Stirling cycle microcoolers that allow the LWIR detectors to be cooled to ~75 K (−189°C). For the most part these two advancements eliminated the need for scanners and cryogenic cooling.

Quantum detectors are mainly semiconductors whose photonic characteristics are selectively modified by suitable doping to achieve specific performance in different camera designs. These devices can operate as either a photovoltaic device (photodiode), where incident photons generate a voltage across a p–n junction in the semiconductor, or as a photoconductor, where the absorbed incident photons change the electrical conductivity (resistivity) of the semiconductor by creating free charge carriers. Table 8.1 gives a few examples of semiconductors that are used as photon detectors in modern thermal imagers.

Photon detectors are designed for optimum response for specific spectral regions (Table 8.1). Indium antimonide (InSb) is an example of a photovoltaic detector with an energy gap confining its use to the MWIR region while mercury-cadmium-telluride (HgCdTe) can be fabricated (with some difficulty) to operate well as either a photovoltaic or a photoconductive device, making them suitable for either LWIR or MWIR. Sensors with 15 and 12 μm pitch are in the development stage and have been demonstrated at both MWIR and LWIR using both HgCdTe and InSb (Coffey, 2011).

There are two other photon detectors that are used in a number of thermal imagers being offered in commercial markets. An MWIR photon detector based on platinum silicide (PtSi), which works as a Schottky barrier photodiode, is used in large detector arrays. The focal plane fabrication process is based on well-developed silicon integrated circuit fabrication technology, which can inexpensively produce very large and uniform arrays. The arrays, however, must

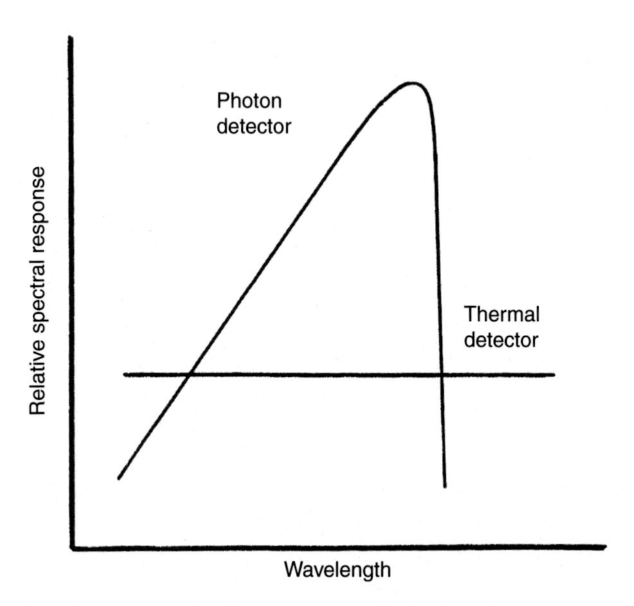

FIGURE 8.1 **The relative spectral response of a thermal detector and a photon detector.** Note the wavelength independence of the thermal detector.

be cooled to 77 K and unfortunately have a relatively poor quantum efficiency (~2%). Another quantum detector that takes advantage of well-established manufacturing techniques and produces very repeatable results is the quantum well infrared photodetector (QWIP) detector. Large staring arrays can be fabricated from layers of gallium arsenide (GaAs) and gallium aluminum arsenide (GaAlAs), which are tailored to create a spatial variation of the conduction and valence band edge. This variation forms alternating quantum wells and barriers for electrons and holes. Slight changes in the architecture allow for adjustments to the spectral dependence of the responsivity so that QWIPs can be tailored to have spectrally narrow or wide ranges of response at a particular wavelength.

Perhaps one of the most fundamental metrics and the best to use when comparing detectors is obtained by normalizing the detectivity $D = 1/\text{NEP}$ where NEP is the noise equivalent power $\text{NEP} = V_{\text{RMS}}/R_{\text{d}}$ or the root mean square noise V_{RMS} divided by the detector responsivity R_{d}. Jones (1957) defined the quantity D^* (D-star) as $D^* = [(\text{NEP})^{-1}(A_{\text{d}}\,\Delta f)^{1/2}]$, which is just the NEP (the signal level that produces a S/N of unity) normalized to the detector active area A_{d} and the electronic bandwidth Δf. The relative spectral response for a typical photonic detector and a thermal detector are compared in Figure 8.1. The response of a thermal detector depends on the radiant power incident on the detector (or rate of change in radiant power for pyroelectric detectors) but not on the spectral characteristics of the incident radiation. As a result the response of a thermal detector generally does not exhibit spectral features as photon detectors do. The broad flat responses exhibited by thermal detectors make them candidates for a wide range of applications that do not require high detectivities or high speed. They operate at room temperature and are low-cost devices as compared to photonic devices. The spectral response of photon detectors on the other hand is strongly wavelength-dependant, which is tailored during fabrication to maximize D^* in either the LWIR or the MWIR spectral bands. To achieve high D^*s the detectors must be operated at low temperatures (see Table 8.1).

To summarize, applications that are primarily surveillance oriented, such as animal surveys and censusing, which require high detectabilities at relatively long ranges, would benefit from the time response and detectivities offered by thermal imagers equipped with photon detectors. For applications that are conducted using shorter ranges typical of those used in laboratory studies or when dealing with penned or captive animals for thermoregulation or physiological studies (specific temperatures within some thermal image must be determined) then cameras fitted with microbolometers would be the imagers of choice. For applications monitoring nests, dens, tree cavities, or other aspects of animal ecology that are not critically dependant on imagery collected at long ranges either photon or thermal detectors would be appropriate. Thermal cameras are uncooled and therefore much less expensive. Those interested in reading more about the comparative performances, including D^* and the spectral response for photon detectors, how they are fabricated, how they are cooled, and how their performance parameters are tailored, should see Wolfe and Kruse (1995), Wolfe (1965), Fink and Christiansen (1982), Kozlowski and Kosonocky (1995), Kruse et al. (1962), and Rogalski (2003a, 2003b).

SELECTING AN IR IMAGER

Applications, more than anything else, will drive the imager selection process. Many applications require special performance capabilities and features in an imager to enable success. Some of the more important requirements are range and whether the application requires collecting data in the dark. Other applications might need very high thermal sensitivity or spatial resolution. Others may require fast time response. In fact, all of these may be needed for a particular application. It is recommended that when selecting an imager, ensure the specific requirements are met before considering other features that may be nice to have but are not required. Today most cameras can be used for a variety of different applications.

Application Requirements

The detector incorporated into a thermal imager will have a significant influence on the type of application that falls within the capability of the imager, but it is not the only criterion of importance. Basically, imagers are designed for measuring temperatures or temperature differences (spatial distributions of temperatures or temperature profiles) or for use as a surveillance tool. A good imager could be used for either task but particular camera features may be better suited for one or the other. For example, for most surveillance applications accurate temperatures are generally not needed so the requirement for accurate temperature scale and emissivity compensation is relaxed. On the other hand cameras intended for surveillance applications need to have high spatial resolutions and the capability to detect very small temperature differences. The atmospheric conditions will have the greatest influence on the choice of MWIR or LWIR systems if the imagers are to be used primarily outdoors. In most cases fieldwork must be scheduled when the atmospheric conditions are optimized for the task planned. Many times this will simply require waiting for the right opportunity.

Some imaging applications must be undertaken in the dark or during the nighttime hours when there is no radiation to illuminate the objects of interest so the objects themselves must provide the radiation used to generate the thermal image. This type of passive imaging provides a means of seeing under conditions of poor illumination, in the dark or at nighttime, or where artificial lighting and solar radiation are absent. In these applications the true strength of thermal imaging is demonstrated because it offers vision through the complete diurnal cycle. Warm or heated objects such as machinery, vehicles, and biotic homeotherms are clearly imaged with respect to their background.

If an application is being undertaken that requires visibility through smoke, dust, or mist the longer wavelengths of the MWIR and LWIR imagers offer a particular benefit. Because the wavelengths used are approximately an order of magnitude longer than the wavelengths in the visible part of the spectrum there is less scattering of the radiation by particulates in the atmosphere. This advantage is of great importance to firefighting applications where the necessity of "seeing" through smoke-filled buildings is commonplace. In general the ability to acquire longer ranges through poor atmospheric conditions is a significant benefit provided by thermal imagers.

There are numerous applications of importance stemming from the fact that the radiation emitted from any surface is a function of the temperature of the surface. As a result, thermal imaging can be used to determine the temperature of the surface or to provide an image showing the spatial distribution of the temperature differences across the image. This particular feature has been embraced by medical professionals as a supplementary diagnostic tool and to monitor various diseases and injuries that lead to changes in body surface temperature. We have already mentioned several factors that can affect the thermal mapping outcome of surfaces such as viewing angle, emissivity, reflections, distance to surface, and shielding effects. These are basic physical phenomena that affect the output imagery but much more is needed for the standardization of thermographic imaging for medical applications (Cockburn, 2006).

Obtaining values from radiometric data (i.e., a simple temperature measurement or perhaps an image of a temperature distribution) collected at some range from the source of the radiation has been examined by comparing experimental data with atmospheric modeling using the computer model MODTRAN (Moderate Resolution Atmospheric Transmission) (Richards and Johnson, 2005). They found that for both the MWIR and LWIR, using two major attenuating species, CO_2 and H_2O vapor, that it was unlikely that radiometric measurements would be possible for ranges >100 m and certainly not out to a kilometer.

For surveillance applications an imager is used over long distances and therefore long path lengths through the atmosphere. The transmission characteristics become more important the longer the ranges become. In addition, climatic conditions vary and some will favor the MWIR (warm moist atmospheres) or the LWIR (cooler atmospheres, atmospheres with particulates such as smoke or haze).

Wavelength Selection

So far the discussions of thermal imaging systems apply to both the MWIR and LWIR spectral regions. Figure 7.2 suggests that one factor to consider when selecting the wavelength band of an IR imager (MWIR or LWIR) would be the path length or range over which the thermal images would be collected. While there is an overall reduction in the atmospheric transmittance as the path length is increased (as it would be for aerial surveying and censusing activities), it is more severe for the LWIR region. As the path length is increased the reduction in the atmospheric transmission for the LWIR bands is more significant than for the MWIR, as can be seen in Figure 7.2. This reduction is due primarily to absorption by water vapor or absolute humidity (water in gram per cubic meter in the atmosphere). The absolute humidity increases with both the ambient temperature and relative humidity such that at 40°C and 100% relative humidity (tropical environment) the absolute humidity is approaching 100 g/m^3 and at −25°C and 0% relative humidity (arctic environment) it is near 0 g/m^3.

At times the absorption and scattering losses due to various aerosols (particulates suspended in the atmosphere) can make thermal imaging a difficult task. These are usually dust, smoke, and water droplets of different sizes. The water droplets can vary in size, increasing from light haze to dense fog and raindrops. The ratio of the particle diameter to the wavelength determines the magnitude of the scattering loss. When the particulates are large with respect to the wavelength there is a significant reduction in the transmittance. The aerosol diameter for light haze is ~1 μm so that scattering losses in the visible are significant but there is very little effect in the MWIR and LWIR spectral regions. When the relative humidity increases above ~80% these tiny particulates act as nuclei for condensation, which results in a wet mist consisting of small liquid droplets in the several micrometers range that scatter light more efficiently and reduce the range in the visible spectrum. If the droplets continue to grow to the 10 μm size range they become fog. For light to moderate fog (~8–9 μm particle size) the MWIR and visible regions are degraded and for dense fog and rain or snow (~15 μm) all three regions – visible, MWIR, and LWIR – are degraded.

Care must be taken when comparing visible ranges with those obtained in the MWIR and LWIR because, while the visible range is negatively affected by light haze and the infrared is not, there is not much IR radiation to be detected in the visible. To see long range there must be something radiating in the IR and not just scattered visible radiation. Such conditions require an object that is much warmer than its background.

The possibility of achieving longer ranges in the MWIR seems to suggest that the MWIR would be the better choice for an imager to use in aerial surveys. However, there are other factors that must also be considered, since the thermal sensitivity (NETD) is a function of the background temperature (see Chapter 7). As the ambient temperature decreases the relative NETD increases, and the performance of both MWIR and LWIR detectors are degraded (MWIR more so than LWIR). A thermal imaging system only responds to relatively narrow bands of radiation in the total spectrum of the emitted radiation (Figure 5.2). The fact that the peak of the blackbody radiation curve shifts to longer wavelengths as the temperature drops would seem to favor an LWIR imager. Notice that this suggestion does not include the role of the thermal sensitivity of the imager or the signal-to-noise ratio. Perhaps more important is the lack of photon production at shorter wavelengths as the temperature is lowered below 273 K (0°C), which points strongly toward using an LWIR detector. The sensitivity of an MWIR imager decreases (NETD increases) as the radiating bodies and their surroundings become cool because there are fewer photons to collect in the 3–5 μm range at temperatures below 0°C. Based on these arguments, it is unlikely that thermal imaging in the 3–5 μm range will be useful in an arctic environment. Generally, the NETD becomes excessively high in the MWIR region when the background temperature drops below 273 K (0°C) and excessively high in the LWIR when the temperature drops below 245 K (−28°C) (Holst, 2000, p. 116).

To summarize, there are many factors that can limit the range of infrared cameras, some of which are due to the camera design and the task at hand such as the spectral range, thermal or photon detectors, thermal sensitivity or NETD, optics, size of the animal, temperature of the animal, and the temperature of the background. Another factor affecting the range is the atmospheric transmission and how it is changed from ideal by the following conditions: humidity, haze (aerosol in the atmosphere), and of course weather (fog, rain, snow).

When IR imagers are used on clear days, solar reflections will pose problems for imagers in the MWIR. As seen in Figure 5.1, the peak of the blackbody curve shifts to shorter wavelengths as the temperature of the blackbody increases such that there is 16 times more solar radiation at the surface of the earth in the MWIR than there is in the LWIR. Reflections of this solar radiation and their magnitude depend upon the reflectivity of surfaces in the scene and the viewing angle of the imager. If in a scene a high-emissivity blackbody (an animal, e.g., where $\varepsilon > 0.95$) at ambient temperature (~293 K or ~20°C) is competing with solar reflections then an LWIR imager will give superior performance. This is

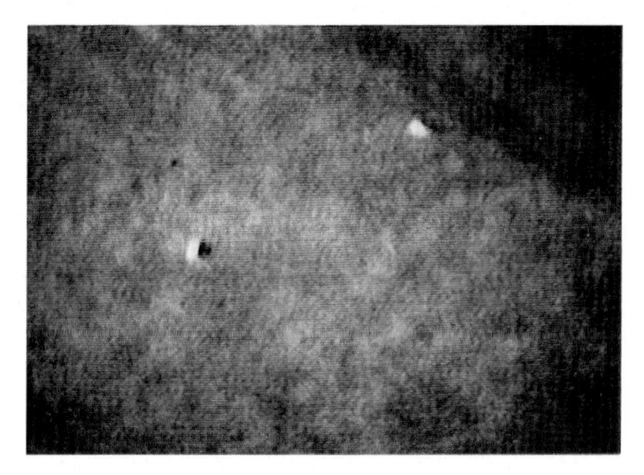

FIGURE 8.2 **A photograph showing the effect of solar radiation on the imagery gathered with an MWIR imager.** The two deer in this image are casting shadows in the direct sunlight, indicating that solar radiation is being reflected from the animals' background.

because the self-emission of a ~293 K blackbody is ~85 times that of reflected solar radiation at the surface of the earth in the LWIR but the self-emission of the same blackbody is ~6 times less than that produced by reflected solar radiation in the MWIR, so the reflected solar radiation could be expected to be more problematic for an MWIR imager. The problems associated with solar reflections occur only during daylight under cloudless conditions. They are not a problem if measurements are made after dark. If the measurements must be made during the daylight hours there are two options to reduce the effects of solar reflections: select a cloudy or overcast day or use a camera fitted with a filter to eliminate radiation below ~3.5 μm. The downside of adding the filter will be an increase in the NETD. Figure 8.2 shows a thermal image of two white-tailed deer taken from a helicopter with a 3–5 μm thermal imager during the daytime under sunny conditions when the background was still at a fairly uniform temperature. Note that there is a shadow to the right side of the deer in this image as a result of being illuminated by direct sunlight. This points once again to the fact that reflected sunlight could be a problem during some imaging tasks.

It is important to develop proper methodology for the use of these imagers and there are many considerations other than performance specifications to be made when selecting an IR imager for field studies. If the situation calls for a survey to be conducted during the day and there is direct solar radiation illuminating the scene then great care must be taken sorting through the imagery with regard to reflected solar radiation (solar glint) and absorbed solar radiation (solar loading). We have already noted that in general the transmittance, absorptance, and reflectance are wavelength dependent. Water has relatively high transmittance in the visible and nearly zero in the infrared. The low transmittance in the infrared is due to both absorption and reflection. Figure 8.3 below shows the thermal image of a beaver (*Castor canadensis*) as it is about to enter the water. This image was taken with a 3–5 μm thermal imaging camera and

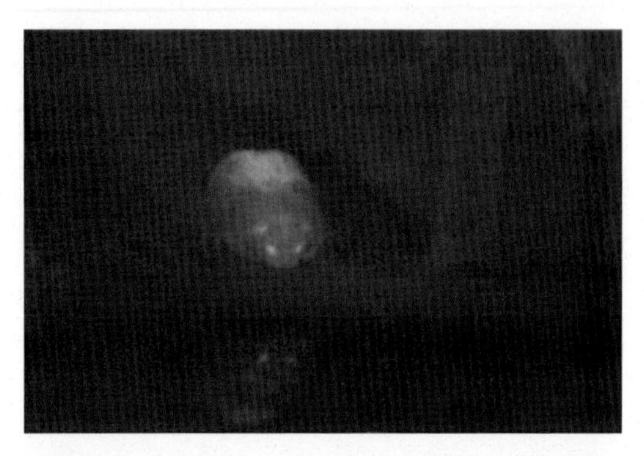

FIGURE 8.3 **A photograph showing a 3–5 μm infrared image of a beaver at night at the water's edge.** The pond surface is reflecting 3–5 μm radiation that is radiating from the beaver's eyes.

the 3–5 μm radiation emitted from the beaver, particularly emanating from its eyes, is clearly reflected from the surface of the pond. If there were ripples on the surface of the pond then the reflected radiation would not be stationary and steady but would appear to be flickering as the radiation is reflected into and out of the camera's FOV as the surface changes in the vicinity of the reflections (i.e., the surface becomes a mirror with periodic distortions).

MWIR and LWIR (Experimental Comparisons)

Several field experiments were conducted where at least one of the main goals was to provide a comparison between MWIR and LWIR thermal imager performance for survey tasks. Early thermal imaging efforts in the aerial survey of animals were conducted with infrared thermal scanners operating in the 8–14 μm or 3–14 μm spectral ranges. The technical specifications of these infrared scanning imagers were not made available because of security regulations so a determination of their full capability was not possible. These thermal imaging systems were large, heavy units that were cryogenically cooled (liquid nitrogen) and used mercury cadmium telluride detector arrays.

McCullough et al. (1969) compared the results of optically filtering a HgCdTe 3–14 μm imager to operate in the 3–5 μm range or in the 8–14 μm range and Graves et al. (1972) compared a scanning 3–5 μm InSb detector with a scanning 3–14 μm HgCdTe LWIR detector. Both of these studies set standards for subsequent animal survey efforts and are discussed in Chapter 10.

Graves et al. (1972) compared the performance of an LWIR and an MWIR thermal imager. They showed that deer emit enough radiation in the 3–5 μm region of the spectrum to allow detection by airborne infrared imagers operated under certain conditions. They suggested that a 3–5 μm imager using an Indium Antimonide (InSb) detector is the best sensor for the detection of large animals in the warmer months when the animal is trying to stay cool by radiating away body heat. They also suggested that a 3–14 μm imager using a Mercury

Cadmium Telluride (MCT) detector is the best choice for the cooler months when the animal's coat reduces radiation of body heat, and the animal is conserving energy. Their conclusions regarding the relationship of seasonal and wavelength sensitivity effects were correctly attributed to differential thermal radiation. We saw in Chapter 5 that the power radiated from a surface above absolute zero depends on the temperature of the surface, the background temperature, and the temperature of the surroundings (ambient temperature or temperature of the environment). It also depends on the emissivity of both the surface and the background. These, however, are not the only influencing factors. A complete explanation must include the fact that the peak of the blackbody radiation curve shifts to longer wavelengths as the temperature drops (which would favor a 3–14 μm imager), or equivalently, photon production at shorter wavelengths is reduced as the temperature is lowered. This means that the sensitivity of a 3–5 μm imager decreases as the radiating bodies become cool because there are fewer photons to collect in the 3–5 μm range at low temperatures.

Both LWIR (HgCdTe) and MWIR (InSb) detectors were used by Best et al. (1982) to conduct cold weather aerial censusing of Canada geese (*Branta canadensis*) on the Missouri River in South Dakota. They reported that they were unable to detect geese against the cold background presented by the water (slightly above 0°C) even at the lowest altitude flown (305 m agl) with the MWIR detector. The spatial (IFOV = 2.5 mrad) and temporal (ΔT = 0.5°C) resolution was only marginal for detecting individual geese ($\leq$1 m) at this altitude. The LWIR detector could detect very low densities of geese resting on open water from 305 m agl. This work affirms that the performance of MWIR detectors is limited by low photon production at cold temperatures. It still needs to be determined if today's MWIR systems using staring focal plane arrays will be useful in detecting animals at low temperatures since the thermal sensitivities would be an order of magnitude better than the scanning units used here.

The methodology used in the work conducted by Dymond et al. (2000) is very good and resulted in successful thermal imaging surveys in the field. They conducted their surveys using the strengths of two forward looking infrared (FLIR) systems in their systematic investigation to determine the optimum conditions for surveying possum (*Trichosurus vulpecula*) populations in New Zealand. Both MWIR (Prism DS) and LWIR (Model 2000) imagers from FLIR® Systems, Inc. were tested. They set their flight conditions based on the imager optics, spatial resolution, and the desired video display that would allow them to discern individual possums with confidence.

They determined that both MWIR and LWIR imagers performed well, that convective heat losses due to wind must be avoided, and that flight conditions be commensurate with the imager optics and desired video display. They indicated that they still had problems with image interpretation due to spurious signals originating from objects in the background that were emitting previously absorbed solar radiation with magnitudes on the order of those associated with possums. This is a problem noted by all users and is the single most important issue that a practical methodology must address.

A review of past work (see Chapter 10) in which IR thermal imagers were used for animal surveys reveals two significant facts. One, the performance of the IR imagers used prior to 1995 was only marginal in actual field studies and surveys. Two, for the most part the methods used to collect the imagery in these efforts did not fully exploit the strengths that the earlier imagers actually offered. There was, however, significant progress made in these early efforts toward pointing out limitations of the technology. Advances made in the performance capability of the cameras and changes in the proposed methodology for using these cameras have raised these limits significantly.

Table 8.2 provides a comparison of the spatial resolution IFOV (mrad) and a temperature differential ΔT (°C), which is probably the NETD or thermal sensitivity as discussed above for the referenced works reviewed in Chapter 10. Unfortunately, the system specifications given in the earlier studies to compare IR imaging and conventional surveying techniques were not provided in many cases and only partially provided in most cases. We assume that the NETD is more appropriate here since the MRTD was not used to describe the temperature differential provided in these references. In most cases the values given for the NETD of the system were only estimated. In either case the value given for the temperature differential (ΔT) is a function of the background or ambient temperature and is quoted as the NETD or MRTD at 290 K or ~23°C.

The evolution in the performance characteristics of thermal imagers used in animal surveys and studies is compared in the following table. The specifications listed in Table 8.3 are representative of a new collection of thermal imagers. There are a number of rugged systems that are now commercially available. These systems were designed for ground base use and are handheld portable units that can be used in a variety of applications and from a variety of platforms. These units can also be fitted to ground-based vehicles and both rotary- and fixed-wing aircraft. The new technology is moderately expensive but has significant improvements in both imagery and data manipulation over the older units used in the past for animal studies.

CAMERA FEATURES

Most imagers intended for use in industrial applications to detect, monitor, and measure thermal abnormalities are handheld and for the most part fitted with uncooled detector arrays. However, when cooling is necessary temperature control is provided by thermoelectric devices, which use very little power. In situations where very high resolution or sensitivity is needed, as in some thermography applications, cooling to liquid nitrogen temperatures may be required and this is usually accomplished with a Stirling cycle cooler that preserves the smallness and portability of the imager, albeit at a higher cost. In a laboratory

TABLE 8.2 Comparison of Spatial Resolution (IFOV) and ΔT (NETD) for Referenced Works in Chapter 10

References	IFOV (mrad)	ΔT (°C)	λ Range
Croon et al. (1968)	3.0	1.0	LWIR
McCullough et al. (1969)	3.0	1.0	LWIR
	3.0	1.0	MWIR
Graves et al. (1972)	3.0	NA	MWIR
	3.0	NA	LWIR
Parker and Driscoll (1972)	2.5	0.2	LWIR
Best et al. (1982)	2.5	0.5	LWIR
Kingsley et al. (1990)	???	NA	LWIR
Kirkwood and Cartwright (1991)	2.0	0.1	LWIR
Klir and Heath (1992)	2.0	0.1	LWIR
Wiggers and Beckerman (1993)	1.4 (WFOV)	0.25	LWIR
	0.25 (NFOV)	0.25	LWIR
Sidle et al. (1993)	0.375	0.2	LWIR
Reynolds et al. (1993)	2.3 (WFOV)	0.2	LWIR
	0.4 (NFOV)	0.2	LWIR
Boonstra et al. (1994)			
Garner et al. (1995)	1.4 (WFOV)	0.16	LWIR
	0.35 (NFOV)	0.23	LWIR
Naugle et al. (1996)	1.4 (WFOV)	0.16	LWIR
	0.35 (NFOV)	0.23	LWIR
Gill et al. (1997)		0.1	LWIR
McCafferty et al. (1998)	1.1	0.1	MWIR
Burn et al. (2009)	0.625	0.12	LWIR
Staring systems			
Havens and Sharp (1998)	0.6	0.05	MWIR
Dymond et al. (2000)	1.4	0.16	LWIR
	1.0	0.10	MWIR
Bernatas and Nelson (2004)	10° WFOV	0.25	LWIR
	3° NFOV		
Bernatas (2010)	20° WFOV	<1.0	LWIR
	5° NFOV		
McCafferty et al. (2010)	1.3	0.08	LWIR
Focardi et al. (2013)	1.3	0.10	LWIR
Franzetti et al. (2012)	0.6	0.06	LWIR

TABLE 8.3 A Listing of the Performance Specifications of Some Representative IR Imagers That are Currently Available From Commercial Sources

Detector	HFOV	VFOV	IFOV	NETD	MRTD (@ 0.5)	λ Range
Handheld Units						
PtSi (256 × 256)	9.0	7.0	0.6	0.07	0.05	MWIR
InSb (256 × 256)	9.0	7.0	0.6	0.025	0.025	MWIR
	4.5	3.5	0.3	0.025		
	1.8	1.6	0.12	0.025		
	0.9	0.8	0.06	0.025		
InSb (320 × 256)	11.0	9.0	0.6	0.025	0.025	MWIR
	5.5	4.5	0.3	0.025		
	2.2	1.75	0.12	0.025		
	1.1	0.9	0.06	0.025		
InSb (640 × 480)	15	11.3	0.3	0.025	0.02	MWIR
	10	8.0	0.15	0.025		
	2.5	1.6	0.06	0.025		
InSb (1024 × 1024)			0.3	0.025		MWIR
Airborne						
HgCdTe (four elements scanned)	4.7	4.0	0.3		0.47@1.5	LWIR
InSb (256×256)	17.6	16.5	1.2	0.03		MWIR
	1.76	1.65	0.12	0.03		MWIR

HFOV and VFOV (degrees); IFOV (milliradians); NETD (°C @ 23°C); MRTD (°C @ spatial frequency in cy/s); Most airborne units are also equipped with a bore-sited CCD camera to give video coverage of the habitat covered by the thermal imager.
With the exception of the HgCdTe detector these imagers use staring focal plane arrays.

setting where size and weight requirements can be relaxed, the cooling demands can be met with bulk cooling using liquid nitrogen or with a Joule–Thompson cooler and a source of high-pressure nitrogen.

For most surveillance applications both the animal and its background are radiating a large amount of energy but the difference between them can be very small. Ideally, one wants to achieve the maximum contrast in the image and still be able to display the imagery so that the full range of temperatures (color or gray scale) is available. This is accomplished by adjusting the total temperature difference (gain) that can be displayed in an image without saturation occurring. The mean value about which this temperature is set is called the level or offset.

Usually there are several temperature ranges built into the camera and on some cameras the operator sets the offset for the intended application. Some cameras such as the FLIR ThermaCAM® SC640 are fitted with an automatic atmospheric transmission correction based on inputs for range, atmospheric temperature, and relative humidity. Other features usually found on imaging cameras that the user should be familiar with include: autofocus, range finding, laser pointer, GPS capability, choices of lenses and FOV, data handling capability (speed), interfaces (Fire–wire, USB, IrDA, SD card), and built-in NUC (nonuniformity correction). Improvements to the performance and updates on features to make the imagers easier to use and to perform additional functions are constantly being developed so it is wise to check with the manufacturers to see if they are available for the unit you select.

There are many other considerations that must be made when deciding what imagers are best for a particular field study, whether it is an aerial survey of large mammals or a ground-based study of insect nesting sites. Important factors to consider are weight, size, air or ground use, power consumption, battery life, recording equipment, monitors, multichannel monitoring capability, robustness to mechanical and weather abuse, data output and recording options, GPS videotape overlay, gyroscopic stabilization, and cost. Care must be taken when choosing subordinate equipment so that the image quality is not compromised by inferior monitors or recording equipment, for example. Most thermal imagers today are small, compact, and robust devices built to be used outdoors.

VERIFYING PERFORMANCE

By comparing the specifications from camera to camera one can get a rough idea of what to expect when the unit is deployed in the field. However, the specifications of the camera are determined in a laboratory setting, so if the images are not collected in a laboratory environment the projected performance of the camera (in the limit) will not be realized. The performance of the camera should be verified prior to every use and tweaked to get optimal imagery. Each field application will have its own set of parameters that can compromise the image quality that the camera is capable of producing. The prevailing atmospheric conditions at the time of use are one of the most important considerations for the choice of an MWIR or LWIR imager. The effects of the atmospheric transmittance, solar reflections, and ambient temperature will be important factors in the determination of image quality, regardless of detector choice. Unfortunately these factors are beyond our control and at best we can only hope to minimize their effect on the imagery by selecting the best opportunity offered for data collection. Note that there is generally some leeway in the time selected for survey and census fieldwork, as opposed to the urgency demanded by military, police, and fire/rescue operations.

Aside from the obvious problems posed by the local environment and adverse atmospheric conditions, we can focus our attention on mastering the factors that we can control that influence the image quality. A correct judgment regarding image quality cannot be made by just observing an image without knowing the details of how it was collected (methodology) and the conditions under which it was collected. An important part of the methodology begins before the imager is actually taken into the field. The user needs to determine what is required to achieve optimum image quality. In this regard, it is necessary to put the imager through a series of tests to measure its performance. All of the performance parameters should be tested and all of the supporting equipment should be optimized for obtaining the best imagery. It should be tested with test targets (different size and intensity), backgrounds, temperature differences, background clutter conditions, ranges, and atmospheric conditions (the limits of the imager should be established for a wide range of atmospheric conditions). Users should be confident in operating the camera and making adjustments to the gain, level or offset, focus (verify that autofocus is operating properly), and above all they should be patient in developing their expertise to successfully operate their camera for the intended applications.

The methodology used to acquire the desired images for each application, regardless of the IR imager selected, is more important than camera specifications. Nonetheless, given a choice we would select the best imager available within our available budget. The important parameters to consider for both LWIR and MWIR imagers are the spectral response (LWIR or MWIR), field-of-view (FOV) and instantaneous field-of-view (IFOV), the thermal sensitivity or the noise equivalent temperature difference (NETD), and the minimum resolvable temperature difference (MRTD). If we consider how each of these parameters affects the performance of the imager and then design our field experiment or survey to completely utilize the benefits provided by each of them then we have taken the first step in optimizing data collection.

TYPICAL MWIR CAMERA

Some of the thermal imagery used in this work was acquired with an Inframetrics, 3–5 μm InfraCAM® Model-A handheld thermal imager and supporting electronics, as shown in Figure 8.4. This camera uses a PtSi (256 × 256) FPA with an approximately 80% fill-factor cooled to 75 K with a self-contained Sterling microcooler capable of 2000 h of maintenance-free operation. For a 50 mm lens and 30 μm detector elements, the camera has an IFOV of 0.6 mrad, an NETD of 0.07°C @ 23°C, an MRTD of 0.05°C @ 0.5 cy/mrad, and a field-of-view (FOV) of 8° (H) × 8° (V). The camera is equipped with an electronic viewfinder consisting of a miniature TV with a 1.8 cm CRT display and an adjustable eyepiece for obtaining sharp focus of the display. The camera is battery-powered, as are the other components shown in the photograph. The gyrostabilizer mounted on the camera prevents jitter in the imagery. The camera weight complete with lens, battery (standard 6V camcorder), viewfinder, and gyrostabilizer is ~1.5 kg, which is easily managed with one hand. While this

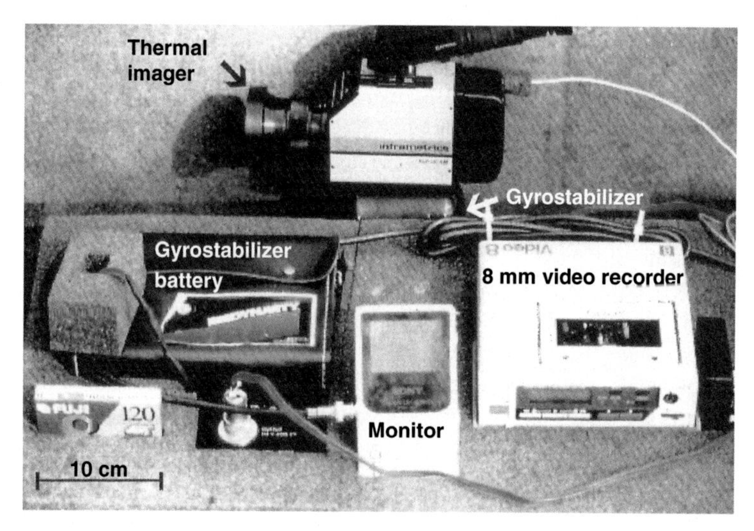

FIGURE 8.4 **A photograph of an early version small portable MWIR thermal imaging system ready for field use.** The entire package fits into a small padded case, which easily slips into a small backpack.

represented a significant improvement in camera capability over earlier imagers there are many new imagers now available with even better characteristics, as noted in Table 8.2.

In the past the single most influential thing in preventing thermal imaging from advancing was the cameras themselves. The methodology was slow to advance because of a general lack of understanding regarding the operation of thermal imagers. In many cases it wasn't clear as to what the imagery should be, based on the specification of the cameras. The image quality was inferior and the root cause of this degradation was difficult to pinpoint. Many times the camera was deemed to be inferior but the methodology also contributed greatly to the loss of image quality.

Most thermographers agree that for applications requiring long ranges, high sensitivity, and the best time response, a camera with a cooled FPA of photon detectors should be used. As far as fieldwork goes for censusing and surveying, particularly on a landscape scale (e.g., Burn et al., 2009), there is a big difference in using thermal imagers that utilize quantum detectors as the sensitive elements as opposed to detectors that rely on thermal effects to generate an image. A number of texts have been written describing the details of using pyroelectric and bolometric imagers for a wide range of applications. On the other hand the use of 3–5 μm and 8–12 μm photon detectors has not been covered and as a result, when researches take these cameras into the field, the cameras are generally not used in the correct fashion to optimize their utility. Vollmer and Mollmann (2010, p. 114) noted: "For more demanding applications requiring higher sensitivities or/and time resolutions such as in R&D, photon detector FPA cameras are used. For the MW and LW spectral region, the FPA is cooled to liquid nitrogen temperature $T = 77$ K. The photon detector FPAs exhibit a much larger responsivity/detectivity and a spectral response varying with wavelength. Cameras using photodiode arrays offer time constants of ~ 1 μs and therefore smaller NETD values and higher frame rates than microbolometers cameras." These features generally cost more but the user must decide if thermal imaging will solve the problems at hand and if it is worth the price.

As we saw in Chapter 7, the performance of thermal imaging cameras can be characterized by a handful of parameters. However, the extraction of useful imagery from a state-of-the-art infrared imager is a complex process. It involves much more than just conducting surveys with the best imager available. It is not enough to simply view an image to determine if it is optimized. For some applications the image quality may not be the most important criterion for successful fieldwork. It may be that only hot spots are needed to make accurate assessments of the presence or absence of a particular species or event. In addition to the performance capability built into the cameras we must also consider a number of other miscellaneous factors that ultimately affect the image quality. There is a clear distinction between seeing a spatial distribution of thermal energy (detection) and seeing the detail in that thermal energy distribution (resolution). Atmospheric transmittance, the scene content, observer experience, subordinate equipment, aircraft type, ambient illumination, ambient temperature, and other factors affect image quality. These factors are not necessarily mutually independent of one another, which contributes to the complexity of interpreting image quality.

Accessories

The availability and longevity of the supporting accessories needed for a day of thermal imaging in the field can often be a limiting factor as to how long one has to collect the data or even where the data can be collected. Days of planning can be wasted by faulty or uncharged batteries, for example. Windows of opportunity provided by excellent atmospheric conditions can be wasted because supporting data storage equipment failed in the field. To operate a thermal imager in the field there are many other considerations that need to be addressed to make sure the imager will survive the day and remain functional for the day during its intended use. Table 8.4 provides a list of suggested items to check prior to deciding on buying an imager for use in the field (specific applications) or if the imager has already been purchased then what will be needed to keep it functional in the field.

If we just consider the performance parameters associated with the imager, we intuitively want the IR camera to be capable of forming thermal images of small objects that have small temperature differences with respect to their backgrounds. Additionally, for animal censusing, we would want to be able to resolve these images from a considerable distance and over a considerable area

TABLE 8.4 A Checklist of Infrared Camera Accessories Which Should be Available Before Going to the Field to Collect Imagery

Size and weight
Handheld unit (weight, portable)
Tripod mounted
Stabilization requirements
Vibrations or oscillations (slow changes in pointing stability) or high- speed changes (jitter)
Vehicle or aircraft mounted
Mounting requirements (adequate space, secure mountings, FAA approval, room for observer and operator, room for supporting equipment)
Power requirements
Battery operated
Quality (lifetime of operation)
Chargeable (charger and power source available in the field)
Viewing platform powered (vehicle or aircraft)
Power requirement available (other equipment needed to provide compatibility)
Cooling requirements
Battery powered
Replacements available (chargeable)
Bulk cooling
Liquid nitrogen supply
Joule–Thompson
High-pressure nitrogen supply
Size and weight of supporting equipment
Portable
Mounting and compatibility with platform
Electronics, monitors, computers, data storage, printers (power sources)
Cost
Entire system (camera, batteries, chargers, computer, software, data storage, interfaces)

(i.e., from an aircraft flying transects). The differences in the performance parameters required to meet the demands of a specific application will, to a large extent, depend on whether the application is for surveillance, where the ability to detect and recognize animals at a long range is the goal, or whether the application is one where measuring temperature or determining the temperature distributions and their relative values in an image is of importance. Williams (2009) points out that manufacturers of thermal imagers do not always provide the data that is relevant to the particular application intended or it is not presented in a manner whereby the user can assess the suitability of an imager for the intended application. To ameliorate this situation Williams (2009) has included in his book a typical datasheet that might be provided by a manufacturer. He refers readers to the appropriate section in his book, which provides more detail on specific parameters of importance.

Chapter 9

Properties of Thermal Signatures

Chapter Outline

In Chapter 7 we described a thermal image, briefly discussed how it was formed, and identified some of its basic properties. Thermal signatures of animals are a part of the thermal image formed by the apparent temperature difference between the animal and its background. In Chapter 5 we showed that the radiation that appears to emanate from the animal depends on its emissivity and any radiation that is reflected from the animal from its surroundings. As such, a signature is a special part of a thermal image and can be classified by its size, shape, intensity, and intensity distribution relative to other parts of the image. Generally, it is a goal to detect the presence of an animal in the thermal image, which means we need to look at the size, shape, and intensity or brightness of any anomalies that are found in the total image. If the anomaly is recognized as something we are searching for, then the perceived anomaly is called a signature. Otherwise, it is an unwanted anomaly called background clutter. Signatures and their quality can change with the prevailing atmospheric conditions so care must be taken to account for these changes when viewing thermal images.

INTRODUCTION

An infrared (IR) camera senses and displays a spatial distribution of thermal (heat) energy instead of visible light. This allows one to see in total darkness, through smoke, and other low visibility or low contrast situations. These cameras can also be used during daylight hours to see heat-generated images when

Thermal Imaging Techniques to Survey and Monitor Animals in the Wild: A Methodology
http://dx.doi.org/10.1016/B978-0-12-803384-5.00009-9

visual observation is inadequate for distinguishing a heat-emitting animal from its background. Generally speaking, the camera provides a display of the spatial distribution of the thermal energy emanating from a scene (thermal image) and the individual animals of interest within the scene are referred to as thermal signatures. Each signature is a composition of the spectral distribution (wavelength band of infrared radiation), spatial distribution (size and shape), and intensity distribution (temperature) that allows it to be distinguished from other signatures and the background of the scene. There is also a temporal component associated with these distributions that can be used to aid in the interpretation of collected imagery (i.e., the animals may be moving). It is instructive to examine a typical thermal image to get a feel for some of these factors. Since thermal images are formed by the radiated intensity difference between animals and their backgrounds, it is important to recognize that the temperature and the surface of the animal play a major role in the recorded energy distribution that the camera captures. In other words, it is more than just a temperature difference; it is an "apparent" temperature difference (see Chapter 5) that produces the image since the surface moderates the emissivity and the resulting energy distribution emanating from the object. The reflected and scattered ambient radiation also contributes to the energy distribution responsible for the imagery.

If we were to examine the side view of a cylinder with a thermal camera, its thermal signature would appear as a rectangular-shaped object with blurred edges along the long axis as a result of the angular dependence of the emissivity on the viewing angle. Its end-on view would be a circular-shaped thermal signature and the edges of the signature would appear sharp with no blurring since the angular dependence of the emissivity would not play a role in the sharpness of the image contrast. Both of these signatures would be detected and recognized by their spatial distribution, one as a rectangular-shaped object and one as a circular-shaped object, but neither of these images viewed alone would allow us to identify the object producing the images as a cylinder. In general we have only the shape of the thermal signature to use for identification and as this example points out, the shape depends on how the object is viewed.

In the spatial domain the size, shape, intensity distribution, and grouping of thermal signatures in a thermal image can contribute to the processes of detection, recognition, and identification. In the thermal image of Figure 9.1, for example, the spatial distribution of 3–5 μm thermal energy emitted by a house cat (Sniper) and his surroundings provides the necessary and sufficient information for detection. The intensity of the signature also provides valuable information but to a lesser degree. If the portion of the image portraying the cat is removed from the photograph then one might have difficulty in determining what the remaining image is about. The reason for this is that we lose all information related to size when the signature of the cat is removed. If we could determine or recognize that the nearest object in the foreground is a small tree then we could regain the size indicator. The shape of the thermal signature presented by Sniper is one that is readily recognized as a cat. The signature of the cat is

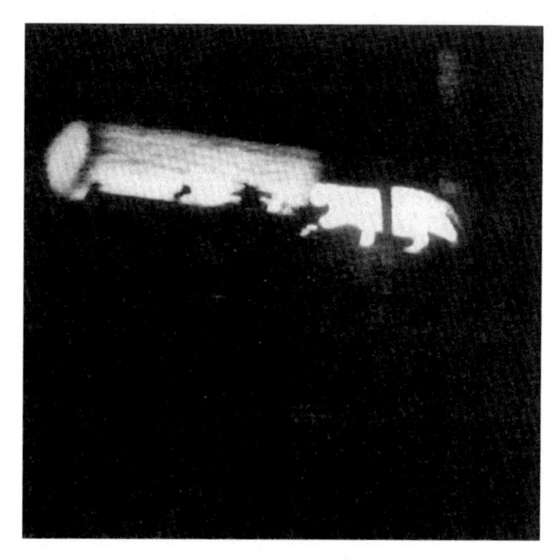

FIGURE 9.1 **Photograph of a thermal image of a house cat in a forested setting just after sunset on a warm evening in coastal Virginia.**

also the brightest object in the scene and this is the preferred situation (objects of interest are the brightest). When the objects of interest are the brightest then the intensity of the signature is very important in the detection scheme since we can adjust the gain and light level of the imager to eliminate all the other bright objects in the scene except the cat.

IMAGE QUALITY

Many factors can influence the decision of when is it best to collect thermal imagery in the field. In the last two chapters the performance parameters and camera features that are needed to get high-quality images were discussed and were basically separated into two separate areas of interest: surveillance and temperature measurement and thermal mapping applications. In both of these missions care must be taken if the data collection is to occur outdoors or in the field. Most surveillance applications require only qualitative measurements and it is generally not necessary to know the absolute temperatures and emissivities of the backgrounds or of the animals being observed. The relative difference in the radiant temperatures within a scene is all that is needed to detect animals of interest.

Signature Saturation

The spatial distribution of an intensity pattern can provide significant information. For example, the heads and legs of turkeys and herons and the eyes and ears of many mammals can be much brighter than the rest of the animal's body. The thermal signature of a heron shown in the photograph below (Figure 9.2)

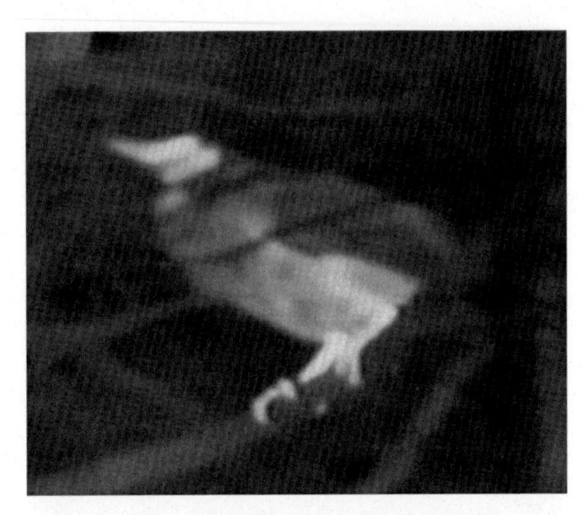

FIGURE 9.2 **Thermal image of a yellow-crowned night heron showing the spatial distribution of radiation across the image as a result of its plumage.**

has a distinct pattern (intensity distribution) associated with the plumage, particularly at this viewing angle. If ground surveys were conducted with multiple species occupying the same roosting site this type of imagery would allow for the identification of the species. These features can provide the uniqueness to the signature of the species being surveyed. By identifying species-specific physiology contained in a thermal signature the identification process can be greatly enhanced, whether it is related to a morphological difference as simple as shape or something more detailed such as pelage or plumage changes across the surface of the animal's body. Animals can easily be detected from their thermal signatures and they can be distinguished by body shape and any regular variations in the spatial distribution of intensity or emission across the signature if the signature is not saturated (see Chapter 8).

Figure 9.2 is a thermal signature of a heron perched in a tree. The image was taken with a 3–5 μm imager in coastal Virginia in the early evening.

In cases where there are very bright signatures often the spatial distribution associated with anatomical features will be lost unless the thermographer has the ability to reduce the gain of the camera, as discussed in Chapter 8. High-gain imagery generally produces saturated signatures that are easy to detect but the intensity may make the signature more difficult to identify because only the shape of the signature will be available to assist with the identification. Every effort should be made to acquire unsaturated imagery unless the goal is just detection or if the shape of the signature will allow unambiguous recognition or identification even when the signature is in a saturated state, as in the case shown in Figure 9.3A, which is a thermal image containing saturated thermal signatures of three deer (a buck and two does). In these signatures there is no information or details observable within the uniformly white signatures except perhaps their cool noses. In most saturated signatures this will be the case; however, in this particular situation the shape of the signatures is adequate for detection, recognition,

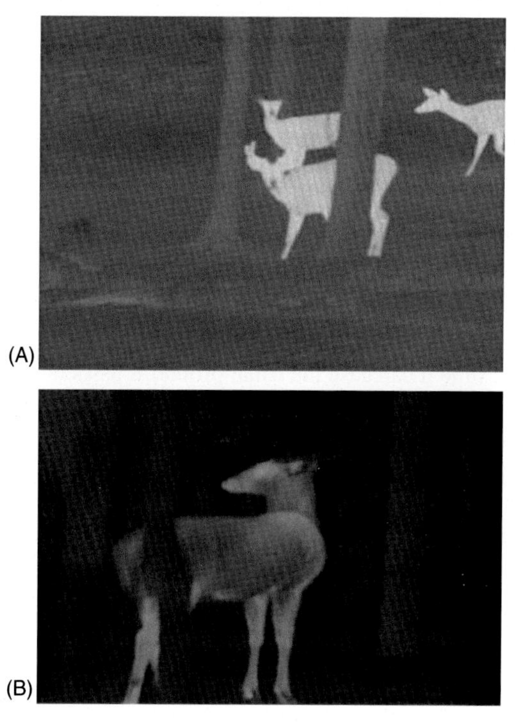

FIGURE 9.3 (A) Thermal image of three deer in a forest. The thermal signatures of the three deer in this image are saturated (spatial distribution of emitted radiation appears to be uniform). (B) Unsaturated thermal signature of a deer obtained by adjusting the gain and light level of the infrared camera.

and identification. Figure 9.3B shows an image with an unsaturated signature of a deer for comparison. If the imagery of the deer in Figure 9.3A was recorded with less gain such that their signatures were unsaturated and more like those shown in Figure 9.3B then identification of individuals within the group could be possible. Normally the saturation of signatures is not a problem for surveying or censusing animals where detection is the primary goal.

Saturation of imagery and signatures is more of a problem for thermographic studies in medicine and physiology than it is for surveillance applications unless identification is an integral part of the surveillance application. The information sought in thermographic observations involves comparing and, in most cases, measuring the variations in the distribution of heat (temperature changes) across the surface of the animal. This requires that the thermographer use gain and level settings that do not saturate the thermal signatures. It is sometimes best to work close to the saturation level for surveillance applications unless there are other targets close to the same apparent temperature difference in the scene. If this is the case then some discrimination might be called for to ensure that the detection rates are not including false signatures.

The quality of thermal images collected in the field depends to a large extent on the imaging cameras (see Chapter 8). In many cases the cameras can be set to collect imagery without additional adjustments, depending on the application. There is much that can be done in the field during the image acquisition phase that will help to secure quality imagery. In the field an observer is typically viewing a monitor or other viewing device that allows the imagery being collected to be reviewed as it is being collected and recorded to video. If during a survey there are significant changes in the apparent temperature differences within the scene because of changes in the camera pointing direction then adjustments can be made in the gain and level to maintain the best contrast and image quality. Since the data is also being recorded to video for comprehensive analysis at a later time it can ease the usual problems associated with visual surveys (observer experience, bias, workload, and fatigue). Nonetheless this phase of data collection will play a role that is influenced strongly by training and experience to ensure that quality imagery is being recorded. The significant difference here is that the decisions related to issues involving image observation/interpretation in the field can be significantly reduced or eliminated.

Once an IR camera has been selected for use then field studies can be designed (finding, monitoring, or counting), realizing that we will be using radiation in a specific region of the infrared spectrum and that the objects of interest can have spectral, spatial, and temporal components. A scene may contain biotic objects of interest that present thermal signatures of varying spatial extent since they may be viewed instantaneously at different angles, be partially obscured, or appear to be of differing sizes. The backgrounds in which these biotic objects of interest are imbedded may include signatures of a similar size and shape but are not of interest (i.e., inanimate objects, rocks, deadfall, standing water, etc.) that are radiating. Some objects in the scene may be moving with respect to one another and with respect to their backgrounds so that the spatial extent of their thermal signature is continually changing in time. After the images are produced an observer can examine them to determine their content. At this phase, when the scene content cannot be changed, the observer must select, monitor, or count objects of interest to the particular field study. To this end, the spectral, spatial, and temporal content of the collected images is all that can be used for accurate interpretations. In practice this is not as difficult as it first appears. The imagery is recorded as video so there is motion as well as changes of viewing angle, animal posture, and groupings. Morphological characteristics can become evident in the signatures, all of which can aid in the image interpretation.

SPECTRAL DOMAIN

As pointed out in the previous section, thermal imaging systems only respond to a small portion of the spectrum. They only detect radiation from the background and animals within the spectral range of the imager being used. Additionally, the background and animals generally only emit a fraction of the

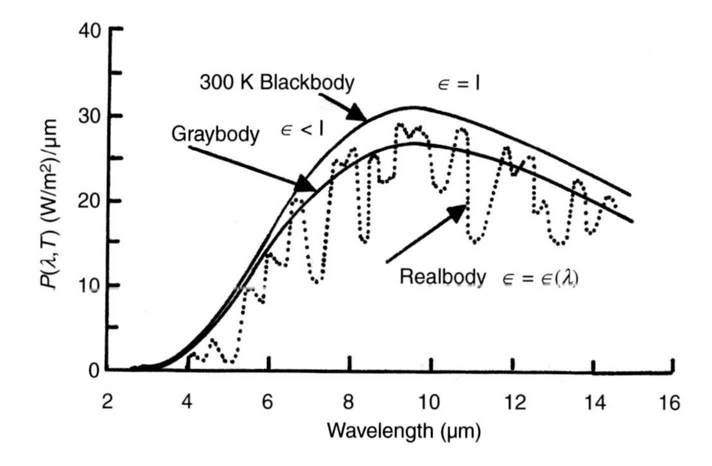

FIGURE 9.4 The emitted power density depends on the emissivity of the animals (real bodies), which are only approximated as graybodies.

radiation emitted by a blackbody. Recall from Chapter 5 that we assumed that Kirchhoff's law equation (5.2) was wavelength independent and that the transmittance (τ), absorptance (α), and reflectance (ρ) had no spectral characteristics, which in general is not true. We can rewrite equation (5.2) showing the spectral dependence as

$$\tau(\lambda)+\alpha(\lambda)+\rho(\lambda)=1 \tag{9.1}$$

Up until now we have treated the emissivity as a constant value $\varepsilon < 1$ for opaque objects $\tau(\lambda) = 0$, such as animals and objects in the background of a scene, as graybodies. The emissivity of real emitters such as animals and objects in a scene is not constant but depends on wavelength (Figure 9.4). This is due to a number of factors, including selective absorption and reflection of the surfaces, local atmospheric conditions, and the dependence of emissivity on the surface conditions of the animals at the time the imagery is collected. As a result, real emitters can only be considered approximations to graybodies. The closer the emissivity is to a constant in the spectral range of the imager the better the approximation will be.

For fieldwork, the sun is the source of heat for creating temperature differences between objects in a scene and for animals with respect to their backgrounds. The bulk of solar loading occurs in the visible and near IR spectral regions $\lambda < 2$ μm (see Figure 9.5). Recall that the sun is a blackbody radiator with a temperature around 6000 K and the earth's ambient temperature (soil, trees, rocks, water, and animals) is about 300 K. The peak of the spectral radiant exitance from the surface of the earth is given by Wien's law ($\lambda_{max} = 2893/T$ μm) or 9.64 μm, which lies approximately in the middle of the LWIR spectral range. It is important to note that the peak radiant exitance for the sun is in the

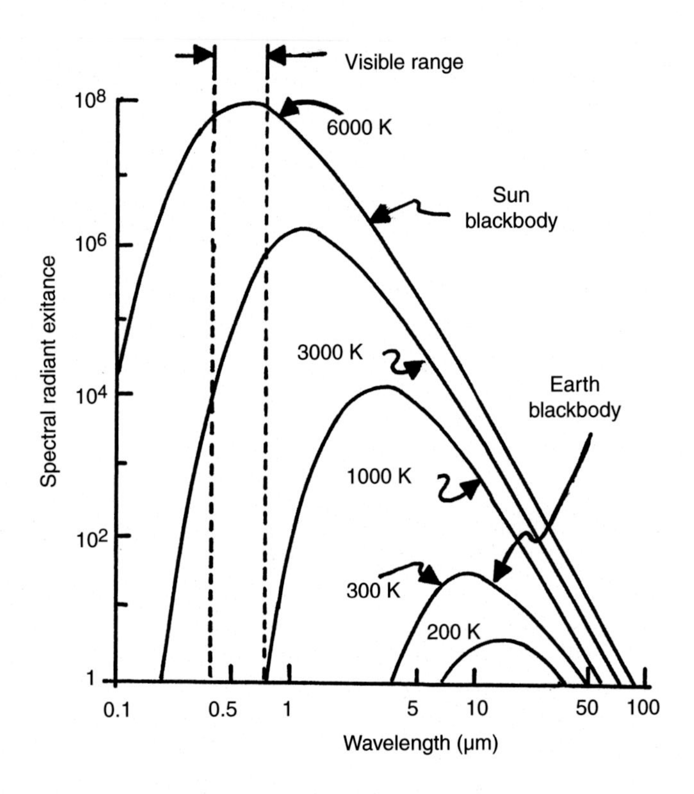

FIGURE 9.5 **The spectral distribution of radiation from blackbodies at various temperatures.** The radiation from the sun peaks in the visible region of the spectrum and the radiation from the earth peaks in the LWIR at 9.64 μm.

visible spectral range at 0.5 μm and that reflected solar radiation in the spectral range from 0.5 μm to approximately 3 μm dominates the radiation leaving the surface of the earth. Beyond 3 μm emission of thermal radiation dominates.

The direction and intensity of solar radiation at the surface of the earth is constantly changing throughout the day and naturally occurring objects such as rocks, earth, trees, and surface vegetation are heated by the absorption of this energy according to their absorption coefficients α_{SOLAR} at these wavelengths. These same objects are constantly remitting this energy in the infrared spectral region with an emissivity ε_{IR} similar to the real emitter in Figure 9.4. Note that this same solar loading will also be occurring for animals but with the caveat that they will also be radiating energy derived from metabolic activity as well. Gates (1980) gives the mean solar reflectance and absorptances for the dorsal and ventral aspects of a wide range of animals including birds, mammals, amphibians, and reptiles. The background and the objects with high absorption in the visible and near infrared ($\lambda < 2$ μm) will all heat up from solar loading. The actual temperatures will depend on the mean absorption coefficients at the solar wavelengths α_{SOLAR} and the emissivity in the infrared ε_{IR} where heat is lost as radiation. The ratio of the absorption to emission determines if an object is heating up or cooling down. For example, if $(\alpha_{SOLAR}/\varepsilon_{IR})$ is large the object will warm up and if it is low the object will cool down. If solar absorption ceases as a result of sunset or by shading effects from trees or passing clouds then the objects will all cool, regardless of their infrared emission characteristics. Emissivities are usually $\varepsilon > 0.90$ for animals in the infrared so they will be cooling or emitting under most conditions in the field.

For the most part thermal images are collected in the far infrared spectral range at either 3–5 μm or 8–12 μm wavelength bands. Wyatt et al. (1980) made a determination from examining the field data gathered by Marble (1967) and Parker (1972) that thermal contrast (ΔT) (the intensity difference between deer and its background) was insufficient to identify to species level. The problem they encountered was that the deer signatures were easily detected (i.e., recorded by the thermal imager) but were imbedded in a background of other thermal signatures of uniform intensity with various shapes that were equally or more intense than those of the deer; this is a situation known as background clutter (see Chapter 11). As a result of a large number of competing signatures of comparable intensity the data was unsuitable for use in an automated data identification system based on the intensity of the signatures gathered but it does not mean that thermal contrast (ΔT) is insufficient to detect, recognize, and identify deer in the field. For the identification of animals, the use of smart and cheap image processing techniques like studying video could work where size and shape might play a role.

The results of these studies formed the basis for the design of a multispectral classification approach for the remote detection of deer (Trivedi et al., 1982, 1984). The effort to develop an automatic recognition system for deer based on spectral reflection data in the near-IR spectral band met with limited success. They reported that when controlled field data collected in a previous study at four wavelengths preselected for a maximum discrimination of deer, green vegetation and dry brush on a snow background was subjected to a multistage pattern recognition algorithm, they obtained a deer detection accuracy of 55.2% with no false counts. The development of a deer detection system is only mentioned here because it was an attempt to improve on a previous method, even though it is not a thermal imaging system but rather a remote sensing system.

Wyatt et al. (1985) determined from emissivity measurements that most biological samples have emission spectra that do not have appreciable wavelength dependent features, whereas reflectance measurements indicated the presence of unique spectral signatures for mule deer (*Odocoileus hemionus*) and some commonly occurring backgrounds like snow and evergreen. The determination that the emission of biological samples shows little spectral dependence indicates that they are fairly good graybodies in the spectral region measured. It appears that the three sands that were measured show the greatest variation in emissivity (~20% change). The other objects that were measured would typically be found on the open ranges where the deer would be surveyed either spectrally or thermally and their emissivities were all found to be approximately the same in value ($\varepsilon > 0.9$). An open range with brush and juniper provide ideal conditions (Trivedi et al., 1982, Figure 3) for thermal imaging survey since the deer will maintain their body temperature during the diurnal cycle and all of the other objects will not. This sets the stage for thermal imaging surveys where large apparent temperature differences will be the norm.

SPATIAL DOMAIN

When examining thermal signatures the properties of interest (size, shape, intensity, and grouping) are useful in making a determination regarding the probability of detecting a species of interest. The actual determination can be influenced by many factors and they are not entirely independent of one another. For example, is the size appropriate for the species? Does the signature have a shape that conforms to what is expected? Is the signature brighter than other objects of similar size? Is it significantly brighter than the background and is there more than one signature with similar properties in the image? The size, shape, intensity, and grouping of signatures in a thermal image creates a natural hierarchy, relating the properties of the spatial distribution of radiated energy in any given thermal image as to their usefulness in the detection phase of image interpretation. Perhaps more importantly, these properties can be enhanced somewhat during the data collection phase of the field studies.

Size

For a given optical system the size of the thermal signature (which can be used for the identification of species, sex, and/or age) will vary for a given range, R_C, depending on the animal's position within the footprint cast by the field-of-view (FOV) of the imager on the landscape. Animals of the same size will appear smaller when located at the most distant extent of the footprint than animals closer to the observer. The actual difference in the observed size will depend on the optics and viewing angle selected for the survey but typically vary by a factor of two or less across the extent of the footprint. If the animals are close together then sex and/or age can be easily determined (e.g., see Figure 9.3A).

The size of recorded thermal signatures is related to camera optics, IFOV, distance to the object (range), object size, and viewing angle. The size and shape of a thermal signature can, in some cases, be intimately coupled such that a change in one changes the other; however, in many cases size alone can be used to achieve ~100% detection rates in the field if proper survey designs and proper methodology are used. For example, the process of determining size can benefit significantly if something in the image is available to provide a

FIGURE 9.6 Photograph of a thermal image containing the signature of an animal, which lacks sufficient information in the background to establish size.

comparison. This can easily be accomplished during the data collection phase of the field studies.

A thermal image is shown in Figure 9.6 that contains a nearly saturated signature of a gray squirrel (*Sciurus carolinensis*), which would only give the viewer an idea of its size if it were already known to be a squirrel. Note that the squirrel is clearly "detected" and "recognized" as a signature of interest but not much more information can be learned from the signature as it is presented. If the thermographer is satisfied that this is the species of interest then a survey could easily be performed with this quality of imagery. If there is doubt that this is the species of interest because of the uncertainty in the size of the signature (there is nothing in the image to allow a determination of relative size) then a closer inspection is warranted. The following image was collected a few seconds later with a slightly lower gain setting on the camera (usually a very quick and easily activated function on most cameras) that allowed features in the background to be included in the imagery (Figure 9.7).

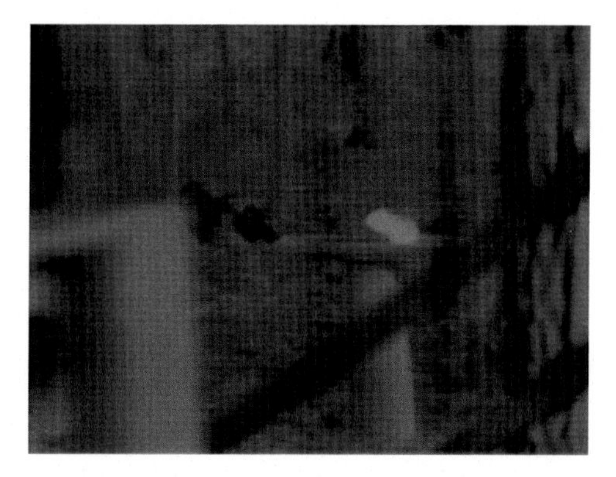

FIGURE 9.7 Photograph of the same animal with sufficient background information to allow a determination of size.

In this image the squirrel is seen sitting on a feeder in a wooded setting, which provides the observer with a way to estimate size. The detected, and now recognized, signature also provides verification of species or "identification".

Shape

Objects of a kind that are to be surveyed will have the same shape when viewed from the same orientation. Typically, aerial views close to the nadir will produce images of the same objects, showing identical shapes albeit with differing orientations (facing different directions) and sizes (as a result of adult/juvenile or male/female). Off-nadir views will present slight changes in the shape of animals of the same species. Animals that are partially shielded by thermally opaque objects (trees, rocks, and deadfall) or atmospheric or weather related shielding (rain, mud, snow, or ice on the pelage or plumage) might also present a different shape. Problems associated with partial obscuration can easily be eliminated. Panning the scene with the thermal imager can provide several different viewing aspects for identification. If animals of similar size, rabbits and fox, for example, are widely dispersed in the same habitat, then the shape of the thermal signature would be key to the success of the survey since identification of the species would be required. In the example above, the shape of the squirrel was formed by the lateral view, which is usually the view of animals that provides the most information when identifying signatures. This brings up a point that is often overlooked when viewing thermal imagery taken from an aerial platform. In aerial surveys most of the imagery shows the dorsal aspect of the animal unless the animal is at the farthest ranges in the survey transect width (swath width provided by the FOV of the imager).

Intensity

The intensity or brightness of the spatial distribution of an object can provide valuable information even though it can vary from animal to animal within the same species, depending on their physical activity and physical attributes such as thickness of coat. Weather conditions, wind, and precipitation can also influence the intensity of the observed signatures. Care must be taken when selecting the time of field surveys or observations since nonbiotic objects in the background that have been subjected to solar loading can present brighter or more intense signatures than the animals being surveyed. Usually the thermal signatures of animals collected in the field are saturated, indicating that there is more than enough signal for the thermal imager to collect useful data. However, when the signatures are saturated it means that the gain is too high and any warm object in the background will also be gain-enhanced, which leads to confounding the imagery. As mentioned before, the spatial distribution within an intensity pattern can provide significant information since the heads and feet of birds and the eyes and ears of mammals are typically much brighter than the rest of the animal. There should be an effort to preserve these features in the collected imagery.

Viewing Angles

From a strictly geometrical point of view, there is an angular dependence on the observed size and shape of any object and its image, whether captured in the visible part of the spectrum or with a thermal imager in the infrared part of the spectrum. A typical thermal signature of an animal in the field at any point in time is a representation of the animal in two dimensions and it can have an infinite array of shapes and intensity distributions (see Monteith and Unsworth, 2008, p. 105). This is because the cross-sectional area of the animal viewed by the camera varies with the viewing angle. As mentioned above, when a typical image is displayed on a monitor for detection purposes, we almost always see a nearly uniform white spatial distribution of radiation (very little contrast within the borders of the image). Under these conditions, imagine a side view and frontal view of a horse. It is safe to guess that nearly 100% of the viewers would recognize the side view as a horse but considerably less would identify the horse from the front view. As we noted above, it is advantageous for thermal signatures to have some distinctive intensity distribution within the border of the signature that can be associated with a species of interest since the problem of recognition and identification can then be significantly mitigated.

Kirchhoff's law $\alpha + \rho + \tau = 1$ for a rough opaque surface or animal (considered graybodies) reduces to $\varepsilon + \rho = 1$. When the viewing angle is normal to the surface of an animal, the emissivity is high and there is very little reflected radiation from the surface of the animal. If the viewing angle is changed to say 45° then both the emissivity and the reflectivity have modest contributions. As the angle is increased even further such that we are viewing the animal at a grazing angle (near 90° with respect to the surface normal), the reflectivity is large and the emissivity is minimal. For angles greater than 45°, the reflectivity begins to increase markedly and a concomitant decease occurs in the emissivity. It should be noted that at angles approaching the grazing angle ($\rho \sim 1$ and $\varepsilon \sim 0$) even rough black surfaces tend to become good reflectors. Fortunately, we won't have to be greatly concerned for most fieldwork with regard to the angular dependence of the emissivity, particularly at the ranges used to conduct most surveys.

The viewing angle is continuously changing when panning with the camera (see Chapter 11) and this affects the size, shape, and intensity (through the emissivity) of observed signatures. Panning will provide different viewing angles of the same individual animal(s), which will give a number of different spatial distributions for identifying species and/or individuals within a species for delineating ratios of adult/juveniles and male/female, for example. Figure 9.8 is a thermal image of a group of deer moving through a forested system. A panning technique used here provided nearly 30 s of video of this group of deer,

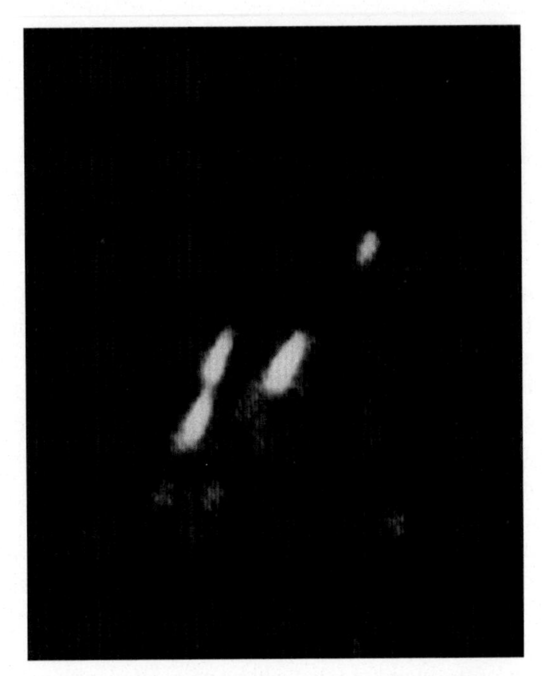

FIGURE 9.8 **Thermal image of a group of four deer in forested habitat.**

allowing the animals to be identified and counted even if the direct view is temporarily obstructed by forest canopy. This image was taken from a helicopter at an elevation of ~180 m above ground level at a speed of 93 km/h.

Grouping

If animals are closely grouped together such as roosting birds, bats, herd/pack animals, or insects their signatures will, under most circumstances, share the above properties (size, intensity, and shape), which can aid significantly in their detection, recognition, and identification. Problems associated with counting the number of individuals comprising large groups can be mitigated by the opportunity to review video of the collected imagery, many times if necessary. The task of counting many animals in a video can also be accomplished by using appropriate software (see, e.g., Melton et al., 2005; Conn et al., 2013; Burn et al., 2009).

TEMPORAL DOMAIN

The temporal domain primarily offers information that can aid in identification but it is also valuable for monitoring, locating, and counting in the field. Reviewing data that was captured on video for later analysis will preserve the information provided in the temporal domain. The information sought is of two types; both involve changes in the observed thermal signatures over time. One occurs in very short timeframes (seconds to minutes) and the other over much longer periods of time (days, months, and years). However, some aspects of both can be recorded on film to aid in detection, recognition, and identification. In the temporal domain we can extract information from motion and the temporal dependence of habitat occupancy.

Motion

When reviewing video of thermal imagery the rapid change in the appearance of a thermal signature can almost always be associated with motion of the object itself. We occasionally see changes in the observed signatures of animals in the field that are not due to the motion of the animal but are due to the motion of the aircraft or vehicle used to gather the data. In these situations the animal's signature is partly obscured by an object (e.g., trees or rocks) and the image of the signature captured on video appears to be moving because its thermal signature is changing. These changes are not rapid changes but occur in the timeframe of minutes, as opposed to seconds. Moving objects in a stationary scene are easily detected, and the nature of the motion clearly identifies the animal in many situations. Figure 9.9 shows two photographs of a black bear (*Ursus americanus*) taken from individual frames of a video only a few seconds apart. The bounding type of locomotion exhibited by the animal is very distinct and clearly identifies the animal as a bear when observed in the video.

Habitat

The periodic use of a specific habitat by certain species can provide invaluable information when determining their identity. The occupation can be daily during certain times of the diurnal cycle or can be for extended periods of time (winter/summer ranges, breeding areas) occurring on a yearly basis. Many times it is important to know when and in what habitat the animals are located. In these cases the gain and light level of the thermal imager can be adjusted periodically to annotate the habitat being surveyed by capturing the background features such as watered areas, fields or meadows, forests, scrub/shrub, or marsh.

VISIBILITY BIAS

Thermal imagery provides a method to observe, count, and study animals throughout the diurnal cycle and there are many times that thermal imagery would be suited for daytime applications. The biggest problem with most visual surveys is that the observers cannot see the animals of interest nor can they pick them out of video or photographs taken during the survey since the visibility problems remain and are just moved from the field to the photographs.

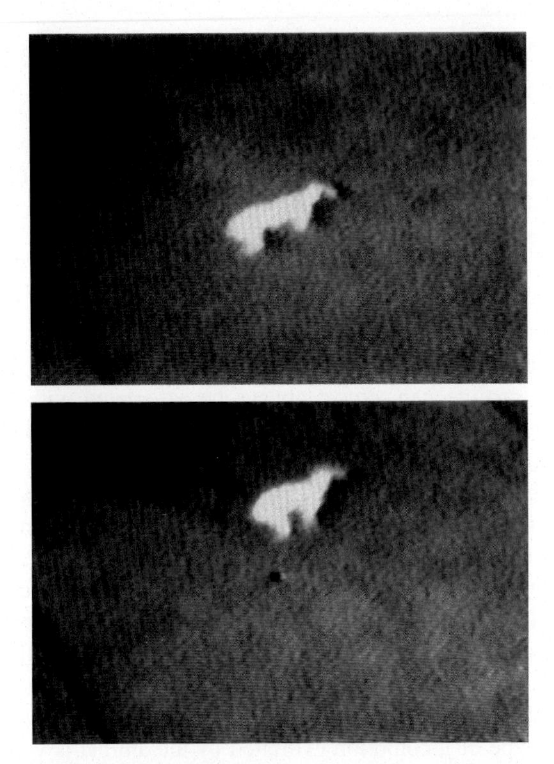

FIGURE 9.9 **Thermal images of a black bear taken from a helicopter at an elevation of ~180 m above ground level at a speed of 93 km/h.** The poor quality of these photos (smearing and blurring at the edges) is due to several artifacts associated with preparing the photographs for presentation as still pictures. The photographs are the result of photographing the images from a video display on a television monitor and subsequently enlarging them to appear as they are shown. If one views the video one sees clear sharp images of a bear moving for the cover of a hammock in southwest Florida.

Thomas (1967) pointed out the difficulty of extracting accurate counts from aerial photographs when there was a lack of snow cover. Likewise, Siniff and Skoog (1964) reported that it was difficult to locate animals on aerial photographs when conditions of heavy vegetative cover prevailed. Garner et al. (1995) concluded that the use of standard VHS video recordings of white-tailed deer and moose were ineffective.

The process of determining the presence of animals in photographic imagery is generally faced with the same problems of visibility bias as those encountered during actual survey conditions in the field except that the observers will have as much time as they need to review the photographs. In the following example try to imagine the amount of information that an observer would be able to extract from the photograph (Figure 9.10A) if the observation time is restricted to a fraction of a second, as it might be in the case of an observation made from a moving platform (e.g., vehicle or plane). By applying the same quick

FIGURE 9.10 (A) 35 mm color photograph of a house cat moving through litter on a forest floor in coastal Virginia during daylight hours. (B) Thermal image of the same cat as in Figure 9.10a taken at the same time as the visible image.

observation to a second photograph (Figure 9.10B) it is easy to gain an appreciation of the capability that thermal imaging provides to an observer, even an unskilled observer. These two photographs were taken from an elevated platform in coastal Virginia in early August when the temperature was ~30 °C in the late afternoon. These photos were taken at the same time and the animal was moving from left to right. The color photo was taken with a 35 mm camera using a zoom lens to closely match the size of the image recorded with the 3–5 μm thermal camera. The thermal camera output was displayed on a TV monitor and photographed with a 35 mm camera and a 50 mm lens. There is very little viewing time during a survey (particularly an aerial survey) and a photograph or video is helpful in extending the viewing time since the photograph or video can be carefully studied after the survey. The problem of visibility bias is still present but with intensive searching of the photos there is a chance that animals missed during the survey will be detected. Now the problem becomes one of deciding how much of the survey needs to be photographed. Would it be of value based on the reports of earlier work using video and photography? (See Chapter 2.)

The most pervasive problem in estimating animal populations stems from the inability to obtain accurate counts of animals in the field during surveys and censuses because they are not detected (see Chapter 2). The lack of detection is directly related to the observer's inability to see the animals in their natural habitats. When thermal imagers are used during the counting phase rather than direct observation, the problems associated with visibility bias are minimized because the coloration differences between the animal and its background are maximized.

Presently there is an ongoing effort by the military to develop an integrated two-color image system by combining a VO_X microbolometer (8–14 μm) and InGaAs (0.7–1.6 μm) detectors into a single focal plane array (Coffey, 2011). Some military imagers operating in conjunction with software provide for a number of specialized tasks such as automatically identifying military targets, performing tracking tasks, controlling camera features such as gain and level, performing image enhancement tasks, and fusing visible images with thermal images. We see from the two images shown in Figure 9.10A and B that even though visible images provide spectral and spatial details, if an animal has the same color and similar spatial characteristics as its background it cannot be readily distinguished from the background. This is due to visibility bias in the visible spectral range; that is, the coloration of most animals blends in with their habitats to the point that they are very difficult to discern in the field. There are a number of efforts to integrate different sensor technologies to move image fusion into the mainstream of remote sensing (Jiang et al., 2013; Sahu and Parsai, 2012). The process of image fusion is one where two or more images are combined such that the combined images reveal more information than is contained in the individual images. Many electro-optic/infrared multispectral systems for intelligence, surveillance, and reconnaissance applications combine at least one reflective band electro-optic sensor with one thermal infrared sensor (Coffey, 2012). Some applications that would benefit from image fusion, in addition to military applications, are remote sensing, animal ecology, robotic vision, and medical imaging. This topic will be revisited in Chapter 12.

SURVEILLANCE

The basic requirements for surveillance applications are considerably different than those for thermoregulation studies or thermal measurement and monitoring applications but all applications require high-quality thermal images to be successful. The image quality is continually judged by the observer on a basis that usually stems from the experience of the observer. Much of the work developed by the military regarding the specific tasks of detection, recognition, and identification based upon the minimum resolvable temperature difference (MRTD) and the criterion developed by Johnson (1958) can be adopted for use in animal ecology studies, which involve surveillance applications. These applications are usually conducted outdoors, where atmospheric changes, time of day, seasonal variations, vegetative cover, large survey areas, animal size, and mixed species might all be contributing factors in any particular surveillance application.

Recall that the MRTD of a system is determined in a laboratory by human observers who are asked to determine contrast threshold values for scaled models of large targets such as tanks and humans. These are then compared with the threshold contrast levels for spatially periodic patterns (standard bar patterns) viewed with the same devices. A correlation is developed between the ability to resolve the spatially periodic bar pattern when compared to the minimum dimension of a target having the same contrast. By varying the spatial frequency (cycles/mrad) or number of bars in the pattern and the temperature of the bar pattern relative to the background temperature, specific tasks are performed such as detecting or identifying a military target (i.e., tank or human). The number of bars for each minimum dimension is increased until the bar pattern is just resolved. The military uses the MRTD to predict the ranges at which targets can be detected, recognized, or identified.

Within the framework developed by Johnson (1958) only the overall features of the military targets are used in these discrimination tests, such as the average temperature difference ΔT_{AVG} between the target and its background and its size. As a result this method of discrimination is based on only a few representative sets of conditions that only apply to certain sets of thermal signatures. The models do not treat a specific target in a specific background. This is done because the alternative is impossible since signatures observed in the field are in a continuous state of change due to the changing environment and changing target temperatures. This means that while the sensor is modeled in detail, the target, background, and observers are treated as ensembles. Nonetheless, these performance metrics provide a bridge between sensor characteristics and task performance (target detection, recognition, and identification).

The minimum dimension, *d* (a linear quantity), of a target used in the early development of this methodology was later converted to a critical dimension (area), which is given by the square root of the target area. (see Night Vision Thermal Imaging Systems, 2001). This change involved testing for both the horizontal and vertical dimensions of the target and is known as the two-dimensional MRTD. These contrast threshold measurements relate the target-angular-subtense and the spatial frequency scale on the MRTD versus spatial frequency plot.

Discrimination Levels

The N_{50} (50% probability) discrimination levels developed by Johnson are shown for both the one- and two-dimensional models in Table 9.1. Here N_{50} refers to the number of cycles that are required for detection, recognition, and identification (e.g., 1, 4, or 8 cycles, respectively) for the one-dimensional model and 0.75, 3, and 6 for the two-dimensional model, which reduced the number of required cycles by 25%.

TABLE 9.1 The 50% Probability of Discrimination for the Tasks of Detection, Recognition, and Identification for Both One- and Two-Dimensional Models

	N_{50} Discrimination Levels		
Model	**Detection**	**Recognition**	**Identification**
One-dimension	1.0	4.0	8.0
Two-dimension	0.75	3.0	6.0

Field tests by Ratches et al. (1975) extended the threshold experiments for discrimination, resulting in the target transfer probability function (TTPF). These results are given in Table 9.2. In this table the multiplier for 50% probability is 1 and to increase the probability the multiplier increases accordingly. The bottom line under each task shows the discrimination levels determined by Johnson. If higher probabilities of discrimination are required for a task then the

TABLE 9.2 The Probability Function Multiplier for Both One- and Two-Dimensional Models for 50, 80, 95, and 100% Discrimination Probabilities

	Probability Function Multiplier		
Discrimination Probability	**Multiplier**	**N_{50} One-Dimension**	**N_{50} Two-Dimension**
Detection			
1.00	3.00	3.00	2.25
0.95	2.00	2.00	1.50
0.80	1.50	1.50	1.12
0.50	1.00	1.00	0.75
Recognition			
1.00	3.00	12.0	9.00
0.95	2.00	8.00	6.00
0.80	1.50	6.00	4.50
0.50	1.00	4.00	3.00
Identification			
1.00	3.00	24.0	18.0
0.95	2.00	16.0	12.0
0.80	1.50	12.0	9.00
0.50	1.00	8.00	6.00

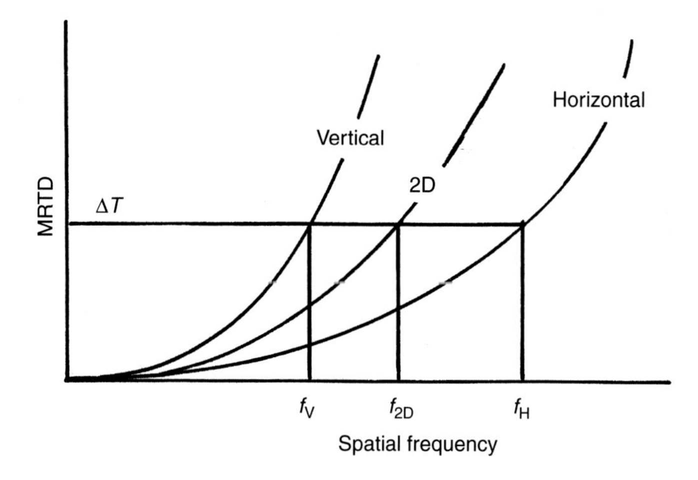

FIGURE 9.11 **Two-dimensional MRTD constructed by taking the geometric average of the vertical and horizontal MRTD frequencies at each value of ΔT.**

one-dimensional or two-dimensional 50% discrimination levels are multiplied by the appropriate multiplier. To obtain a 0.95% probability for identification in the two-dimensional model 12 cycles are needed and 16 cycles are needed in the one-dimensional model.

When the horizontal and vertical MRTDs are combined, as shown in Figure 9.11, the horizontal and vertical frequencies obtained at each thermal contrast value are geometrically averaged to obtain $f_{2D} = (f_V \times f_H)^{1/2}$ at those values. By bringing together the vertical and horizontal MRTD frequencies to form a two- dimensional spatial frequency, there is an approximate response with a radial symmetry associated with the detected object or a "blob" like appearance. When this is done, the bias associated with the one-dimensional model is removed by reducing the one-dimensional discrimination levels by 25%. The procedure for producing a discrimination curve that gives the static probability of detection, recognition, or identification as a function of range requires four parameters: (1) ΔT_A, (2) the characteristic or critical dimension $d_C = (A_T)^{1/2}$, where A_T is the area of the target or animal, (3) an estimate of the atmospheric transmission within the spectral band of interest, and (4) the sensor two-dimensional MRTD. Once a ΔT_A is obtained, the highest spatial frequency that can be resolved by the sensor is determined by finding the spatial frequency on the MRTD curve that matches ΔT_A. The Johnson criterion is important because it connects the spatial frequency scale on the MRTD curves with the angular subtense of the animal being viewed in the field. The apparent temperature difference of the animal is a function of the atmospheric attenuation, which we can write as,

$$\Delta T_A = \beta^R \Delta T \tag{9.2}$$

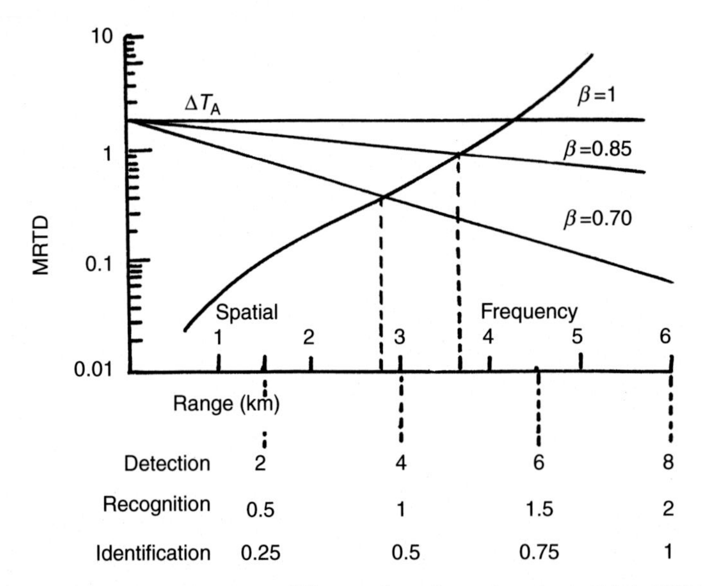

FIGURE 9.12 **Apparent temperature difference for a thermal contrast of ΔT = 2 °C is plotted for three different average atmospheric attenuation coefficients to predict the ranges for detection, recognition, and identification for the two-dimensional model.**

This means that the thermal contrast ΔT is moderated by the atmospheric attenuation β as a function of the range R and allows us to determine the minimum resolvable temperature difference for a particular task such as detection. The target load line is the target contrast modified by the prevailing atmospheric transmission. Three load lines are shown in Figure 9.12, representing the three attenuation conditions imposed by the prevailing atmospheric conditions at the time of data collection. The apparent temperature difference for a thermal contrast of ΔT = 2 °C is plotted for three different average atmospheric attenuation coefficients β = 1/km, 0.85/km, and 0.70/km. Note that when the apparent temperature difference in equation (9.2) is plotted as a function of the range it is a straight line that intersects the MRTD curve and where they cross the MRTD curve they predict the range for the desired tasks of either detection, recognition, or identification.

The number of cycles across the critical dimension d_C that can actually be resolved by the sensor (see Chapter 7) at a particular range determines the probability of discriminating (detecting, recognizing, or identifying) the animal or target at that range.

To convert the values of the spatial frequency to range conditions we need to use a target discrimination value shown in Table 9.1. The spatial frequencies in Figure 9.12 are converted to range values by using the number of cycles that appear across the animal's minimum dimension d_C in meters at a range of R in kilometers. The spatial frequency f_{2D} in cycles/mrad at the sensor can be determined from $N_x = (d_C/R)f_{2D}$.

For the example shown in Figure 9.12 we selected a moose (*Alces alces*) with a $d_C = 2$ m and used the 0.95% probability of discrimination for the two-dimensional model, which means $N_x = 1.5$ for detection (Table 9.2). The conversion from spatial frequency to range is $R = (2/1.5)f_{2D}$ for detection, $R = (1/3)f_{2D}$ for recognition, and $R = (1/6)f_{2D}$ for identification. In this example, with atmospheric attenuation of 0.85/km the moose is detected at 4.9 km, would be recognized at 1.25 km, and identified at 0.625 km, all with 95% probability.

Three levels of discrimination can be extracted from the thermal imagery collected. Detection means we detect and can count warm objects radiating in the 3–5 or 8–12 μm spectral bands, recognition means we can determine if the detected animals are biotic objects of interest, and identification means we can determine what species we have detected. In prior work we have demonstrated that we can identify individuals within a species (Havens and Sharp, 1998). The military devised the methods of using the performance parameters noise equivalent temperature difference (NETD), minimum detectable temperature difference (MDTD), and MRTD (Chapter 7) to meet needs for battlefield tasks and one of the most important is the ability to predict the range at which a target such as a tank, plane, or ship might be detected, recognized, or identified. These methods and concepts have been extended to the use of thermal imagers to predict the performance of all surveillance systems, including those in the field of animal ecology. This is important because it allows one to estimate the ranges that can be expected for a survey or observation when the atmospheric conditions are accounted for during fieldwork.

Surveillance Application

In the example that follows thermal images are shown that were taken in north central Maine about 4 h after dark on a warm evening in July. There was minimal solar loading during the daylight hours due to nearly complete cloud cover. The skies cleared just after sunset and resulted in excellent imaging conditions. The background was uniform after the day of cloud cover with the gravel roadbed showing a little thermal contrast with the vegetative cover due to its higher thermal inertia. The following series of photographs were extracted from a short video and are presented in the order in which they were viewed. Six frames of the video are presented at various stages of a thermal imaging sweep across the scene. The images were taken from a very slow moving vehicle with a 3–5 μm handheld thermal imager. In the first image (Figure 9.13) a deer (*Odocoileus virginianus*) is detected, recognized, and identified and because of the uncluttered background it is the only thermal signature in the scene.

This is good imagery and we also note that the deer is standing in dense low grass and shrubs at the edge of the road. The next photo (Figure 9.14) shows another signature detected by the imager farther down and on the left-hand side the road. The deer in the left of this image is calculated to be approximately 80 m distant from the camera by using the optics and IFOV of the camera.

FIGURE 9.13 **Thermal signature of a deer (*O. virginianus*) standing on an old roadbed in north central Maine.** Range is approximately 80 m.

FIGURE 9.14 **The same deer as in Figure 9.13 and another unrecognizable signature at perhaps 1000 m farther down the road.**

This small signature is one of interest because it is moving and there is no background clutter that could be giving a false signature nor could it be a vehicle because of where it is positioned on the road. The signature is very bright and well within the detection range of the camera but beyond the range of recognition. A conservative estimate based on the rear view of a moose 1.25 m wide and 2 m tall would give a critical dimension of $d_C = 1.5$ m, which would predict a detection range of ~1700 m. However, the animal is probably closer to 1000 m from the camera. In Figure 9.15 the FOV of the camera is now centered on the new signature and the camera is refocused. The vehicle has provided zooming by moving closer to the animal, which reduces the range.

In Figure 9.15 the signature is clearly recognized as that of a large animal that has moved to the center of the track but is still unidentified as to species. At this point we can calculate the range for recognition based on the same value for the critical dimension as used above. The range for recognition is calculated

FIGURE 9.15 **Thermal signature is recognized as a large mammal at ~450 m range.**

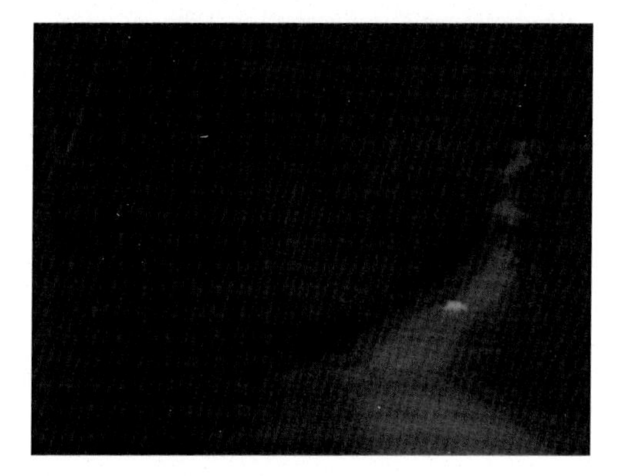

FIGURE 9.16 **More details of the thermal signature can be seen (legs and head) but not well enough for identification.**

from equation (7.10) as $R = (X_{min})/\theta_{IFOV}$ where $d_C/3 = 1.58/6$ m is used as the minimum detectible size X_{min} for 0.95% recognition using the two-dimensonial model (see Table 9.2). The calculation gives the range for 95% recognition as ~450 m. The focusing and zooming continue and Figure 9.16 shows more detail of the animal but not quite enough for identification.

The final frame, Figure 9.17, was taken well within the identification range of ~ 250 m and clearly shows the signature of a moose (*A. alces*) calf. The example above illustrates the concept of using a thermal imager to find or detect a significant difference in the intensity distribution during a scan of a particular scene and shows how a subsequent interrogation of the scene with the imager leads to the recognition and identification of species.

Adams et al. (1997) conducted a series of aerial moose surveys in northern New Hampshire in January, 1995, using a Westinghouse WesCam DS® 10× infrared sensor and a >10× Sony® zoom lens color television camera. They conducted the surveys at approximately 1000 m agl in a fixed-wing Cessna

FIGURE 9.17 **This image clearly shows the thermal signature of a young moose (*A. alces*).**

337g. In the example above we demonstrate that the range for detection is about 1700 m and at 1000 m Adams et al. (1997) reported that they achieved 88% detectability for their surveys (based on 100% detectability during intensive searches) and, perhaps more importantly, this detectability was influenced by the terrain and elevation. Compared to the visible camera they noted that many moose positively identified with the IR sensor could not be viewed with the video camera because they were obscured by trees or dense canopy. They stated that these individuals would probably have been missed by an observer during a traditional aerial survey. If a helicopter were used instead of a fixed-wing aircraft, the flight altitude and air speed would have been reduced and higher detectability would have been achieved.

There is also a safety issue with low-altitude flights and high air speed. Kufeld et al. (1980) pointed out that for animal surveys in rugged mountainous terrain a helicopter was better suited for meeting the survey requirements for several reasons. There are obvious dangers associated with flying fixed-wing aircrafts in mountainous terrain where low altitudes are required for surveying applications. Even when using helicopters, they must be powerful enough to quickly handle erratic wind conditions and large changes in elevation.

Some survey applications can be greatly simplified if the goal is to simply count animals in an area of single-species habitats. Since everything detected is an animal of interest only detection is required and when the background is uniform the data collection can be performed at much longer ranges than those surveillance applications requiring recognition or identification. There are many applications that fall into the realm of surveillance that can be quickly, thoroughly, and efficiently carried out with very little disruption to the animals involved (e.g., locating nests, tree cavities, and dens). It is not always necessary to count animals that are detected; observing them at a distance can be sufficient for studying behavior, such as nocturnal behavioral patterns over time or any application that involves seeking out animals in their natural habitats. Note that this is distinctly different from industrial inspections, thermoregulation studies, or thermography for medical science. The main difference between these two types of applications is that surveillance is generally qualitative and those involving temperature measurements or thermography are more quantitative (McCafferty, 2007; Cilulko et al., 2013).

It is important to realize that there is a lot the thermographer can do in the field to assure that the best signature possible is being obtained at any given time. For example, the temperature of an animal in the field is pretty much constant yet the background can present a wide range of thermal backdrops with which the animal can be compared. The background in which the animal is situated is constantly changing and it does so without affecting the temperature of the animal. It is up to the thermographer to determine how the best apparent temperature difference can be obtained in each situation. The background is typically water (ocean, lakes, ponds, or streams), sky, clouds, forest, rocks, sand, or soil (dry, wet, or snow covered) and the character of these environments changes with each season. There can be a variation in the temperature of the animal as well due to thermoregulation. The backgrounds presented to the thermographers will depend on the viewing platform being used, that is, airborne platforms may have bare soil, clouds, sky, water, and many different types of vegetated surfaces. The goal of the thermographer is to use these varying backgrounds to obtain the best apparent temperature difference between the animal and the background.

Chapter 10

Thermal Imaging Applications and Experiments

Chapter Outline

BACKGROUND

This chapter provides a review of the past efforts using thermal imagers to observe and count animals in the wild. The review is chronological for the most part and as a result it closely follows developments in the technology from both an imager improvement standpoint and from advancements in user skills and understanding. Here we review many past efforts described in the literature to find, monitor, and count animals in the wild using thermal imagers. Most of the early works were experiments to test the feasibility of using thermal imagers as a tool in wildlife monitoring applications and large ungulates were generally the animals of choice for these efforts. In most of these comparisons thermal imaging proved to be superior even though there was very little effort made to use the thermal imagers in ways that optimized their performance. Remarkably, in some cases, researchers refused to accept the results of their own work when thermal imaging provided improved detectability performance. The strengths and weaknesses of the techniques used in those efforts are examined and suggestions are offered for improvement through the use of remote thermal imaging as a technique. From the outset we see that comparing the results obtained with thermal imagers with that of data collected with other methods, such as spotlighting and visual surveys, must necessarily be skewed and these efforts, while commendable, do not allow for a fair comparison of the data collection capability of the compared techniques. In most cases, when the goal of these efforts was to determine the superiority of one technique over another there was no reasonable explanation provided as to why the results favored a particular technique. Thermal cameras are suitable for surveys and counts throughout the

Thermal Imaging Techniques to Survey and Monitor Animals in the Wild: A Methodology
http://dx.doi.org/10.1016/B978-0-12-803384-5.00010-5

24-h diurnal cycle while other methods are not. These studies by their nature and design mean that the results of data collected with a thermal imager will be compared with data collected using an alternative method that was optimized for the conditions of the survey at hand. For example, consider the comparison of data collected during a visual survey and data collected via thermal imagery using the same temporal and spatial conditions for both. By selecting the same temporal and spatial conditions the survey must be conducted during the daylight hours because the visual spotters need daylight to see the animals of interest. It is interesting to note that there are no reports of nighttime aerial surveys comparing the results of data collected visually with that collected with a thermal imager. Thermal cameras can also detect the animals of interest during the daylight hours (e.g., see Figure 9.10,) but there are concomitant conditions required for the optimization of the thermal survey if it is conducted during the daylight hours. These conditions can be met in a relatively easy manner but were generally ignored during these comparative studies so the results reported were skewed and in some cases grossly inaccurate. We review many of these comparisons and offer alternatives.

LITERATURE REVIEWS

Many applications such as surveys and censuses involve counting animals or habitat features relating to specific animals and do not necessarily have to involve long ranges. We will review work devoted to observing and counting mammals (both large and small), birds, and insects as well as nests, dens, lairs, and tree cavities. It should be noted that these reviews are by no means intended to be all-inclusive. Generally there is a point made regarding some facet of the work being reviewed that enhances the overall understanding of using thermal imagers as a tool to improve the detectability of wildlife in the field.

Mammals

Introduction

During the past 35 years a significant number of experiments, studies, and surveys have been conducted in the field using thermal imagers. Infrared (IR) imagers have been used to observe animal behavior, locate dens, lairs, nests, and burrows, and conduct censuses for a wide range of animals. These investigations and surveys were carried out for a wide range of conditions that tested the limits of the imaging systems. These experiments were conducted during different seasons, at different times of the day, under a variety of weather conditions and terrain, and in both temperate and arctic climates. Collectively, the result of this work has moved the technology to a reasonably advanced state. As with any new technology introduced to enhance or replace methods that have become ingrained as standard (through general acceptance and wide use), thermal imaging has been met with the skepticism that usually accompanies new technology and ideas. For the most part these problems have been worked out in one piece of work or another but the results were not readily recognized. In many cases the researchers themselves would propose invalid justifications to explain away the superior performance in censusing experiments afforded by thermal imaging. At other times the methodology used would be flawed in some way and would result in some facet of the experiment that would actually limit the performance that the IR imagers were capable of achieving. In this section we review some of the strengths and weaknesses of past work that has been done in the field using thermal infrared imagery for locating, monitoring, and censusing heat-generating animals.

Important factors that we will highlight during this review will be first and foremost the need for detection to locate animals, to monitor their activity without disturbing them, and to facilitate counting so that population estimates can be made. The counting process should be capable of giving accurate population densities, whether the animals are clustered together in high numbers dispersed in groups or individually dispersed over large areas. The versatility of IR imaging as a noninvasive technique for conducting aerial- and ground-based surveys, behavioral studies, determining the age and sex characteristics of animal populations, and locating dens, lairs, and nests occupied by animals at any time during a diurnal cycle is pointed out.

Mammalian Reviews

The first attempts to use thermal imaging as a technique to survey animals took place in the late 1960s. Since that time attempts to use thermal infrared sensing systems (thermal imagers) to conduct aerial surveys and locate and count large animals such as deer, antelope, bear, and panther are becoming more widespread. It is surprising that the use of thermal imaging is not more widely used for animal surveys since it eliminates completely one of the major sources of inaccurate surveys: the problem of visibility bias. One possibility is that the use of thermal imagers has been slowed by misconceptions related to their potential and the lack of a prescribed methodology for their use. The major drawbacks of the early thermal imaging efforts for surveys were inadequate sensor capabilities for the conditions under which they were used, a lack of understanding of what the thermal imager was actually detecting, and perhaps more importantly the unwillingness of researchers to embrace a new technology even if it could enhance the accuracy and precision of previous work.

In one of the first aerial animal detection experiments conducted using a thermal imaging system Croon et al. (1968) demonstrated that the problem of visibility bias could easily be negated by using thermal imaging to detect animals in the field. In an experiment where penned white-tailed deer (*Odocoileus virginianus*) exhibited a 7°C differential in apparent temperature (i.e., ΔT the thermal contrast or the temperature difference between the animal and its background), they were able to detect ~97% of penned deer. They observed that under the right conditions, thermal imaging was capable of giving better counts over larger areas than any other technique available at that time (Croon et al., 1968, p. 759). In essence they reduced the problem of getting accurate counts to simply establishing the right set of conditions for the actual survey. Even at this early date they pointed out some of the many difficulties associated with the use of thermal imaging, as a technique, for conducting animal surveys. Included in this list were the difficulties of infrared in penetrating green leaf canopy, the variability of animal and background apparent temperatures, which are affected by weather and other factors (Marble, 1967; Parker, 1972), the difficulty of distinguishing between species of animals, and the high initial cost of the imaging system. These difficulties were encountered for the most part because of the relatively poor spatial and thermal resolution of the line-scanned imager used in this work (Table 8.2).

The conditions under which the Croon et al. (1968) experiment was conducted were not optimum but clearly showed that the technique had very strong potential as an aerial survey tool. For example, they flew the imager at midday in January 1967, over a natural reserve area the Edwin S. George Reserve (ESGR) in Michigan with a DC-3 fixed-wing aircraft at an altitude of 1000 ft. above mean terrain that was blanketed by 6–8 in. of snow cover. The ambient temperature during the flight was −4°C and a high overcast cloud cover was present. Based on the spatial resolution (3 mrad) of the imager and the altitude used during the tests, only animals larger than 3–5 ft. would be detected, yet they reported that they easily detected deer in grassland and oak wood (during leaf-off) and when the location of deer under conifer canopy was known they too were detected.

The fact that these early experiments were conducted at 1000 ft. agl with a low-resolution detector (IFOV = 3 mrad) and lead to remarkable results should have started a revolution in the use of thermal imaging equipment in aerial censusing. This did not occur primarily because the availability of IR thermal scanners was limited due to classification issues and because units capable of better performance would not be developed or available until much later. The issue surrounding complete snow cover is discussed below.

This group continued their work and McCullough et al. (1969) tested an imager by filtering the device to accept radiation in three separate wavelength bands: 3.5–5.5 μm (MWIR), 8–14 μm (LWIR), and the combined ranges 3.5–14 μm. Unfortunately, they conducted their experiments using the 3.5–5.5 μm and 3.5–14 μm setups on cold sunny days when the ground was snow covered and they were unable to detect any deer. These results were to be expected since any reflected, scattered, and re-emitted sunlight from nonbiotic objects (rocks, deadfall, standing water, and bare patches of soil) would all be recorded as strong thermal signatures when compared to the thermal signature of a deer, especially in the wavelength band used for these particular conditions (see Chapter 8). In fact, any thermal signatures of deer would be completely lost in the strong thermal background clutter that would be typical of sunlight-illuminated snow cover. Their 8–14 μm detector census was conducted on a day with high overcast skies, and they concluded that, for flights at 1000 ft., the best thermal images were obtained at 8–14 μm in the far infrared spectral region. Their conclusion that 8–14 μm imagery is superior to 3.5–5.5 μm and 3.5–14 μm imagery is somewhat misleading since they were used during different sunlight conditions and perhaps ambient temperatures as well. Furthermore, the conditions for these tests were perhaps the worst possible for MWIR imaging (cold, sunny, and snow covered). Winter surveys will be better facilitated with 8–14 mm imagers than MWIR imagers because of the spectral shift of the blackbody curves to longer wavelengths as the temperatures become cooler (see Figure 9.5).

Their concern regarding the variability in the apparent temperature of both the animals and their backgrounds was the result of not understanding the basic principles that produce these changes. They examined the data of Marble (1967) and pointed out the erratic nature of the data but failed to put it in the proper perspective. Data collected during the daylight hours in full sunlight should be erratic (Parker, 1972). The important issue to consider is that only two sources of heat (solar energy and the metabolic process) supply the radiant energy emitted by an animal. The rare exception to this would occur if the ambient temperature were greater than the temperature of the animal and warming of the animal occurred through convection, as might occur for poikilotherms. The amount radiated from an animal is dependant upon its emissivity and surface temperature, which are in turn moderated by a number of factors (hair, feathers, precipitation, mud/dust/dirt, and wind), radiation from the surrounding environment that is reflected from the animal, and the rate of metabolic activity. For sensing tasks with a thermal imager it is only necessary that there be an apparent temperature difference that is within the sensitivity range of the thermal imager, which is ~1.0°C in the present case. Snow cover is not required.

This early work by McCullough et al. (1969) pointed out for the first time a few important things: (1) the apparent temperature difference (ΔT) would be sufficient to permit the separation of species occupying the same habitat; (2) animals bedded in exposed positions are recorded as well as actively feeding ones and vacated beds may be warm enough to produce false counts; and (3) animals smaller than deer, elk, or domestic cattle could not be detected with the imager available to them at that time. They suggested that the improvements in imagers would need to have much smaller IFOVs leading to improved minimum resolvable temperature differences (MRTDs) while maintaining the thermal sensitivity or noise equivalent temperature difference (NETD). Their predictions for the future of IR censusing were not encouraging based on the apparent interdependence of IFOV and MRTD specific to scanning imagers. However, they still concluded that "In those situations where peculiar requirements of the <u>method</u> are met, infrared will give a better count of animals over large areas than any other method presently available" (McCullough et al., 1969, p. 146). Even at this early date the importance of methodology was noted.

The conclusion that good snow cover is essential to the accuracy of big game counts when conducting visual aerial surveys (Evans et al., 1966; Gilbert and Grieb, 1957) does not apply when carried over to aerial surveys using thermal imagers. Conducting visual aerial surveys over complete snow cover is a fundamentally sound procedure because it enhances the visual contrast between the animal and its background, which is easily detected by the human eye. However, maximizing the visual contrast for the eye does not take into account in any way the thermal aspects of the situation. The perception that this same uniform white background in the visible spectrum would provide a uniform thermal background (because it is cold) in the infrared spectral region has been treated as essential to obtaining accurate animal counts when using thermal imagers. A thermal imager does not need color contrast to form images; it requires thermal contrast. To assume that thermal contrast will be improved by a cold layer of snow providing a high-contrast situation for warm animals is fundamentally wrong from a thermal imaging point of view. The condition of complete snow cover appears to have been carried over from the visual methodology and was included in thermal imaging surveys because of a misunderstanding regarding the relationships of the environment and the functioning of a thermal imager. (See the arguments in Chapters 5, 6, and 7.) Additionally, it does not appear that much thought has gone into the role of thermoregulation and the resulting thermal differences between the mammals and their environments. These misconceptions, maybe more than any other single factor, have slowed the progress in developing a comprehensive methodology for thermal imaging surveys.

An experiment to detect and record the presence of polar bear (*Thalarctos maritimus*) on Arctic ice pack with a thermal imager was conducted by Brooks (1970). He used both MWIR and LWIR detectors to record imagery and compared the two. The Daedalus Enterprises, Inc. unit used for collecting the images had an IFOV = 2.5 mrad and 120 degree FOV. They flew at an altitude of 175 m on a clear day at noon and collected imagery with both detectors. The polar bear's tracks were followed until the animal was visually located and then several passes were flown over the bear with each detector to record the imagery. Ice cover on the Chukchi Sea where the bear was found was complete and covered with 1.5 in. of new snow. The wind was about 5 knots and the ambient temperature was −20°C. The imagery obtained with the MWIR (InSb) detector was of poor quality. The LWIR (HgCdTe) detector provided good resolution and contrast and when the imagery was displayed in an enlarged format on a screen or on photographic paper the bear was not discernibly different from the background. However, the motion of the bear was clearly detectable due to a warm trail that would be recorded showing the polar bear passing. The thermal signature of the bear tracks persisted for several minutes after the passing of the bear. The probable cause for the poor performance of the InSb detector was attributed to scattering of MWIR radiation by the exhaust gases of the aircraft, which would not affect the LWIR because of the small size of the exhaust particulates. This may have been a contributing factor but the lack of IR photons at very low ambient temperatures probably was the chief cause for the poor performance in the 3–5 μm spectral range.

This report may mark the first observation of animal tracks being recorded with a thermal imager. Brooks concluded that in spite of the high cost of infrared scanners the initial testing indicates that this method is probably superior to any other and could have application in line transect sampling systems aimed at yielding estimates of bear numbers over fairly large regions.

Extensive experiments using a scanning thermal imager to detect wild mule deer (*Odocoileus hemionus*) were conducted by Parker (1972). At that time he concluded that thermal scanning for wild deer detection remained a feasible technique. He suggested that improvements would be needed in the performance of infrared scanners in order for them to be suitable for use over rough terrain where a nonuniform background exists. These improvements turned out to be comparable with the performance specifications of today's imagers that use FPAs. That is, an adequate imager should at least have a spatial resolution of ~6.0 in. at 1000 ft. or IFOV = 0.5 mrad and a thermal sensitivity or NETD of 0.5°C @ 23°C.

In his dissertation Parker referred to an effective radiant temperature (ERT), which is just the apparent temperature (T_A) (see Chapter 5), and he made extensive measurements of this parameter as a function of the time of day, solar radiation, and air temperature for deer and objects in the background. Recall that T_A includes all the radiation emanating from an object or from an animal and that it includes the emitted radiation from metabolic activity, reradiated previously absorbed solar radiation, radiation scattered from the background, and reflected solar radiation. The plots of ERT versus time of day show a distinct stabilization of the ERT after sunset, which clearly indicates that the proper time to collect thermal data would be after sunset or before sunrise.

The most important outcome of his observations was that it led him to question the validity of the assumption that complete snow cover was required for meaningful surveys. He attributed the success of his tests to the uniform background temperature (without snow cover) and the large thermal contrast (6–9°C) between the penned deer and their background and suggested that greater thermal contrast may exist in summer, despite warmer background temperatures. He made the important distinction that the uniformity of the background temperature was much more important than the magnitude of the temperature.

The thermal imaging experiments by Parker and Driscoll (1972) were important for several reasons. They collected aerial infrared imagery (0.8–13.0 μm) of 55 deer and 11 antelope (*Antilocapra americana*) confined to the same pens during daylight hours before sunrise during August (air temperature, 56°F or 13.3°C) at Fort Collins, Colorado. Even though they were using a thermal imager with poor spatial (2.5 mrad) and poor thermal (0.2°C) resolution by today's standards, their resulting imagery produced very accurate counts of the total number of animals in the pens. They were able to get very good imagery from a fixed-wing aircraft at 300 and 500 ft. agl but were unable to detect deer at 1000 ft. The imagery was interpreted by three interpreters unfamiliar with imagery and without knowledge of the number of animals present. The average animal count of the three interpreters was greater than 96.5%. The low spatial resolution of their system led to difficulties in distinguishing between deer and antelope confined to the same pen but clearly demonstrated that IR imaging was capable of improving detection capability even with the early imaging systems.

Here the spatial resolution of 2.5 mrad can be used to estimate the range for detection by using equation (7.10) to calculate the maximum range for detecting a deer with 50% probability. The critical dimension $d_C = (1 \times h)^{1/2}$ is estimated as 2.5 by using 4 ft. long and 1.5 ft. wide as typical dimensions for the dorsal view of a mule deer. From $X_{min} = R\ \theta_{IFOV}$ using the critical dimension as X_{min} and an IFOV = 2.5 mrad the range is estimated to be 1000 ft. for a 50% probability for detection. This calculation does not account for atmospheric attenuation in the 8–14 μm spectral range, which could easily be 20–30%, so the imagery shown in this work (p. 490, Figure 1) is quite good. Even though these early experiments clearly pointed out that predawn flight during warm weather (no snow cover) would produce imagery suitable for very accurate animal counts, the (predawn/warm weather) criterion has not been readily adopted by other workers.

Graves et al. (1972) compared the performance of an LWIR and an MWIR thermal imager. They showed that deer emit enough radiation in the 3–5 μm region of the spectrum to allow detection by airborne infrared imagers operating under certain conditions. They suggested that a 3–5 μm MWIR imager using an Indium Antimonide (InSb) detector is the best sensor for the detection of large animals in the warmer months when the animal is trying to stay cool by radiating away body heat. They also suggested that a 3–14 μm imager using a Mercury Cadmium Telluride (MCT) detector is the best choice for the cooler months when the animal's coat reduces radiation of body heat, and the animal is conserving energy. Their conclusions regarding the relationship of seasonal and wavelength sensitivity effects were correctly attributed to differential thermal radiation, but it is not the complete explanation.

We saw in Chapter 5 that the power radiated from a surface above absolute zero depends on the temperature of the surface, the background temperature, and the temperature of the surroundings (ambient temperature or temperature of the environment). It also depends on the emissivity of both the surface and the background. These, however, are not the only influencing factors contributing to their results and conclusions. A complete explanation must include the fact that the peak of the blackbody radiation curve shifts to longer wavelengths as the temperature drops (which would favor a 3–14 μm imager), or equivalently, photon production at shorter wavelengths is reduced as the temperature is lowered, meaning that the sensitivity of a 3–5 μm imager decreases as the radiating bodies become cooler.

The results of two different surveys reported by Graves et al. (1972) contrasted the detectability of deer obtained using spotlighting techniques to that obtained using thermal imagery. For the survey taken when both techniques were used essentially at the same time, between 2237 h and 2253 h (spotlighting detected 21 deer) and between 2242 h and 2245 h (thermal imaging detected 29 deer), two important conclusions can be drawn. The thermal imagery was collected in 3–4 min and the spotlight data took 14–15 min and the thermal imaging technique detected eight more deer than the spotlight count. The second surveys taken in a different location (the two locations are separated by interstate highway I-80 in Pennsylvania) used approximately the same time intervals to conduct the counts but were separated by approximately 1 h. The thermal imagery was again collected in 3–4 min (32 deer counted) about an hour before the spotlight count was made (36 deer). In the first case the spatial and temporal conditions of the survey were the same for the two techniques but for the second survey they were the same spatially but not the same temporally. To compare the two counts in the second survey and make judgments on the performance in a comparative fashion is not recommended since the field conditions were not the same. In the first case such a comparison can be fairly made since the conditions were the same both spatially and temporally. This observation should have been noted as a major endorsement of thermal imagery as a superior technique (since ~26% more deer were detected in one-quarter of the time it took for the spotlight count), but it was not.

There were a number of observations that did advance the prospects of using thermal imagery for animal censuses. They reported, and correctly so, that the detectability from the air was related to time of day, season, altitude, and wavelength sensitivity of the IR detectors. They were able to distinguish (identify) deer, bear, and livestock. We can also conclude from this work that a uniform background temperature, not necessarily a cold one, will provide better results for both MWIR and LWIR imagers. If the ambient temperature is cold then an LWIR detector would provide better results. They (Graves et al., 1972, p. 875) also offer the observation that "during winter, the deer's hair surface is more nearly equal to the average background temperature than it is during summer; this would be adaptive in that body heat is apparently freely radiated during the summer and retained during the winter".

A complete count was conducted in an aerial census of ungulates within the fenced confines of Elk Island Park, Alberta, Canada by Wride and Baker (1977). They used an LWIR Daedalus line-scanning (8–14 μm) thermal imager with a modest spatial resolution of IFOV = 1.7 mrad. By comparison, note that in the work described below, the imager used by Wiggers and Beckerman (1993) had a spatial resolution of IFOV = 0.25 mrad in the NFOV and 1.4 mrad in the WFOV mode (today's imagers typically have IFOV $\leq$ 1). The entire 270 square miles of the park were surveyed in a light single engine aircraft during the winter in morning and midday (overcast conditions) and it was concluded that IR thermal scanning was much more accurate and precise for a census of ungulates than visual techniques. Generally it was concluded that the advantages over visual survey methods included no technician effect; no observer fatigue; operation at night, early morning, and without snow cover; detection of camouflage or neutral-colored animals; accurate results; repeatable results; and habitat permanently recorded. The disadvantages noted were an inability to distinguish the sex of animals and species identification was limited to differentiation by animal size.

They reached these conclusions by examining the processed film to count the animal images that were collected. They estimated that there were a total of 2175 ungulates made up of moose (*Alces alces*), bison (*Bison bison*), and elk (*Cervus canadensis*) in the park. All of the bison (356) were confined to a pen. Two visual surveys conducted the same winter produced ungulate counts of 1010 in December and 1231 in March. The March number of 1231 is 57% of the thermal imaging result (2175). Parks usually added 25% to raw visual aerial survey data because it was considered common knowledge that counts are consistently below actual population figures. This would raise the park estimate to 1450 or 67% of the thermal scan number. This large difference between the estimates from the visual aerial survey and the thermal scan was then analyzed by examining all possibilities of probable error in the visual and thermal counts. Taking into account the experience of the visual spotters, lack of complete snow cover, and the large variation in the detectability coefficients for the three different species (actually two since the bison were isolated and penned), they concluded that the visual estimate is probably on the order of 60% of the actual numbers and that the thermal figure for total ungulates is a much more accurate estimate. Thermal scanning methods provide an important advantage over visual methods because they offer the ability to conduct scans when there is little or no snow cover in the fall or late winter. The contrast between ungulates and their backgrounds provided by uniform snow cover is required for visual methods.

We point out that when they examined the possibilities for overinflating the numbers obtained in the thermal scans, such as solar heated objects, vacant beds, and other animals, they determined that any inflation would be insignificant for their data and likewise that any omitted animal (say behind a tree) was equally insignificant. From these conclusions we infer that for all practical purposes Wride and Baker (1977) achieved near 100% detection using thermal imaging as a survey technique for a census. It should also be noted that these results could be improved upon by working with a more uniform thermal background, using a more sensitive thermal imager, and by using a helicopter for the census.

Several experiments were conducted to test the utility of LWIR thermal imaging in the cold environment of the Canadian high Arctic. Barber et al. (1989) were successful in locating walrus (*Odobenus rosmarus*) hauled-out on pans of sea ice at altitudes of 5.5 miles. Their analysis shows that thermal imagery can be used to detect walrus hauled-out on sea ice at a scale sufficiently coarse to allow complete coverage of a large survey area in a single flight. In addition, they show that they can obtain a walrus abundance estimate from this complete coverage survey. The walrus thermal signatures obtained at high attitudes (and therefore over large expanses of ice and water) appear as bright blobs on a relatively homogeneous background and the relationships between altitude, the maximum and minimum pixel brightness, and pixel frequency within a particular blob are compared with walrus numbers gained from ground truthing using photographs obtained by low-altitude flights over the detected blobs. This method allowed them to conclude that IR imaging presented a method for avoiding the fatigue and boredom of visually surveying large expanses of ice and water devoid of walrus only to encounter difficulty in counting a large number of walrus clustered in tightly packed groups. If the walrus are hauled-out on land then visibility bias also hampers the accuracy of visual counting methods.

This work was continued by Barber et al. (1991) and resulted in refining this technique, leading to the following recommendation. They suggest a sampling scheme using a high-altitude survey (between 2438 m and 3048 m) to collect a stratification variable. They noted from earlier work that walrus tend to concentrate near the interface between floating ice and open water and that complete coverage of this habitat could be achieved in a single 3-h flight. Once the stratification data were obtained, a mid-altitude survey (e.g., 1219 m) would be conducted, where sampling effort is apportioned according to the stratification survey. Finally, photo passes would be conducted at 457 m using systematic transects placed over the walrus concentration areas, both as a means of obtaining a systematic photo estimate and for forward-looking infrared (FLIR) calibration. Further, they suggest a FLIR pass at the 1219-m altitude because a larger coverage can be realized with the same precision available at the 457-m altitude. This work marks a significant step in expanding the possibilities of thermal imaging being used for complete coverage of large survey areas.

Counts of red deer (*Cervus elaphus*) from videotaped thermal infrared imagery collected in extensive open habitats in Scotland were found to be the same as or 6–12% higher than ground counts (Reynolds et al., 1993). The aerial censusing with an LWIR having modest spatial resolution (2.3 mrad) and thermal resolution (<0.2°C) reduced the time required to census deer by 50–75% when compared to the ground counts. They reported that the thermal signatures of red deer in open habitat were very distinctive and could, in most cases, be readily separated from the thermal signatures produced by other species (sheep). For example, the results of the surveys conducted at two different sites indicate that of the 602 animals detected in open habitat and counted from the thermal images, species could not be determined for only four animals (0.7%) by one interpreter and 11 animals (1.8%) by another. It was pointed out that while sheep also had a very distinctive signature they were not always detected with the thermal imager because their weak thermal contrast (well-insulated) was probably taxing the thermal resolution of the camera for some individuals. The superiority of thermal imagery for censusing in extensive open habitat was clearly demonstrated in this work but was not fully recognized at the time.

Penned white-tailed deer (*O. virginianus*) were identified from thermal imaging scans of deer pens acquired from a fixed-wing aircraft during night flights in warm weather (i.e., no snow cover) on the Louisiana State University Idlewild Research Station and a privately owned plantation. This important experiment, carried out by Wiggers and Beckerman (1993), showed that a scanning 8–12 μm thermal imager provided the resolution necessary to discern the morphological characteristics of the penned deer. The thermal imager provided adequate resolution to detect all deer in the open pens and permitted the accurate determination of the age and sex structure of the deer present. They used a FLIR Systems®, Inc. 2000G LWIR imaging system that could detect temperature differences of 0.25°C and had a spatial resolution of IFOV = 0.25 mrad in the NFOV and 1.4 mrad in the WFOV mode. For their test studies of image interpretation they used only the NFOV optics. They subjected their collected imagery to interpretation by five biologists. This experiment showed that a skilled interpreter could correctly identify the sex and age for 100% of the deer scanned.

Important relationships between the altitude, speed, and flight pattern of the fixed-wing aircraft used in these experiments were also brought to light. The imager was mounted under the wing of the aircraft and was aimed and focused by the scanner operator. Flights were flown at three different altitude ranges: low (170–270 m), medium (271–370 m), and high (371–450 m). These are above ground level (agl) altitudes and they represent the minimum distances from the aircraft to the animals since the plane was circling above the pens and scanning the animals at oblique angles. The most accurate interpretation of the imagery was that collected for the medium altitudes. The aircraft speed and the pointing stability of the imager (vibrations and jitter) reduced the image quality at low altitudes while the thermal and spatial resolution degraded the imagery at the highest altitudes. Nonetheless, the best imagery allowed inexperienced image interpreters to identify species correctly for 93% of the animals and to identify the sex correctly for 87% of the animals. These numbers should not be confused with detection. All animals in the experiments were detected. With three to four passes they were able to detect all deer in a pen with 73% canopy closure. These ground truth experiments point to the fact that thermal imagers having superior performance capabilities to those found in older designs (see Table 8.2) can easily provide near 100% detection rates. While there were difficulties encountered with imaging areas having a high percentage of canopy closure, this work points out that changes in census methodology can circumvent these difficulties as well. The use of a rotary-winged aircraft and the technique of panning with the IR camera will greatly reduce these difficulties.

It is clear that an imager with a little better sensitivity, a gyroscopic mounting to reduce vibrations and jitter, and flights using a helicopter would have allowed Wiggers and Beckerman (1993) to determine whether investigators could visually identify deer, discriminate between adult and juvenile, and distinguish between sex classes when males are growing antlers using airborne thermal infrared sensing. Perhaps the most remarkable achievement was not the 100% detectivity, recognition, and identification (including age and sex discrimination) achieved by a skilled interpreter but rather that inexperienced image interpreters correctly identified species (93%) and sexual discrimination (87%) from imagery collected under less than ideal conditions. Further, they obtained 100% detectivity with a 73% canopy closure by circling and panning with the imager. The impracticality of multiple passes during field applications could be greatly reduced by using a helicopter, which is a more versatile aircraft. All in all, this work showed advancements in performance of the imagers and advancements in technique to improve detection recognition, identification, discrimination, and ways to mitigate the problems associated with green leaf canopies. Unfortunately, many subsequent survey efforts by researchers utilizing thermal imagers did not use or learn from the teachings of this work.

Garner et al. (1995) used a commercially available LWIR thermal imager mounted to a strut on a Cessna 210 aircraft to conduct surveys of moose (*A. alces*), white-tailed deer (*O. virginianus*), and wild turkeys (*Meleagris gallopavo*) in the field. The imager had two instantaneous fields-of-view available. They point out that an NETD $< 0.3°C$ is typical for their state-of-the-art imager. The imagery collected during the three surveys was subjected to analysis via a computer program. This program compared a reference ΔT collected from individual animals at the start of each survey to the ΔTs of signatures that were captured on the infrared imagery. VHS video was recorded concomitantly with the infrared imagery for comparison. A trained image interpreter determined the final classification of the objects of interest. Untrained observers were also used to identify potential objects of interest in the imagery for the purpose of comparison.

These studies were conducted at three different sites: (1) the Saratoga National Historic Park, located in eastern New York state, where the largest vertebrates are white-tailed deer (*O. virginianus*), coyotes (*Canis latrans*), and wild turkeys (*M. gallopavo*); (2) the western portion of Algonquin Provincial Park in south central Ontario, where the largest resident vertebrates are moose (*A. alces*), deer, wolves (*Canis lupus*), and black bears (*Ursus americanus*); and (3) portions of western Chenango and eastern Cortland counties in south central New York, where the largest nondomestic vertebrates are deer, coyotes, and wild turkeys. Thermal imagery was collected under clear to partly cloudy skies in November 1991 with ambient temperatures near 0°C and no snow cover at sites 1 and 2 during the afternoon hours. The third site was surveyed in March 1992 under overcast skies with 15 cm of snow cover and ambient temperatures a −3°C between 0800 h and 1130 h. The imager used in all three surveys was a 2000A-B FLIR operating in the 8–14 μm spectral band. Visible video was also recorded of the same footprint as the thermal imager for direct comparison. Air speed of 70–75 knots and 154 m agl altitude were used for moose and 70–75 knots and 398 m agl for deer and turkey.

They point out that using VHS video for censusing just shifts the problems encountered in visual aerial surveys to film. The problems are still there but they can be reviewed many times. They make the suggestion that snow cover may be needed for thermal imaging surveys to provide adequate thermal contrast yet they also point out that thermal loading of the scene (background clutter) was responsible for the poor counts obtained. Both the deer and moose surveys were conducted under clear or partly cloudy skies near midday (the worst possible time to conduct thermal imaging surveys). In addition to the false signatures resulting from the reradiation of objects warmed by direct exposure to the sun, the solar melting of snow will act as an IR mirror, reflecting infrared energy into the detectors. The conditions of overcast skies in the morning for the turkey survey were much better than those used in the deer and moose counts. Because they were able to view the flocks of turkeys from many different oblique angles as the airplane circled, they were confident that they achieved 100% detection of all the birds in each flock. By flying concentric circles and using telephoto capability detailed counts of turkeys and deer were possible. This technique – panning with the camera while circling areas of interest in the aircraft, which aids in the detection capability – was also used successfully by Stoll et al. (1991) for a visual aerial survey of deer using a helicopter.

The difficulties associated with counting schemes based upon the magnitude of the thermal contrast between objects of interest and their backgrounds were noted by Garner et al. (1995, p. 236). If the magnitude of ΔT (intensity of the thermal signature above its background) is going to be used for the purpose of identification of species then particular care must be taken during the data collection phase of the experiment, and field surveys would almost certainly be excluded. This is because most biotic objects do not have significantly different emissivities. Furthermore, any change in their metabolic rates through activity, foraging, or exposure to solar radiation will significantly alter their emissivity. There can be any number of objects in the field on clear days that can produce signatures with ΔTs equal to or significantly greater than biotic objects.

This work identifies several problems that can limit the success of thermal imaging surveys if care is not taken to "prepare" the background, so to speak. That is, if the surveys are planned correctly, the imagery will be scheduled for the early morning hours when the background (e.g., vegetative cover, soil, and the litter on the forest floor) has come to thermal equilibrium. It would be even more beneficial if this followed several days of minimal solar loading in the survey areas. This precaution will allow the contrast between the animals of interest and their backgrounds to be maximized and remain constant through the survey period. It will also eliminate spurious signatures resulting from nonbiotic objects that appear in the imagery, confounding its interpretation. The technique of circling and panning the flock of turkeys is a good one for achieving high detectabilities, leading ultimately to accurate counts (i.e., approaching 100%).

We have already seen a number of successful demonstrations regarding the detection capability provided by IR imaging. Naugle et al. (1996) estimated the density of white-tailed deer using spotlight surveys and thermal infrared aerial surveys at Sand Lake National Wildlife Refuge in South Dakota. The objectives of these surveys were as follows: (1) estimate the deer detection rate using IR sensing, (2) develop an aerial transect method to minimize sampling biases, (3) estimate deer density using IR sensing and spotlighting surveys, and (4) compare the density estimates from IR sensing and spotlight surveys.

An FLIR 2000AB filtered for use in the 8–14 μm spectral region with a wide FOV (28 × 15°) and a narrow FOV (7 × 3.25°) was mounted under the wing of a Cessna 210 fixed-wing aircraft for the aerial surveys. For the wide FOV the imager had an NETD = 0.16°C and an IFOV = 1.4 mrad, or the spatial resolution was 1.4 m at 1000 m. Thermal imagery and real-time video were recorded on standard (VHS) tape. Before each flight a signature of a group of deer was obtained to determine contrast ΔT. The dominant vegetation (45%) in the survey area was cattail and common reed. The total area was 99% open with less than 1% scattered deciduous tree belts and had a topographical relief of 3 m. From this area four deciduous tree belts and four areas of emergent vegetation that deer frequented were selected as sample areas for estimating the deer detection rate.

They reported an 88.2% detection rate using FLIR technology and ground truthing was established via deer drives from the sampled areas after each aerial survey. They found that spotlight counts conducted before spring farming underestimated deer density by 38%. This is attributed to the fact that deer were probably dispersed more uniformly throughout the refuge before seasonal plowing began. More deer were observed along the spotlight transects after the farming season started because the deer were attracted to the undisturbed foraging opportunities available within the refuge. Based on the high detectivities and increased estimates achieved with IR sensing they concluded that IR sensing was a more reliable density estimator than spotlight surveys. A number of adjustments were made for the spotlight surveys: (1) spotlight survey data were not included in analyses when the weather restricted observer visibility, and (2) an observability factor was used to account for the loss of detectivity when deer were unequally dispersed or lost in dense vegetation (e.g., emergent vegetation), which comprised 45% of the habitat surveyed in 1993 census data. Photographic examples of IR imagery collected readily showed that deer were detected in the emergent vegetation yet these deer were inaccessible for spotlight counting.

Even though they report that thermal IR sensing detected 88.2% of deer counted by ground personnel and that spotlight counts associated with IR sensing flights underestimated deer density by 38%, it is still important to note that the two methods of sampling/counting (aerial IR imaging and ground-based spotlighting) are very different, with entirely different sets of conditions regarding their successful implementation. For comparisons of counting methods to be truly meaningful it is necessary to conduct the counts in the same place and at the same time. If this is not done then they should not be compared because the results can be interpreted in a variety of different ways, none of which may be correct. For example, what does it mean when data is excluded because weather restricted the observer visibility during a test to compare two counting techniques? Furthermore, what does it mean when an observability factor is used to modify the counts because the observer cannot see deer in emergent vegetation and/or adjust for a nonuniform distribution of animals in the survey area? Figures 2, 3, and 4 in this paper clearly show thermal signatures of deer in corn fields and in emergent vegetation. The detectivity achieved with the thermal imager was determined to be 88.2% in four deciduous tree belts, which was verified by ground truthing. The aerial surveys were conducted at 70–75 knots with a fixed-wing plane and the survey area was snow covered except for the May 1 survey, which was after snow melt. Using a thermal imager from fixed-wing aircraft over snow cover are not good conditions for thermal imaging surveys. An important outcome of the work by Naugle et al. (1996) shows that better results are obtained with thermal imagery when there is no snow cover and when the ambient temperature is warmer.

There are also a number of efforts that have obtained very high detectability for surveys of open populations. Adams et al. (1997) conducted a series of aerial moose (*A. alces*) surveys in northern New Hampshire in January 1995, using a Westinghouse WesCam DS® infrared sensor with a spectral range of 8–12 μm and a >10× Sony® zoom lens color television camera. The two cameras were mounted on a fixed-wing Cessna 337G aircraft to allow both thermal and visible imagery to be collected at the same time. They noted that many moose positively identified with the thermal camera could not be viewed with the video camera because they were obscured by trees or dense canopy and they suggested that they would probably have been missed by an observer during a traditional aerial survey. Two study sites were used that contained extensive deciduous and mixed-wood forests, which were known moose wintering areas. They conducted the surveys at approximately 1000 m agl and they reported an 89% average sightability (based on 100% detectability during intensive searches) and, perhaps more importantly, this detectability was influenced by the terrain and elevation. For their surveys ground coverage of 1.0 min/km^2 was used. They concluded that when using an IR imager to replace visual observation techniques the method should provide an acceptably accurate and precise moose population density estimate applied over a large area, as in a Gasaway et al. (1986) style survey.

The conditions and implementations used to collect data in this study were not optimized, yet the results are encouraging. They flew fixed-wing surveys at 1000 m agl, with patchy snow cover, between 0800 h and 1500 h. These hours corresponded to acceptable flying conditions rather than desired peak periods of moose (*A. alces*) activity at dawn and dusk. The operators of the IR imaging cameras were confident that they detected all moose during the intensive searches so the 89% detectivity is based essentially on a flyby at 1000 m agl. These surveys would have benefited from flights providing slower air speed and survey time schedules set for dawn when the moose are most active and thermal contrast is the greatest. A helicopter would allow a lower flight altitude and reduced air speed so higher detectability would be achieved. It may also have helped with maintaining the flight path during the windy conditions encountered during the surveys.

A battery-powered LWIR handheld thermal imager (NETD, 0.1°C, and an IFOV of ~0.5 mrad) was used for distance sampling by Gill et al. (1997) to estimate deer populations in seven forest areas in mainland Britain. The sites varied substantially in habitat, terrain, and deer species composition: roe (*Capreolus capreolus*), muntjac (*Muntiacus reevesi*), red (*C. elaphus*), fallow (*Dama dama*), and sika (*Cervus nippon*). Distance sampling is an efficient estimation method for habitat, with widely varying visibility conditions imposed by tree type, age, and understory development. Transects were both walked and driven where roadways or trails were available. Density estimations were made following the procedure outlined by Buckland et al. (1993).

This ground-based survey technique using an IR imager produced a number of predictable results. Substantially more deer were detected at night (4.5 deer/km) than during the day (0.85 deer/km), detection distances were related to vegetation density, and detectivity declined as perpendicular distance from the transect increased. They point out that the imager had sufficient thermal and spatial resolution to display images of natural scenes in some detail as well as making homeotherms very conspicuous. When unobstructed by vegetation, red deer could be detected at distances of up to 2000 m, with species, age class (adult/juvenile), and sex differences becoming increasingly discernible at closer range. Since environmental features such as water, sky, trees, vegetation, and dead fall typically differ in surface emissivity, detecting and locating the position of deer in the landscape is relatively easy, even if much of the body surface area is concealed. However, in general, the amount of detail visible through the imager only decreases with increasing distance. "Although vegetation limits detection, it is only likely to limit accuracy where a substantial portion of the forest includes dense vegetation and the animals do not exhibit a tendency to select more open areas at night" (Gill et al., 1997, p. 1285). This conclusion is simply stating that the sampling method (distance sampling) clearly requires a means of detection and that it must be good enough to acquire large sample sizes if accurate population estimates are desired.

The fact that many more deer were detected at night using a thermal imager than along the same transect routes in daytime is sufficient to warrant the use of IR imaging technology, regardless of the sampling method used. Additionally, even if much of the body surface of the animal was concealed by vegetation, the position of the deer in the landscape was easily determined and bedding sites found with the thermal imager were also used to determine the location of deer. This would not be the case with binoculars or other visual locating techniques, including spotlighting. Using distance sampling methods, where disturbing the animals, or surveying when they are bedded down or inactive results in fewer animals being detected and poorer results. A wise choice would be to use a thermal imager. Even though the relevant assumptions of the distance sampling method (Buckland et al., 1993) can be met by other counting techniques they are much easier to verify when using thermal imaging techniques.

Havens and Sharp (1998) conducted a large-scale field survey of deer (*O. virginianus*) in southwest Florida over the Florida Panther National Wildlife Refuge and Big Cypress National Preserve. We flew transects in a helicopter from a half an hour before sunrise to 1 h after sunrise and recorded thermal imagery to 8-mm videotape with an MWIR Model A Infracam® thermal imager. These experiments represented the first time that a thermal imager with a staring focal plane array was used in the field for censusing work. The 256 × 256 platinum silicide array and 50 mm optics provided a spatial resolution of 0.6 mrad, NETD of 0.07°C at 23°C, and an MRTD of 0.05°C @ 0.5 cy/mrad, which represented approximately an order of magnitude improvement over scanning systems used in past fieldwork. We incorporated a number of techniques in these aerial surveys to improve the detection capability such as panning the camera to penetrate the tree canopy, gyrostabilization of the camera to reduce pointing instabilities, and soft-mounting the recorders to reduce recorded image jitter. The combination of improved camera performance and improved techniques coupled with ideal survey times (early morning or overcast sky) led to ~100% detection rates. We also found that the detectability of thermal contrast (ΔT) between biological objects and their backgrounds was sufficient to permit species identification.

For example, in work conducted in the Florida Everglades using 3–5 μm thermal imagers, numerous species were identified in the field, including deer (*O. virginianus*), bear (*U. americanus*), feral pigs, cows, fox squirrels (*Sciurus spp.*), different species of birds, and panther (*Puma concolor coryi*). A preliminary inventory of deer using a rotary-winged aircraft flown along transects covering approximately 800 ha in three survey transects found a deer density of one per every 10 ha, which did not differ significantly among transects (Havens and Sharp, 1998). For large survey areas, stratified random transects coupled with Monte Carlo statistics can be used to determine the optimum amount of sampling without sacrificing accuracy (Havens and Sharp, 1995).

A flight protocol for counting deer on snow with a helicopter was conducted by Beringer et al. (1998). They used a fenced study site (794 ha) dominated by mature hardwoods with steep topography. Two experienced observers and a pilot searched for 69 previously marked deer by using a census (complete count) protocol that consisted of flying 200 m transects at an altitude of 60 m and an air speed of 23 knots. They were confident that their transects did not overlap >50 m. They report that the 60 m altitude and slow air speed did not cause deer to flush >100 m. They searched in the early morning and late afternoon over ≥10 cm of uniform snow cover and obtained an average detection rate of 78.5%. They commented that it was difficult to see deer beyond 100 m at the altitude (60 m) that they were flying, but visibility was good except for one large cedar patch. They commented that they could easily see the forest floor through the oak–hickory cover types.

They also reported that they had used a scanning thermal imager mounted on a helicopter for surveying (unpublished data) over oak–hickory hardwoods in Missouri and obtained variable detection rates, often <50%, but they do not provide the details of the survey conditions other than results of similar flights using fixed-wing aircraft. In these flights they recorded as little as 25% of estimated deer.

The average detection rate of 78.5% is computed to the nearest 0.5% but the control over the swath width (and the deer therein) during this census work could have varied by as much as 75% at any given time (deer flush ~100 m and transect overlap ~50 m on a 200 m swath). It is unclear why the authors report on unpublished thermal IR imaging survey results unless it is offered as a means of comparison with the present report for visible aerial census work.

A comparison of three devices for observing white-tailed deer at night was conducted by Belant and Seamans (2000) at the National Aeronautic and Space Administration's Plum Brook Station (PBS), Erie County, Ohio. PBS is an enclosed (2.4 m high chain-link fence) 2200 ha facility with mixed vegetation. The deer population was estimated at 475 ($\geq$21 km^{-2}) at the time of the surveys. They established a 10 km survey route along roadways in the enclosure and categorized vegetative associations along the left side of the route (one side survey) based on the dominant vegetation observed every 0.2 km as grassland (39%), shrub (30%), and wooded (31%).

The first of the three devices used was an FLIR P100 Nightsight® (Texas Instruments) mounted on the cab roof of a pickup truck. The P100 is an LWIR imager with a (28 × 18°) FOV. The pointing direction of the FLIR was controlled from within the cab. Next, the image intensifier used was an AN/PVS-7B night-vision goggles (NVG) fitted with a GEN II^{PLUS} image intensifier tube with 1× magnification and a 40° FOV. The final observation devices were two different brands of 1,000,000 candle power handheld spotlights. They conducted two trials of 12 consecutive nights each in winter (leaf-off) and summer (leaf-on) with surveys starting at 30 min after sunset. Each night they conducted three surveys, once with each device. The order in which the devices were used each night was rotated such that each device was used an equal number of times.

In winter they observed 825 deer with the FLIR, 716 with spotlight, and 243 with NVG. The percentage of deer observed among vegetation associations with FLIR and spotlight were similar with about 82% of deer observed in grasslands and 18% observed in shrub or woodlands. For NVG 99% were observed in grasslands.

In summer they observed 570 with FLIR, 445 with spotlight, and 152 with NVG. In summer they observed the most deer in grasslands: 87% with FLIR, 82% with spotlight, and 93% with NVG. They observed only 2% of all deer in shrub associations with each device and for the wooded areas: 16% for spotlight, 10% for FLIR, and 5% for NVG. The numbers of deer observed during inclement weather (fog, rain, snow) on a 10 km route (80 km surveyed/per device) were: ~580 for FLIR, ~440 for spotlight, and ~190 for NVG.

It was noted that each device was limited by vegetation blocking the line of sight, particularly in the shrub and wooded areas. The FLIR was effective because only a portion of the infrared image of a deer was necessary to identify the animal and detection was not dependent upon deer looking at the device, as it was for spotlighting at distances >50 m. They felt it was easier to detect deer in poor weather with an FLIR because of increased scattering with spotlights. They do not recommend using NVG under the conditions tested during their study.

This study was undertaken to evaluate the three techniques of FLIR, NVG, and spotlighting to detect deer in habitats typical of airport facilities. Deer-aircraft collisions are common and occur typically at night so there is a need to determine the abundance of deer at an airport and maximize the effectiveness of deer harassment or removal efforts at night. It is unclear what population densities would be tolerated or acceptable on airport facilities. Their concerns of using spotlights on or near airport facilities are well-founded as is their conclusion regarding the use of NVG for this application. NVG require ambient near-infrared radiation from the moon and stars to be effective as surveillance applications (see Chapter 3).

The methodology used in the work conducted by Dymond et al. (2000) is very good and verifies that sound criterion can be developed that will allow successful thermal imaging surveys in the field when an effort is made to optimize the detection rate. They developed a methodology for their experiments consistent with exploiting the strengths of the two FLIR systems they used in their systematic investigation to determine the optimum conditions for surveying possum (*Trichosurus vulpecula*) populations in New Zealand. Both MWIR (Prism DS) and LWIR (Model 2000) imagers from FLIR Systems, Inc. were tested. They set their flight conditions based on the imager optics, spatial resolution, and the desired video display that would allow them to discern individual possums with confidence.

A theoretical calculation of the apparent temperature difference was found to be in close agreement with their field measurement of caged possums. This was fortuitous in that the temperature called the "brightness temperature" of the environment, T_e, was approximately equal to the temperature of the background when the measurements were made. It is important to note that when a thermal imager records a scene the ambient temperature (temperature of the environment) is generally not the same as the background temperature against which the apparent temperature of the animal is compared in any particular image. It is, however, possible to arrange the conditions to be such that the processes of conduction, convection, evaporation, and radiation coupled with time will aid in this task.

They determined that both MWIR and LWIR imagers performed well, that convective heat losses due to wind must be avoided, and that flight conditions be commensurate with the imager optics and desired video display. They indicated that they still had problems with image interpretation due to spurious

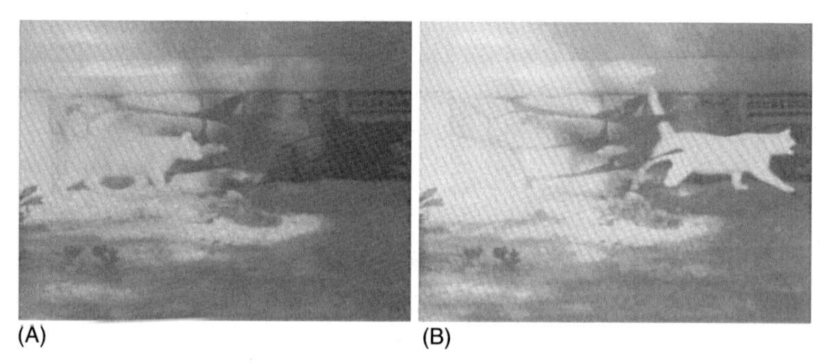

FIGURE 10.1 (A) This photograph of a thermal image depicts a cat lost in the background clutter formed by a strongly reradiating background that was previously heated from solar radiation during the afternoon hours. (B) This photograph is of the thermal image after the cat moved to the right into a region of the background that was not subjected to the solar loading of the afternoon sun.

signals originating from objects in the background that were emitting previously absorbed solar radiation with magnitudes on the order of those associated with possums. This is a problem noted by all users and is the single most important issue that a practical methodology must address. The information provided in their Figure 1 regarding the specifications and performance characteristics of the imagers they used is a feature often neglected in most papers on thermal imaging surveillance applications, which forces a reader to guess these properties or search elsewhere to find the needed imager specifications. This work provides instructive insight with regard to properly using thermal imaging for applications involving the surveillance of animals.

Consider the following example to illustrate the difference between the ambient (environment) temperature and the background temperature (Figure 10.1). Imagery of an animal taken (after sunset) in a location that has been shaded during the diurnal hours will be completely different from that taken in a location that experienced full solar exposure, yet the ambient temperature is nominally the same in the two cases. Note that the two cats could be standing side by side in this example. Knowing the actual value of the temperature difference is not as important as is maintaining the value throughout a survey. The left side of these images has been exposed to afternoon solar loading and the right side was shaded by a tree during the same time. When the cat is compared to the side that experienced solar loading it is difficult to extract the thermal signature due to the strong radiative component of the background. In the photograph (Figure 10.1B) the cat is shown compared to a cooler background, which is radiating much less than the side that experienced the solar loading and it is easily detected and identified as a cat. The resulting improvement is due to optimizing the apparent temperature difference by using a more advantageous background yet the ambient temperature is the same in both photographs.

The possibility of using thermal imaging to determine deer abundance in New Zealand forests was undertaken because the only methods available at the time were counting fecal pellet groups or using hunter kill returns, and both methods are imprecise and inaccurate (Wilde, 2000). A FLIR Systems Prism DS MWIR thermal infrared imager with an IFOV = 1 mrad, FOV (17° × 13°), and an NETD = 0.1°C was used to census penned fallow deer in a 15 ha paddock that contained both pasture and open indigenous forest. They aligned a set of flight lines for each of three flight altitudes (550, 1000, and 1650 ft. agl) to completely cover the paddock without overlap of transects. They videotaped the thermal imagery in early summer from three separate flights: at dawn, late morning, and dusk, using a helicopter fitted with a differential GPS. They determined that while the technique was good for detecting the presence of animals and for showing trends related to time of survey and flight altitudes, it would be unlikely to provide a reliable count index except under some well-defined situations because of the high level of variability between counts. They found that the best time to count deer using thermal imaging was shortly after dawn when both the ground and vegetation were still cold and there existed a significant temperature difference between the deer and their background. They also found that the best flying altitude to maximize deer sightings for their imager with an IFOV of 1 mrad was 550 ft. agl. They conclude by saying that if animal detection and counting with thermal imaging is to be successful within forest or scrub lands, the canopy must be sufficiently open for a good portion of the deer to be detected.

Surveying or censusing with an imager of this resolution is capable of detecting more deer than was realized in this effort. If the flight lines are aligned so that the footprint on the ground (swath width) is to one side of the aircraft at an angle slightly over 45° with respect to the nadir and the imager is panned slowly back and forth in a direction parallel to the flight line (dithered), the problem posed by tree canopy can be mitigated or perhaps negated.

Focardi et al. (2001) conducted a study in the field to compare animal detectability achieved with spotlighting and that achieved with thermal imaging as a function of animal size, species, detection range, season, group size, and composition. They used a cryogenically cooled thermal imager with a thermal resolution of 0.2°C equipped with two lenses having 5× and 30× magnification. The imager was mounted on a van and the imagery was recorded to VHS tape. The spotlight surveys were conducted using two observers in a motor vehicle traveling at 5 km/h equipped with a high-intensity spotlight. Binoculars (8 × 30) and (10 × 40) were used to aid in animal identification. Both the spotlight and thermal imager surveys were conducted along a 19.3 km circuit on park roads. Data were collected on red deer (*C. elaphus*), fallow deer (*D. dama*), wild boar (*Sus scrofa*), red fox (*Vulpes vulpes*), European rabbit (*Oryctolagus cuniculus*), and brown hare (*Lepus europaeus*).

Unfortunately, an evaluation of the performance of the two sighting techniques (spotlighting and thermal imaging) was not carried out, although the experiment was ideally suited for distance sampling. Distance sampling (Buckland et al., 1993) is a powerful tool for minimizing undercounting and/or changes in environmental conditions that influence detectability during sampling but the method depends strongly on sample size. This work does, however, provide valuable insight relative to the improvement in detectability that can be expected from thermal imaging. On average, spotlighting underestimated the number of animals/species recorded in the field by 53.8%. Thermal infrared imaging was significantly more efficient than spotlighting for all body size classes. They found that spotlighting underestimated the small animals (*V. vulpes, O. cuniculus*, and *L. europaeus*) by 32.7%, the medium animals (*S. scrofa*) by 92.1%, and the large animals (*C. elaphus* and *D. dama*) by 41.0%.

Another important observation was made for the large animal class. The two species of deer showed a different relative detectability with spotlighting and thermal imaging. Both the red deer and fallow deer were underestimated by spotlighting, but the amount was significantly different (42.7% for red deer and 9.3% for fallow deer). Focardi et al. (2001, p. 137) point out that this result is due to the fact that "Fallow deer on our area were observed much closer to the roads than red deer, possibly due to their being more comfortable with roads because they are from semidomestic stock. Under these specific conditions, spotlighting performs quite well for fallow deer, while its performance decreases with the wilder red deer mainly observed in the more remote areas of the park". As a result of the significant undercounting of wild boar (*S. scrofa*) by spotlighting (38 animals by spotlighting as compared to 481 by thermal imaging), an astonishing 92.1%, they recommend that wild boar should only be counted with thermal imagers.

A very important consequence of recording the imagery is that it becomes a permanent record that can be reviewed many times after the data is collected and decisions relative to the significance of the images recorded can be thoroughly examined. The imagery provided by Focardi et al. (2001) is excellent, and one needs to imagine watching the video in real time as opposed to viewing the images shown in this publication, which is only a single frame of the video. The added benefit of video is that it provides a continual scan of the landscape containing the animals as the vehicle moves along the transect. This in turn provides a continuum of viewing angles of the same animal or group of animals that will aid significantly in the task of counting and sizing individual animals. This work shows that thermal imaging is much more efficient than spotlighting for detecting wild boars, which are well-known to be difficult to count because of their elusive behavior and preference for densely vegetated habitats. Furthermore, this work served as the impetus for the advancement of thermal imaging techniques to serve as a primary survey tool for nocturnal distance sampling.

Haroldson (1999) and Haroldson et al. (2003) estimated deer abundance during replicated thermal imaging surveys, evaluated the precision and accuracy of these estimates relative to an independent mark–resight population estimate, and identified sources of variation in thermal imaging estimates. The mark–resight surveys were conducted using a helicopter with one pilot and the same two observers for all surveys. They flew at 38 m agl at 60 km/h during all flights. Each observer covered a 100 m swath width on their side of the flight line for a total swath width of 200 m for each flight.

For the thermal surveys they used a Micro-FLIR WF-160DS® Airborne Surveillance System mounted under the wing of a Cessna 337 twin-engine aircraft. Imagery was recorded with a Hi8 video cassette recorder. This imager is an 8–12 μm LWIR with an NEDT of 0.9°C, a WFOV of 10° × 7.3°, and an NFOV of 3° × 2.2°. They attempted to maintain a flight altitude of 610 m agl and a ground speed of 160 km/h during all flights. They could visually differentiate deer from background objects with similar thermal characteristics (e.g., cattle, water, rocks, stumps, etc.) when viewed using the NFOV. All flights were flown in March 1996, under dry weather conditions (absence of rain, snow, and fog). The 1400 ha study site was divided into nine 203 ha circular quadrats (1.6 km diameter). Sensor operators searched for deer during three consecutive revolutions around each quadrat by systematically panning the sensor from the quadrat perimeter to the center. The thermal imaging surveys were conducted twice per night (midnight and 1–2 h before sunrise) for five nights. Portions of eight of the nine quadrats were outside the study area and were not searched. In general, when a deer group was sighted, operators made no attempt to enumerate animals but obtained sufficient video footage for an accurate postflight video tally. They were unable to directly calculate ground coverage within each quadrat. The results obtained from these surveys were considered to be highly variable by the authors, ranging from 31% to 89% based on the mark–resight population estimate of 311 deer. They attributed the variability in detection to observer (flight crew) bias and inconsistent thermal contrast between deer and the background, and they acknowledge that the procedure for interrogating the circular quadrats was difficult to do in a repetitive manner for both flight crews.

This effort suffered from several factors, including those pointed out in retrospect by the authors. The same survey design might have produced better counts if a helicopter was used to conduct the thermal imaging surveys. The flights could have been conducted at lower altitudes and much slower speeds, which would have made the panning much more stable and effective. It would have made the goal of surveying the entire quadrats significantly easier and perhaps a reality. As generally indicated by the thermal contrast measurements, the predawn flights were better than the midnight flights because the background had more time to move toward thermal equilibrium. The work done here was conducted at midnight and in the predawn hours, as it should be, to minimize the effects of solar loading during the daylight hours. A typical 8–12 μm thermal imager does not detect visible light, but it responds to radiation in the form of heat in the 8–14 μm spectral range. This occurs regardless of the source of that radiation (see Chapter 9). If, for example, this was treated as a visual survey, a snow background would be used to maximize contrast and hence detectability. Similarly, the apparent thermal contrast between the deer and its background must be maximized. To do this the apparent temperature difference between the deer and its background must be "adjusted," a task that is easily accomplished but which is intimately tied to the local atmospheric conditions prior to and at the time of the survey. These conditions are much easier to meet in warmer weather because in warmer weather the apparent temperature of a deer is larger due to thermoregulatory processes.

Since the study area was 1400 ha and the nine quadrats were 203 ha each, part of the area encompassed by the quadrats was outside the study area, which caused some implementation problems for the operators. The area of the quadrats totaled 1827 ha and the study area was 1400 ha. It was conceivable that the entire study area could have been surveyed using a helicopter flying linear transects while panning the imager in a slow cycle of fore and aft along the flight line (slower in the forward direction and faster in the aft direction), which would be tied to the ground speed of the helicopter. Additionally, this would have been more consistent with the flight protocol used for the mark–resight surveys to determine the estimated population.

Their suggestion that the detection of deer with thermal imaging is highly variable is most likely due to the survey design and implementation coupled with the fact that imagery was collected before thermal equilibrium conditions were reached in the background, rather than detection inadequacies of the thermal imager.

A study was designed by Dunn et al. (2002) to compare the effectiveness of visual aerial surveying techniques with that of thermal infrared sensing to count elk on five different ranges in southwestern United States. This work reinforces many of the previous reports of deleterious effects resulting from the possible misuse of thermal imagers in the field. They used an Inframetrics IRTV-445G MK II FLIR capable of detecting temperature differences of $\Delta T = 0.3°C$. This LWIR thermal imager was flown at 300 m agl to collect imagery from a series of transects for comparison with visual surveys that were flown at 60–90 m agl along the same transects 20–30 min after the FLIR surveys. It is difficult to fully understand the reason why such a method was selected if the goal was to compare survey techniques to determine which would produce superior detection rates. A fair comparison would, at the minimum, require that the two surveys be conducted under the same spatial and temporal conditions.

Assuming that the two methods actually surveyed the same ground coverage during transect flights there is still the 20–30 min variance in temporal uniformity between the two surveys that may have skewed the results. The assumption for this methodology was that the same elk would be available for counting during both surveys because the flights were close enough that environmental conditions were similar and the elk were sedentary during the FLIR surveys. Nonetheless, they point out that "The Timber Mesa FLIR count was higher because we observed a group of 243 elk during the FLIR but not during the visual survey".

All surveys (visual and FLIR) were conducted from 0630 h to 0900 h on mornings with calm winds (<16 km/h) and clear skies. Visual surveys were carried out at 60–90 m agl and the FLIR surveys at 300 m agl. The aircraft typically used by the agencies managing these ranges were also used in these surveys. "We mounted the FLIR detector to provide a 200–300 m field-of-view on each side of the flight line and set it at 1.5× magnification during surveys".

This work verifies the observations of previous workers in that if the FLIR is used in a way that does not exploit its strengths then the results will be poor. However, when comparisons are made between different survey techniques such as thermal imaging and visual surveys they should be done, at a minimum, under the same set of conditions with regard to the spatial and temporal extent of the survey. If the animals are confined spatially then the temporal condition might be waived (i.e., FILR could be done at predawn and visual at midday).

It is difficult to gain an understanding of the results presented here because important information regarding the flight conditions are not stipulated and where they are stipulated the conditions are too variable to make valid comparisons between the two techniques being tested. For example, the FOV of the thermal imager was not included nor was the transect width for the visual surveys so comparing counts is somewhat difficult. If the FLIR was mounted to provide a range of transect widths varying between 200 m and 300 m on each side of the flight line (a total of 400–600 m swath width or survey width) then there is an uncertainty of 33% in the surveyed area for the FLIR. From the information presented in this paper we cannot determine the area that was visually surveyed. Flying the FLIR at 300 m agl implies that the FLIR must have had a total VFOV of 65–90°, depending on whether a 400 m or 600 m swath width was used. The total FOV (sum of the two spotters) used during the visual surveys would have to be 156° for 60 m agl and 146° for the 90 m agl flights if a similar 600 m swath width was used.

Apparently there was some confusion about the FOV of the imager they were using. The IRTV-445G-MKII® FLIR has an 18.6° (WFOV) and an IFOV of 1.2 mrad. The largest FOV (footprint that extends outward from the flight line) that can be accommodated by this FOV at 300 m agl is a 100 m swath width (50 m on each side of the flight line). If this is to be compared to the visual at a 400–600 m swath width then the estimates obtained from the FLIR survey are grossly underestimated. This is a factor of 4 or 6 and would mean that they should have counted four to six times more elk with the FLIR than they reported (temporal error aside). If we take the Reserve data as an example (550 by visual to 210 by FLIR) then we should have at least 840 elk counted by FLIR.

Surveying with a thermal imager during periods of peak solar loading must be done with caution, and special attention must be given to the actual background that the subject animal will be compared with thermally if one expects to get usable imagery. Difficulties encountered due to reradiated solar energy from tree canopy were noted (Dunn et al., 2002, p. 966) and the effects were reported as abruptly changing thermal signatures in the imagery from medium gray to bright white (saturated) with very little solar illumination, which points out, once again, that the timing of thermal imaging surveys is just as important as it is for visible surveys.

Reviewing these survey conditions we find that the surveys were conducted in a manner and with a methodology that precluded the full utilization of the thermal imager, which is usually the cause of poor results. If the survey was carried out such that the thermal imager was used to its full capacity and the visual survey was conducted under the same conditions, then the visual survey would be ineffective. This result would arise from the methodology used when comparing the two survey techniques, rather than the strengths or weaknesses of either technique. For example, if the FLIR and visual surveys were conducted simultaneously (querying the same spatial and temporal fields) before daylight without snow cover, then the visual count would be zero. Researchers using thermal imagers need to be aware that they are not looking at visual images formed by scattered and reflected solar radiation (light) but are looking at images formed by apparent temperature differences (see Chapter 5). The condition required for optimized visual counts includes the use of snow cover to improve visual contrast between the animal and its background. The same high-contrast conditions between the apparent temperature of the background and the apparent temperature of an animal must be sought when acquiring thermal images.

A forward-looking infrared thermal imager was used to locate and verify California bighorn sheep (*Ovis canadensis californiana*) in rugged canyon habitat located in Owyhee County in southwestern Idaho (Bernatas and Nelson, 2004). Elevations in the study area ranged from 1380 m to 1660 m and included gentle rolling uplands and steep canyons ranging from 30 m to 300 m deep with widths ranging from 300 m to 1500 m. A commercially available LWIR Westinghouse WesCam DS16® FLIR with an NEDT = 0.25°C, a wide (10°) FOV, and narrow (3°) FOV) was mounted on a fixed-wing aircraft. Flights were made at 600 m agl looking straight down along the nadir so that the footprint on the ground was 105 m for the wide FOV and 31 m for the narrow FOV. They obtained paired observations using 30 radiocollared bighorn sheep during surveys conducted in March of 3 consecutive years: 1998–2000. Two planes were used during the surveys; the telemetry crew (pilot and biologist) flew and maintained altitudes $\leq$150 m agl and the FLIR crew (pilot, sensor operator, and biologist) flew and maintained an altitude of 600 m agl. The telemetry crew would locate groups of sheep and identify the subunit location and other demographics to the FLIR crew biologist for recording. The sensor operator would then conduct a search for the bighorn sheep. A series of three overlapping orbits (1.1 km radius) were flown over the plots to acquire the thermal imagery in the steep linear canyon habitat, which was recorded to videotape. Surveys started 30–60 min before sunrise and continued until thermal contrast was lost due to solar loading of the background. The detectabilities for each of the three yearly surveys were 85.2, 89.4, and 88.9%, respectively.

The survey work here was conducted with a good understanding of the problems associated with collecting thermal imagery and dealing with them (i.e., ceasing data collection when the thermal contrast was compromised by solar loading, establishing a flight protocol that allowed for high detectivities in steep canyon habitat, and proper sensor operator training). With regard to the choice of fixed-wing versus helicopter as aerial platforms we point out that many visual surveys using helicopters to count ungulates are designed to flush the animals and very low agl flight patterns (typically 30–90 m) are used. Kufeld et al. (1980) (see Chapter 3) noted that the advantages of helicopters over fixed-wing planes are as follows: (1) with a pilot and two observers sitting three abreast there are three observers searching for deer, (2) terrain and cover problems are minimized since the helicopter can turn and ascend/descend rapidly and it can hover over thick patches of vegetation to flush animals, (3) helicopters are capable of operating at air speeds much lower than that of fixed-wing aircraft, which increases observation time, (4) the sound of the rotor blades tends to flush deer, increasing their detectability, (5) forward visibility is superior to fixed-wing aircraft, (6) helicopters are less likely to subject observers to forces that are capable of producing disorientation and airsickness, and (7) a helicopter can fly safely at lower altitudes than fixed-wing aircraft.

The work by Bernatas and Nelson (2004) points out that the type of aircraft used in FLIR surveys must be selected based on the species and terrain to be surveyed and when there is doubt with regard to the reaction of the species to the aircraft selected it must be determined prior to survey implementation or it will probably be a wasted effort. Clearly there are advantages offered by helicopter surveys with FLIR (slow speed, low altitude, more maneuverability, and possibly safer conditions for the crew) as long as the survey goals can be met. Surveys should not be stressful for the species studied.

An interesting and informative experiment was conducted by Kissell and Tappe (2004) to test the detection capabilities of thermal infrared imaging techniques using humans as white-tailed deer surrogates. They conducted a complete survey (a census) of a 2.56 km^2 site that had a resident deer population and they added 20 randomly placed people who assumed reclining postures to mimic deer signatures during an aerial survey of the study area. Four of the 20 people were located in water and 16 on dry land. The survey was conducted 1.5 h after sunset from 2000 h to 2200 h but the degree of solar loading during the daylight hours preceding the flight was not provided. The maximum ambient temperature during the day was 18.3°C and declined from 12.2°C at the start of the flight to 10.6°C at the conclusion of the flight. They flew a Mitsubishi IR-M700® MWIR thermal imager mounted on the belly of a Cessna 182 fixed-wing aircraft. They noted from past experience that a compromise needed to be made regarding the altitude (agl) of the flight and aircraft speed (~130 km/h) because at 305 m agl the signatures would be too hard to distinguish and at higher altitudes such as 610 m the sensitivity of the imager (range) was taxed, so they settled for a flight altitude of 457 m agl for the survey. The imager signal was GPS referenced to a digital video recorder and the video was reviewed using a high-resolution video monitor. The resulting signatures were recorded as people, deer, possible deer, possible people, or unknown. The known locations of people were correlated with the video signatures and the detectivity associated with the people was determined.

The four people located in water were not detected and one person on dry land was not detected so a detectivity of 75% was achieved for this experiment. If the effect of the water is removed then 1 of 16 people was not detected or the detectivity without water would be 93.8%. During their analysis of the experimental results they mention a number of things that may have introduced bias and suggest a number of things they might have done differently to obtain higher detectivities. The inclusion of people into the survey plan to serve as deer surrogates has shed light on a number of problems associated with thermal imaging surveys. These problems are pointed out by Kissell and Tappe (2004) and are discussed below in the context of adjusting portions of the methodology to mitigate or remove completely the problematic issues.

The logic of this type of test is excellent and will provide operators with the opportunity to determine proper camera settings for the habitat features and the prevailing atmospheric conditions, which should always be taken into account when planning surveys and collecting imagery. This type of preparation is easy to implement and will usually provide valuable information for the sensor operators about how to best proceed with the planned survey (see Chapter 8).

With regard to the excellent work by Kissell and Tappe, it is clear that they have identified and specifically noted ways to improve their results. For example, they noted that it would be better to survey at 305 m but fly much slower than 130 km/h to exploit the sensitivity and spatial resolution of the imager. This would fix the problem of signature identification (deer/person). The survey should be done at the lowest possible altitudes at the slowest possible speeds unless there are safety issues or the resolution is adequate at higher agl levels and lower resolutions. The same area could be surveyed at a cost of a longer flight time.

The people could be detected in water if they were not completely submerged or soaked if the survey were conducted just before dawn (this could be easily tested with surrogates). Since water has a higher thermal inertia than vegetative material and soil, the signatures associated with standing water will be problematic during the time period just after sunset. If the survey were conducted just before dawn the water would have more time to radiate heat to the cold night sky (assuming a cloudless night). The objective is to bring the thermal background (not to be confused with the ambient temperature or air temperatures) into thermal equilibrium such that the imager sees a uniformly radiating background, regardless of its actual temperature or contents (i.e., water, rocks, deadfall, trees, and shrubs). When this is achieved the radiation emanating from animals and deer will be brightly contrasted against the radiation emanating from the background (see Chapter 11).

They correctly point out the difficulties of flying circular transects or surveying circular plots using oblique angles. The panning should be done to interrogate the ground under tree canopy but the obliqueness of the angle from nadir should not exceed 45°. Further information on panning is presented in Chapter 11. We have used this technique to image deer in forested habitat during leaf off and in dense mixed-canopy situations.

Drake et al. (2005) point out that counting suburban deer populations is difficult for more reasons than the technical aspects associated with counting methodologies and the resulting detectabilities. For example, they point out that spotlighting may not be suitable in residential areas due to the intrusive nature of bright lights. In addition, hunting is generally not permitted in suburban areas, ruling out density estimates based on biological data from hunter check stations. Other methods may suffer from a lack of access to private property in suburban areas. For these reasons they proposed to compare two techniques that may prove to be suitable for determining the size of a suburban white-tailed deer population: road counts and counts obtained by a FLIR imager.

Their study area consisted of a 324 ha suburban community in Somerset County, New Jersey that was fenced (2.4 m high), which ensured nominal deer immigration and emigration. They divided the study area into five relatively even-sized regions and drove 2–3 transects per region, which were 4 km in length and followed an established road. Five observers were trained to count deer from the road while driving at 8–16 km/h. Four road counts were conducted between June 2001 and January 2002, starting 1 h before sunset and requiring approximately 20 min to complete.

Three FLIR flights were conducted in January 2002, between 2000 h and 2330 h. Each flight required about 45 min to complete and the environmental conditions and deer behavior changed throughout the flight period, with temperatures becoming colder and deer becoming less active as the evening progressed. The imager used was an FLIR Series 2000F® LWIR camera with a WFOV of (28 × 15°) and a NFOV of (7 × 3.75°). They flew the imager with a helicopter at 80 km/h at 152 m agl and covered the area using a swath width of 129 m.

If the FLIR deer counts of January 9 are considered as independent then they should be unaffected by any preceding flights, which may not be the case here. The marked decrease in the counts after the first flight suggests that the deer may have been flushed to cover by the disturbance produced by night helicopter flights. This possibility should also be considered when evaluating the results of this survey comparison since it is one that can be mitigated during future work. A close examination of the video should provide information regarding the number of deer occupying open areas on subsequent survey efforts. The footprint or swath width used for the surveys was 129 m and apparently was designed for complete coverage of the enclosure, which means that the imager was pointing in the forward direction at an angle of ~54° with respect to the nadir. Slow panning with the imager along the flight line affords a range of viewing angles to help with tree canopy.

The road surveys had approximately 80% of the study area observable from vehicles while driving transects, which means that the entire area was not completely covered. The results show that the total deer counted on January 10 from the road was 222 by the five trained drivers. The FLIR, on the other hand, was 214 on average for the three counts of January 9. The first flight recorded the largest and most uniform count at 251 deer, but there may have been double counting, as was possible for the first road counts in June and August 2001. All in all, the results are very close and the authors recommended using FLIR in suburban areas dominated by private property where ground access is restricted.

An attempt was made to compare two aerial survey techniques on deer in fenced enclosures. Potvin and Breton (2005) tested a visual double-count method (used by the Province of Quebec to survey deer over its entire range) and thermal infrared sensing. Six surveys were used for the double-count method and two for the thermal surveys. Four large enclosures (A through D) varying in size from the smallest at 6 km^2 to the largest at 29.4 km^2 were used to evaluate the accuracy of the two techniques. The actual number of deer in the enclosures was estimated by reconstructing the population based on registered hunting harvest and winter mortality data. They point out that the evaluation of the accuracy of aerial survey techniques was dependent on the validity of the reconstruction of deer populations, a task that is more difficult for larger enclosures than it is for smaller ones. The bulk of the work in this paper was devoted to the reconstruction of the deer populations so there would be a way to make estimates. The actual times of the surveys and the populations being surveyed were not the same and in fact only one of the thermal surveys was carried out over one of the same enclosures as the double-count method (A at 15.7 km^2). None of the double counts were taken on B (6.8 km^2), which had the densest canopy cover from the residual forest. The only survey conducted on B was a thermal survey in 2001 based on an assumed density at that time and which was used only once in a data analysis that stretched over a 2-year time frame. The double-count method was used three times on enclosure A (81, 83, and 37%), twice on enclosure C (64 and 75%), and once on enclosure D (45%).

There were two surveys conducted with an FLIR 2000A/B filtered for use in the 8–14 μm spectral region with a wide FOV (28 × 15°) and a narrow FOV (7 × 3.25°) with an NETD = 0.16°C and a spatial resolution of IFOV = 1.4 mrad. The imager was mounted to the floor of an AS350B2 helicopter for the aerial surveys. Only the wide FOV was used for the surveys and the spatial resolution would allow the detection of a 0.14 m target at 100 m. The survey of enclosure A gave 89% accuracy and the survey of enclosure B gave an accuracy of 54%. The sensor was flown at an altitude of 91 m agl at 70–100 km/h. The 28° horizontal FOV created a swath width of 57.2 m at a 30° forward-looking angle from the nadir. Presuming that the altitude was held constant, the area surveyed would be very accurate (57.2 m × lengths of transect) and only animals within the FOV were counted.

For visual observers this is not the case. Animals must be estimated as to their location and subsequently it must be determined whether or not they are within the transect by using tape markers on the side window and on the bottom of a rod extending perpendicular to the aircraft. A third observer (sitting behind the pilot) with front and lateral vision accommodated by a bubble window would discriminate and record the observed deer. Additionally, the observers must contend with animals that are moving in or out of the transect as they are flushed. They must adhere to a strict protocol regarding the direction and timing of flushed deer to minimize the possibility of double counting. There is no permanent record to review in the case of miscues with this procedure.

If the goal of this work was to evaluate the effectiveness or the accuracy of the two counting techniques they should have been carried out at the same time (i.e., using the same reconstructed population estimate) over the smallest of the enclosures, which presumably would have been the most accurate according to the authors. As it is, the only commonality in these series of surveys is tied to enclosure A. To make a reasonable comparison one would use the counts from the two techniques obtained for enclosure A and assume that the updated herd reconstruction numbers are accurate for this enclosure from October 2001 to August 2003. A somewhat valid comparison can be made between October 26, 2001 data from thermal imaging and the three surveys taken over enclosure A using the double-count technique in January and August of 2002 and the survey in August of 2003. The results of these surveys are reported in Table 5 of this paper and show that the thermal survey had the highest accuracy (89%) for enclosure A and the three double-count surveys gave 81, 83, and 37%. A legitimate question might be, "Why is double visual counting in the same enclosure so disparate?".

The statement "our results on thermal infrared sensing are limited and contradictory" (p. 324) is only partially correct; that is, they are limited but they are not contradictory. The suggestion that they are contradictory is offered because the results differ from double counts over different enclosures than those surveyed using the thermal imager or that the accuracies obtained in enclosures A and B via thermal imaging are thought to be too disparate in accuracy. In fact it may be proper to suggest that the double-count technology is contradictory in that one survey (August 2003 37%) is unaccountably significantly different from the two other double-count surveys of 80 and 82% in the same enclosure.

Finally, the last sentence in the abstract of this paper is without foundation based on the work presented: "Because of closed forest canopy, thermal infrared sensing of deer along systematic survey lines was not a reliable technique". There is absolutely no data in this paper to suggest that this is a valid statement. They did two surveys with a thermal imager and the performance was significantly better where a valid comparison can be made with double-counting results (enclosure A). For the results obtained in enclosure B there is no comparable double-count survey nor is the reconstructed population for this enclosure considered a very firm estimate (see Potvin and Breton, 2005, p. 324). The FLIR survey conducted on A was carried out in the morning from 0700 h to 0902 h (good methodology) and that of B was carried out in the afternoon from 1545 h to 1703 h after a full day of solar loading (bad methodology). The result for the FLIR survey conducted in enclosure B (highest residual forest canopy cover at 72%) was conducted at the worst possible time for preserving a suitable thermal contrast between the deer and their backgrounds and yet it still produced a reasonable 54% accuracy. We point out that vegetation is very quick to warm and reradiate under the influence of the sun (even on cloudy days) so that the apparent temperature of the tree canopy at those times can be very similar to the apparent temperature of the deer being surveyed.

A closed mark–recapture design was used to explore the efficiency of spotlights for detecting white-tailed deer by operating thermal imagers and spotlights simultaneously (Collier et al., 2007). The study was conducted at Brosnan Forest (5830 ha), a lower coastal plains habitat in Dorchester County, South Carolina, which was 93% forested and contained 330 km of navigable roads. Four management units (1134–1316 ha) were surveyed using a multiple observer approach where two observers surveyed from each side of the vehicle; each pair had one observer using a thermal imager and one observer using a spotlight. The thermal imaging observer and spotlight observer were separated from one another by an opaque partition, ensuring independent observations. The thermal imagers used were handheld LWIR Raytheon Palm IR 250 Digital® and Raytheon Palm IR 250 Analog®, fitted with a Barium Strontium Titanate (320 × 240) focal plane array. The spotlights used were 1,000,000 candle power handheld Lightforce SL240® spotlights. Transects ranged in length from 12.6 km to 14.8 km and were each surveyed twice during the study.

In all they identified 883 individual deer during the surveys (total of both detection techniques) and the thermal imager detected 92.3% of the total white-tailed deer seen, whereas the spotlight detected 54.4% of the total. The classification of deer by age and sex was 11% better with the spotlighting than the thermal imaging.

Again we must caution against using results of studies that compare one counting method versus another when the best features of each are not exploited during the comparison. Their results show that detection with thermal imaging was greater than that obtained by spotlighting – 92.3% compared to 54.4% – yet they determined that counts made with thermal imagers cannot be considered a census. They referred work by others in support of this conclusion, specifically the work of Potvin and Breton (2005), whose results showed bias due to incomplete detection, which was actually due to the improper use of the thermal imager during that effort (see comments above). If a more sensitive imager was used in this work it would improve the classification strength.

An improved procedure for detection and enumeration of Pacific walrus (*Odobenus rosmarus divergens*) signatures in airborne thermal imagery was demonstrated by Burn et al. (2009). In earlier work Burn et al. (2006) showed that matching digital photographs concomitantly taken with collected thermal images at spatial resolutions of 1–4 m per pixel indicated there was a linear relationship between the number of walruses in a group and the amount of heat that they produced. Furthermore, this relationship existed for all spatial resolutions tested, indicating that the number of walruses in a group hauled-out on sea ice could be estimated using their thermal signatures. Walruses use floating pack ice as a substrate for birthing, nursing, resting, and for passive transport to new feeding areas. The surveys in this work were carried out in US territorial waters (Bering Strait) where pack conditions ranged from 50% to 100% complete. The first aircraft used to conduct the survey was an Aero Commander 690B turbine engine aircraft fitted with an LWIR infrared scanner with an IFOV of 0.625 mrad and an NETD of 0.12°C; it was equipped to georeference the thermal imagery.

North–south transects ranging in length from 60 km to 225 km were flown at 6400 m and at 3200 m agl, producing imagery with 4 m and 2 m pixel size, respectively. A second aircraft, an Aero Commander 680 piston engine aircraft equipped with a vertical camera port, was used to photograph walrus groups with a high-resolution image-stabilized digital SLR 12.4 megapixel camera fitted with a 200 mm, F/2.8, and a 1.4× teleconverter producing images of 4288 × 2848 pixels. The photographs were collected within an hour of the thermal scans to minimize the effect of changes in group size over time. The same procedures were used to process both the 2 m and 4m resolution images and the new methods presented here are contrasted with the methods used by Burn et al. (2006).

They developed calibration models to estimate the number of walruses in each group based on the group's thermal index since it was detected in the thermal image. For calibration they only used the observations of photographed groups that were detected by the thermal imager because the calibration models are conditional on the group being detected. The new procedure for detection and classification of walrus groups in thermal imagery described in the current work is a significant improvement over the methods of Burn et al. (2006), as smaller groups were detectable in both 4 and 2 m imagery using the new procedure. They postulate that improvements can be had by surveying when the ambient temperature is warmer and point out the walrus skin is known to vary with ambient temperatures. At extremely cold temperatures walruses vasoconstrict to reduce blood flow to the skin, which serves as a mechanism to conserve body heat. False positive signatures (background clutter; see Chapter 11) were open leads that extended only slightly into a tile that otherwise contained only ice and bare rock faces, which were reradiating previously absorbed solar radiation, both of which were easily reconciled by comparing with adjacent tiles or by overlaying GIS data layers for land onto the thermal images (Burn et al., 2009).

This important work demonstrates that the use of airborne thermal imagery to survey marine mammal populations has the potential to sample considerably larger areas per unit time than visual photographic surveys and is applicable to a number of other marine mammals. Here the thermal camera is used to cover vast areas of open sea and ice to locate (detect) hauled-out walruses, followed by visual high-resolution photography and data processing to count the walrus.

Three methods of population estimation for woodland caribou (*Rangifer tarandus caribou*), which is listed as a threatened species in Canada, were compared with regard to effectiveness and cost (Carr et al., 2010). The study site consisted of the Slate Islands, located on the north shore of Lake Superior, which provided a protected and closed population of woodland caribou with very limited predator influence. The total size of the protected area is 47.3 km^2, and population surveys have been conducted every year since 1974. Bergerud et al. (2007) suggest that the population has entered a 5-year cycle where the population fluctuates between 100 and 600 animals with major mortality rates occurring at 5-year intervals. The survey methods compared by Carr et al. (2010) included FLIR technology to count caribou on regular-spaced transects flown by fixed-wing aircraft, observers to count the number of caribou seen or heard while walking random transects in the spring, and mark–recapture sampling of caribou pellets using DNA analysis.

An Agema Thermovision® 1000 (FLIR Systems) operating in the LWIR had an NETD $< 1°C$, a WFOV of 20°, and an NFOV of 5° so that at 305 m agl looking downward along the nadir the footprint of the sensor was 110 m wide $\times$ 71 m long for the WFOV and 27 m $\times$ 18 m for the NFOV. Survey times were between 1000 h and 1400 h on January 29–30, 2009, but the weather conditions during the survey were not provided. The look angle used during the surveys relative to the nadir was 30°, which results in a different footprint than those specified above for looking down along the nadir. The transects were spaced 200 m apart but the width of the transect apparently was variable, depending on the operator's panning technique. All imagery taken within the study area was recorded to video for analysis. The total length of the transect lines for the FLIR survey was 284.4 km. After the survey all video was reviewed frame by frame, forward and backward, and in slow motion to confirm caribou sightings.

Walking surveys were conducted over a 30-day period in July 2008, when 11 transects were used that averaged 4.2 km each for a total survey length of 63.4 km. Fecal pellets were collected on January 30 and February 27 in 2007 for mark–recapture analysis (2-window approach). The eight sampling sites were chosen by randomly selecting lakes and/or sheltered bays within the study area that were appropriate for landing a helicopter to collect samples. Samples were again collected in 2009 on three occasions in January and February (3-window approach) to allow for more sophisticated mark–recapture modeling.

The results of these surveys were used to estimate the population size of caribou on Slate Islands with large overlapping confidence intervals. They feel that the population estimate from the FLIR survey represents a minimal value that may have been limited by the rugged landscape of the Slate Islands and possibly dense conifer forest cover. Subsequent FLIR surveys for moose and caribou in a conifer-dominated landscape on the mainland north of the Slate

Islands indicated that this forest type does not severely limit the detectivity. The genetic sampling method using the 3-window approach provided results similar to those obtained by FLIR but genetic sampling was twice as expensive ($5.8K compared to $13.9K) and it must be conducted during the winter in a closed system.

There are a number of critical omissions regarding the details of the thermal imaging flights that collected the imagery. The vertical FOVs for the imager were not provided but they can be derived from the footprints described above as 13° for the WFOV and 3.3° for the NFOV. They used a 30° survey angle with respect to the nadir accompanied by side-to-side scanning. We assume that the viewing angle of 30° with respect to the nadir is in the forward direction, however, this is not indicated in the survey geometry presented in this paper. As a result it is unclear if the side-to-side scanning resulted in moving the footprint back and forth across the flight line, which could possibly lead to double counting, or if the scanning was parallel with the flight line, which would result in a larger footprint than that stated above. For the WFOV (20° × 13°) footprint used during these flights we calculate the footprint on the ground as 133.5 m × 93 m, conforming to the geometry of a 30° angle with respect to the nadir, which amounts to a 59% increase in the area of the footprint when compared to the footprint obtained by viewing along the nadir at the same altitude.

The speed of the aircraft was not indicated for these transect flights. Slower flight speeds are recommended for surveillance applications and helicopters would be more appropriate in this regard. The surveys were conducted during daylight hours under peak solar loading conditions (100–1400 h), which are the worst possible conditions unless there have been overcast skies since sunrise.

One of the features offered by FLIR technology is the ability to maintain strict transect widths throughout the survey by simply maintaining the same altitude agl. A word of clarification regarding the reference on page 214 to the work of Thompson (2004). When discussing counting techniques, Thompson refers to the inability of thermal imaging to obtain complete counts over the size of areas typically surveyed, a comment that actually applies to all counting techniques. Furthermore, these comments are in regard to rare and elusive species so to infer biases due to incomplete and variable detection rates is inappropriate if one is in fact not looking at the entire landscape. There are, however, numerous results reported of ~100% detectabilities when complete coverage of surveyed areas was undertaken and when the population was actually known and not estimated.

There is a misconception regarding conducting surveys with a thermal imager. Animals are only detected if they are in the FOV of the camera. If the camera is never pointed at the animal it will not be detected. If you are using thermal imaging to survey along transects then you know you will not see animals that are off the transect and nothing can fill that in for you at the detection level. The survey model selected at the outset is used to determine the population estimate, and the thermal camera is used to detect and count during the survey. You must look to see. Bias comes into play when you look and do not see the animals that are present or when the survey design itself is biased.

A technique called "vertical-looking infrared imagery" for estimating deer density that uses distance sampling in conjunction with airborne infrared thermal imagery integrated with GPS and GIS data (for computing distance calculations for individually detected deer) has been demonstrated by Kissell and Nimmo (2011). They estimated the deer density in four bottomland hardwood sites during leaf-off in Arkansas. The smallest of the four sites was 2030 ha and the largest was 3650 ha. The topography (elevation ranges between 30 m and 70 m maximum) and vegetative cover were similar in the four sites, which were dominated by agriculture fields and food plots interspersed with bottomland hardwood forests. Nocturnal flights were carried out along transects at each site between 2300 h and 0600 h during February 2004. Transects were spaced approximately 400 m apart. Based on the altitude (457 m) agl and the imager's FOV (14° × 11°), the swath width at ground level was 110 m, extending 55 m on either side of the flight line. The imager, a NIR-MWIR (1.2–5.9 μm) Mitsubishi IR-M700 with a 50 mm lens, was mounted in the belly of a fixed-wing Cessna 182 with the collection optics directed in a fixed position along the nadir perpendicular to the flight line. They refer to this arrangement as vertical-looking.

Thermal imagery was collected at flight speeds of ~120 km/h and deer were identified by their unique shape and brightness relative to the background. There were no other species that had similar thermal signatures at any of the four sites. They calculated that the probability of observing deer in the imagery was 1.00 for three of the sites and 0.82 for the fourth site. They argue that viewing with the imager in a fixed vertical position relative to the flight line maintains a more accurate transect width, allowing for more accurate calculations of area and distance. They also argue that it maximizes the detectability because oblique angles introduce more obscuration from vegetation, which could become problematic at the far edges of the viewing footprint (see Chapter 11). They feel that they met the four basic assumptions required for distance sampling with this new method: that g(0) = 1, that deer were detected at their original location, that the measurements were exact due to GPS and GIS computations, and that 400 m between transects was sufficient to eliminate the possibility of double counting.

The detectivities obtained in this work (three sites at 100% and one at 82%) are due to careful selection of transect widths, imager sensitivity, surveying at the right times for optimum apparent temperature differences, and good data processing techniques.

The dependence of visual aerial surveys on complete snow cover has been reported by many observers as essential for successful survey counts (Gilbert and Grieb, 1957; Evans et al., 1966; LeResche and Rausch, 1974; Rice and Harder, 1977; Ludwig, 1981; Gasaway et al., 1985; Stoll et al., 1991; Beringer et al., 1998) (see Chapter 2).

These reported observations motivated Storm et al. (2011) to consider thermal imaging as a potential alternative to helicopter surveys for estimating small-scale deer abundance (at ~2.59 km^2) over an extensive area (>500 km^2). The goal of their study was to determine if thermal imaging could provide less-biased counts for both snow covered and nonsnow covered landscapes than is possible with visual observation from helicopters. They wanted to compare fixed-wing thermal surveys and helicopters in terms of (1) relative bias, (2) relative influence of snow cover, and (3) cost and evaluate the precision of fixed-wing thermal imaging surveys. The study areas consisted of four (12.9 × 3.2 km) plots near Madison, Wisconsin and the University of Michigan's Edwin S. George Reserve (ESGR), which is 5.3 km^2. They surveyed plots with both fixed-wing thermal and helicopter visuals with two observers during winter (>20 cm snow) and during spring with no snow and before leaf emergence. The helicopter visual surveys were conducted at 30 m agl at speeds between 65 km/h and 72 km/h. The fixed-wing thermal surveys were conducted at 300 m agl at a speed of 120 km/h and the transect widths were not provided.

The visual surveys were conducted using a previously adopted protocol that was not completely described, but they were conducted during daylight hours between late morning and early afternoon to avoid shadows that might have limited the ability of the crew to detect deer. The low altitude of the flight often caused deer to flush but the observers were still able to record the deer without double counting. The transect widths were not given other than that they were tightly spaced.

The thermal imaging system used was a Poly Tech Kelvin 350 II fit with an (8–12 μm) LWIR sensor with an NETD of <1°C. Each plot was comprehensively surveyed in transects spaced 150 m apart. When the operator detected a potential deer, based on thermal contrast, the aircraft would circle around the deer to confirm the sighting. As a result of these deviations from a transect to circle, suspected deer transect overlap occurred, which was sorted out later from imagery recorded to videotape. They reported that precipitation prevented thermal surveys on three days, low ceiling clouds <300 m agl prevented thermal surveys on seven days, and one day was lost due to the plane being covered in ice during a storm. As a result the thermal surveys were conducted opportunistically, as weather allowed, and they were therefore not restricted to any particular time of day.

The authors state in their discussion of results that the inconsistency between thermal and visual helicopter surveys over snow suggests that the detection rates of one or both of the methods were variable. When snow was absent thermal counts were always higher than helicopter visual counts, indicating that thermal provides less-biased estimates of deer abundance under these conditions than visual helicopter counts. The authors are keenly aware that results show visual helicopter counts in their study area to be more effective with snow cover (they were always >2.5 times higher than counts without snow). They expected this result because without snow on the ground, there is little visual contrast between deer and their environmental surroundings. What is surprising, however, is that they were not aware that thermal imagers require thermal contrast to be effective and that the biggest factor affecting thermal contrast is the prevailing atmospheric conditions preceding and during the thermal survey. Thermal surveys cannot be expected to produce high detectivities unless the thermal contrast criterion is met. Conducting thermal surveys opportunistically does not guarantee that the thermal contrast between the deer and their backgrounds is optimized; in fact, it would be surprising if it were. The thermal flights over ESGR on February 11–12 took place at night (more uniform background temperature) and the counts from this survey are two times that of the visual count on ESGR. None of the flight times are given for the other thermal surveys.

The thermal surveys should be conducted during overcast days from a helicopter flown at altitudes of 100 m or at altitudes comparable to those used during visual surveys. If conducted at night or in early morning/predawn, flight altitudes can be increased for safety. Making comparisons between FLIR and visual surveys should be done only when the conditions for both of the surveys are optimized.

A covariate analysis of game species monitoring using road-based distance sampling in association with thermal imaging was conducted by Morelle et al. (2012) at five sites in Condroz, a natural region in Belgium located between the Ardennes and the Meuse River. Condroz is a mosaic of woods and farmland with 55% consisting of grassland and crops and ~25% patchy forest. The five study sites varied in size from 4600 ha to 6400 ha and had varying forest cover ranging between 14% and 46%. Randomly selected survey routes ranged from 40 km to ~64 km and were designed to cover the habitat availability and also allow completion of the count within the charge time of the thermal imaging batteries. Four nocturnal road counts were conducted from a vehicle at each site (two between 2000 h and 0100 h, and two between 0100 h and 0700 h) by a driver and two observers, one on each side of the vehicle, equipped with a handheld thermal imager.

Two different thermal imagers were used to detect the animals of interest: roe deer (*C. capreolus*), wild boar (*S. scrofa*), and red fox (*V. vulpes*). A FLIR ThermaCAM® HS-324 with a 320 × 240 VO_x FPA and a JENOPTIK Vario Cam® with a 640 × 480 FPA were used and found to be similar in performance for the surveys and the conditions under which they were conducted (essentially searching for "hot spots"). Information concerning the FOV used in these surveys was not provided.

This work points out a very important and useful function that infrared imaging can fulfill that other counting technologies cannot. If the object of a survey is to count a species that is the only animal available on a range or resident in a particular habitat then the imager can be used to detect hot spots. This type of survey is very easy to implement with inexpensive imagers and can provide excellent detectivities with ~100% probabilities (see Chapter 9). Clearing the background of residual hot spots is the only preparation needed to conduct such a survey. In Figure 6B of this work the detection probability (N_{50}) for the forest edge category is at 250 m for the searched animals. The specifications for the FLIR ThermaCAM HS-324 indicate that a man can be detected (N_{50}) at 440 m and recognized at 110 m. These detectabilities can easily be adapted for a wide array of survey and census work when the survey is concentrated on detecting hot spots. This work also points out that there are considerable differences in the cost of imagers available for these applications. For example, the JENOPTIK imager was seven times more expensive than the FLIR imager.

A 10-year study evaluating nocturnal line transect sampling was carried out by Franzetti et al. (2012) in a fenced Mediterranean forest near Rome, Italy. Night surveys of wild boar (*S. Scrofa*) were conducted in the autumn of 2001–2010, using portable infrared cameras to detect animals. The entire study area was sampled on foot by using transects along existing forest roads and paths to limit animal disturbance and to ensure the safety of field workers. The accuracy of these nocturnal line transect studies in monitoring density changes was compared with independent estimations obtained by capture–mark–resight performed on counts at feeding sites.

Two different thermal cameras were used in this work and both were equipped with digital laser range finders and electronic compasses. The ThermaCAM B 640® has a 640 × 480 pixel uncooled FPA microbolometer detector operating in the LWIR spectral range with a FOV of 24° × 18° and an NETD = 0.06°C at 30°C. The spatial resolution was 0.65 mrad. A second camera, ThermaCAM PM 545® with a 320 × 240 pixel microbolometer FPA and an FOV of 12° × 9°, was also used to detect and locate wild boars. They reported that detection and locating of wild boars in the landscape was straightforward with an infrared camera, even in dense vegetation. They recorded the radial distance and compass bearing to the center of the group detected, group size, and features of the animals' behaviors (feeding, lying or standing still, moving, and fleeing response to the presence of the observer). Observations of flushing animals were ignored unless their original location was recorded or was evident from the heat radiating from the ground where they had been lying (see Figure 4.2b for an example of observing heat transfer from an animal to the ground as a result of bedding or resting). They report that flushing response can be largely limited during night sampling by using well-trained observers and proper precautions during data collection in the field (keep noise and stray light sources to a minimum). In this regard thermal imaging helped in detecting animals before they flushed out or moved in response to the presence of an observer.

Franzetti et al. (2012) point out that nocturnal line transect sampling carried out on foot using portable infrared cameras can be used efficiently to estimate the population size of a nocturnal and elusive species (such as wild boar) even in dense habitats but great attention has to be paid to the survey design and the field protocol for data collection. The precautions used in this work were implemented because they were keenly aware that convenience sampling must be avoided or great care taken to assure that the distribution of wild boars around the transects was consistent with line transect survey assumptions and that any observed flushing response was random and not directly related with the presence of observers. Unfortunately, the precautions used here are largely ignored by many researchers.

Nocturnal distance sampling surveys were conducted on foot using thermal imagery once a year from 2001 to 2005 to obtain population estimates for fallow deer (*D. dama*), a nonnative ungulate in the survey area (evergreen Mediterranean woods) near Rome, Italy (Focardi et al., 2013). Great care was taken to comply with the four relevant assumptions required by the distance sampling methodology (Buckland et al., 2001): (1) objects on the line are always detected; (2) objects are detected at their initial location, before any movement in response to the observer; (3) distances and angles are measured accurately; and (4) transects are distributed randomly with respect to the local distribution of animals. The flushing behavior of many species as a result of spotlighting limits the ability of spotlighting to meet the requirements of assumption 2. However, nocturnal distance sampling with thermal imagers can easily satisfy both assumptions 1 and 2.

In the present work, the adherence to distance sampling assumption 4 was assessed by using a set of radio-tagged deer, while assumptions 1, 2, and 3 were facilitated through the use of a handheld thermal imager ThermaCAM PM 545 FLIR using an X2 lens. This camera was fitted with a coaxial laser rangefinder and an electronic compass to record the position of the detected deer with respect to the transect. Additionally, the camera would record photographs of the detected animal, which were associated with an appropriate voice memo. The actual details of their methodology for data collection were designed to meet each of the four assumptions. Their results confirmed that nocturnal distance sampling can provide precise estimates of ungulates in a heavily forested area, reaffirming and supporting earlier work by Gill et al. (1997) and Franzetti et al. (2012). The potential source of bias due to animal flushing before detection (a violation of assumption 2) was investigated by Marini et al. (2009). They studied the response of fallow deer (*D. dama*) and wild boar (*S. scrofa*) to human presence during nocturnal line transect surveys and lessons learned were applied in the present work, suggesting that the potential problem may have been overcome by the use of thermal imagers as a data collection tool. Specifically, during nocturnal surveys with a thermal imager it is more likely that animals are observed before flushing so that distance sampling estimates are more reliable.

The present work shows that if care is taken to properly use thermal imagers with modest ranges and sensitivities nocturnal distance sampling can give reliable population estimates in forested habitats. They point out that they were able to cover the entire study area in less than a week and used only a single imager and two observers. They recommended that the observers be properly trained and that a careful survey design be used that is selected and tailored to minimize potential biases through both the data collection aspects of the survey and through the proper placement and use of transects within the study area. The role of thermal imaging in satisfying the critical assumptions of distance sampling methodology and its contribution to the success of this work and that of Franzetti et al. (2012) should not be overlooked.

Aerial vertical-looking infrared thermal imaging was used to evaluate bias of distance sampling techniques for white-tailed deer on the Security Area (1489 ha) of Arnold Air Force Base, located 112 km southeast of Nashville, Tennessee (Beaver et al., 2014). The basic idea was that, since distance sampling with spotlights and handheld thermal imagers are usually conducted along roads, the assumptions associated with distance sampling may be violated and result in estimates that are biased high. It is always a goal to use monitoring techniques that produce estimates with precision and low bias. Since aerial imaging is not restricted to roads, the sampling and detection area can be randomized by conducting counts along systematic strip transects that are not associated with roads or features preferred or avoided by deer. This will minimize density estimates that may be biased high or low based on feature preferences of resident deer.

Approximately 814 ha of the Security Area was forested. Grasslands and open rights-of-way occupied 197 ha and the remaining 478 ha was occupied by open area, including water, buildings and structures, mowed areas, wildlife food plots, and other open areas. For the ground-based surveys a continuous route of 31.25 km was surveyed by dividing the route into 29 transects (each approximately 1 km in length) and only data collected from alternating transects was used to provide spatial independence. The route was covered at 8–16 km/h on four separate occasions between 1800 h and 2300 h and between 0200 h and 0700 h over three days in January. Only the right sides of the transects were surveyed. A Raytheon Thermal-Eye 250D® LWIR handheld thermal imager with an uncooled bolometric FPA (320 × 240) was used to collect the imagery and with the aid of a spotlight and rangefinder the distance and direction to located deer were recorded to the nearest meter. Four spotlight surveys beginning at 1900 h were conducted from February 8–10, 2010, 2 weeks after the aerial imaging and ground imaging surveys. The weather conditions used for spotlighting were similar to the ground imaging surveys and followed standard protocol for road spotlight surveys of white-tailed deer.

The aerial thermal imaging sampling and equipment used followed a methodology similar to that of Kissell and Tappe (2004) and Kissell and Nimmo (2011) (see above). They used a Mitsubishi IR-M500 NIR-MWIR thermal imager with a 14° FOV mounted on the belly of a Cessna 182 fixed-wing aircraft flying 457 m agl at approximately 120 km/h. The altitude and FOV resulted in ~112 m swath width when the imager was directed down along the nadir. They established 10 test locations outside the perimeter where ground personnel observed the number of deer using handheld thermal imagers immediately after flyover as an independent measure of detection probability. The same altitude and strip width used for aerial imagery was used for the test locations. They assumed that the ground counts would be near perfect, so the proportion of this count detected in the aerial survey would represent the aerial survey detectivity.

The density estimates for the Security Area differed among the three techniques. Furthermore, the results of the aerial surveys indicated that deer tended to use areas close to the roads, and this behavior likely biased the road-based ground surveys.

This work points out the difficulties encountered when trying to meet all the assumptions associated with distance sampling from the ground, regardless of whether it is conducted with thermal imagers or by spotlighting. They point out that aerial thermal imaging estimates were collected randomly across the landscape with minimal animal disturbance and further revealed a tendency for deer to select areas close to roads within the Security Area. They also observed that deer avoided areas immediately adjacent to the roads by troughs in the detection probability curves for the ground-based surveys. They point out that sampling along roads that follow natural habitat features that can structure the deer population violates critical assumptions of the experimental design and could result in biasing the detection curves, giving an inflated density estimate. They conclude that since they detected more deer from the air than were detected by ground observers (which prevented an estimate of the aerial detectivity) they could assume that very few deer were missed during the aerial thermal imaging surveys. They noted that aerial thermal imaging enables representative and unbiased sampling across the landscape over short time periods without the assumptions of distance sampling.

They caution that the continued use of nonrandom road-based surveys as a method for estimating white-tailed deer populations should be discouraged and that in areas that are relatively open with a flat to moderately rolling terrain, aerial thermal imaging is likely to yield more reliable and precise results than road-based approaches. Note that other species may not be road-shy, and if very careful attention is given to meeting the assumptions of distance sampling successful population estimates can be obtained using ground-based thermal imaging techniques (Franzetti et al., 2012; Focardi et al., 2013).

Avian Species (Birds)

Introduction

In Chapter 3 we discussed several surveys that incorporated the use of radar, thermal imaging, and moonwatching and we will mention these again here to emphasize several important points with regard to collecting good imagery of birds in flight or roosting. When birds are in flight it is usually easy to get imagery with the cold sky as a very uniform background. The background of the sky with clouds, however, will not be as uniform as a result of the radiation emitted from the clouds, that is, cloud shine (Holst, 2000). During a study to monitor bird migration with a fixed-beam radar and a thermal imager, Gauthreaux

FIGURE 10.2 **A thermal image of a flock of small birds roosting in a tree during the afternoon hours on a warm summer day in central Virginia.** The cold clear sky provides a good thermal background for the image of both the tree and the birds.

and Livingston (2006) made the observation that "the thermal imager worked equally well during daytime and nighttime observations and best when skies were clear because thermal radiance from cloud heat often obscured targets".

Other ideal backgrounds for collecting thermal imagery of birds are water and ice. Best et al. (1982) found that aerial LWIR thermal imaging was better suited for cold water or ice backgrounds than imagery collected with an MWIR imager. In general, the thermal signature associated with a bird in flight is much brighter than that of a roosting bird because of the sustained metabolic activity. The photograph of resting birds in a small tree (Figure 10.2) shows the details of the thermal signature associated with these birds. The cold cloudless sky is an ideal background for collecting thermal imagery of birds.

The head is the brightest spatial area of the signature and the wings and tail the darkest. The head, feet, and areas under the wings are typically the points of major heat loss through radiation experienced by avian species but are modified somewhat by the type and density of plumage (see Chapter 9). Note that many birds will roost with their heads under their wing so the signatures will lose the feature presented by the head of the birds in those situations. Nonetheless, these birds would still be easily detected when the apparent temperature of the bird is compared with the apparent temperature of the cold sky.

For birds roosting or feeding on water the apparent temperature of the background will be determined to a large extent by the degree of solar loading prior to the observations made with the thermal imager. Additionally, the thermal inertia of water is high so it is slow to heat and slow to cool, however, it is usually uniform. Figure 10.3 shows a photograph of a thermal image of a great blue heron (*Ardea herodias*) against a warm water background that was taken on a sunny day with little cloud cover. A group of mallard ducks (*Anas platyrhnchos*) is shown on the water in Figure 10.4.

This thermal image was taken with a 3–5 μm imager on a warm fall day in the late afternoon in central Virginia. Thermal imagery of birds roosting in a

FIGURE 10.3 **A thermal image of a great blue heron feeding at the edge of a pond on a summer day.** The uniform temperature of the water provides an excellent background for the image. The thermal loading of vegetation in the foreground of this image is evident.

FIGURE 10.4 **A group of mallards in this thermal image are easily detected and recognized against the uniform background of the water.**

tree requires a little more planning than those collected when the apparent temperature of birds are compared to the apparent temperature of the sky or a body of water. In these situations it normally means that the thermographer must wait until the atmospheric conditions come together to form a uniform thermal background within the surroundings of the birds. If the birds are roosting in the canopy of a tree, delay until after sunset (when most birds will be on the roost) to let the tree canopy come to equilibrium before imagery is collected. When this criterion is established good imagery can be collected either from the air or from the ground, as shown in Figure 10.5.

Reviews (Avian Species Surveys)

Both LWIR (HgCdTe) and MWIR (InSb) detectors were used by Best et al. (1982) to conduct cold weather aerial censusing of Canada geese (*Branta*

FIGURE 10.5 **A thermal image of herons roosting in a tree.** This image was collected with an MWIR thermal imager at night.

canadensis) on the Missouri River in South Dakota. They reported that there was ~11% difference between estimates of goose numbers from thermal imagery and the number of geese determined from counts made on aerial photographs. The method used for goose counts made from enlargements of aerial photographs followed a technique developed by Chatten (1952) that gives accuracy to within 15% or less. Geese were undetectable against the cold background presented by the water (slightly above 0°C) even at the lowest altitude flown (305 m agl) with the MWIR detector. The spatial resolution (IFOV = 2.5 mrad) and sensitivity (NETD = 0.5°C) of the LWIR detector was only marginal for detecting individual geese (~2.5 ft.) at this altitude. The LWIR detector could detect very low densities of geese resting on open water from 305 m agl. Density estimates were made by determining the areal extent of goose concentrations and estimating total numbers as the product of area and goose density, as determined from aerial photographs. It is unclear if the thermal imagery provided a higher or lower count than that of the Chatten method.

This work points out once again that the performance of MWIR photon detectors is limited by low photon production at cold temperatures. It still needs to be determined if today's MWIR systems using staring focal plane arrays will be useful in detecting animals at lower temperatures since the thermal sensitivities would be an order of magnitude better than the scanning units used here.

Aerial surveys were conducted by Sidle et al. (1993) to count sandhill cranes (*Grus canadensis*) roosting along the central segment of the Platte River, Nebraska. Counts were made on thermal infrared images collected with an AN/AAD-5 imaging system on board a reconnaissance F-4 phantom jet traveling at ~300 knots just after sunset. The imager was an LWIR system with a spatial resolution of IFOV = 0.375 mrad in the narrow field of view and an NETD < 0.2°C. This field survey was successful in using thermal imagery to count animals in the wild. There was sufficient contrast between the emissivity of the river and the temperature of the cranes to distinguish them from their backgrounds and the spatial resolution was sufficient to detect cranes in small and large groups. Individual cranes roosting at night were readily visible on the imagery. Improvements in the methodology would include flying at slower speeds (helicopter) so the aircraft can preserve the same viewing angle and ground swath width (no banking on the turns by the aircraft) and flying after dark when all the birds have returned to roost.

Radar, passive thermal imaging, and moonwatching studies were carried out in southern Israel by Liechti et al. (1995) to compare the three methods and to shed light on the potential and limitations of the moonwatch method. The main advantage of radar is that it can determine distance. However, on the downside, it cannot distinguish between birds and insects, especially at short distances (< 1 km). They determined that for distances up to 3 km all birds crossing the FOV of the infrared imager were detected. This should be expected since the warm active birds are compared against the background of a cold night sky. About one-third of the birds were missed by moonwatchers for elevations below 3 km and about half were missed at distances below 1.5 km. They showed for the first time that a good LWIR thermal imager, Inframetrics IRTV-445L® with an NETD = 0.1°C and an IFOV = 2 mrad, can detect even small nocturnal migrants up to at least 3 km away and many birds were detected even beyond 3 km. The maximum distance that they could observe a bird with a 40× telescope was 3.5 km, although radar showed many birds beyond this distance. They argue that the findings improve the application of the moonwatch technique, which is still the least expensive and easiest available tool for observing nocturnal bird migration, and they postulate that general guidelines for the method would help to make moonwatching results comparable all over the world.

There are numerous advantages offered by the availability of reliable small radars and the tremendous amount of data from large networks of Doppler radars designed for monitoring the weather. Important applications include monitoring populations of both threatened and overabundant species, locating roosting sites and critical stopover areas necessary for migratory birds, and following movements of flocks that depredate crops. A singular but important drawback of radar is that it does not provide information on species identification; however, this situation has been mitigated somewhat by using thermal imagers to augment data of bird migrations. Gauthreaux and Livingston (2006) devised a technique to accurately enumerate and determine the flight altitude of migrating birds. They combined a vertically pointed stationary radar beam (marine radar with a 61 cm parabolic antenna producing a beam width of 4°) and a vertically pointed thermal imaging camera (100 mm lens and a 5.5° × 4° FOV) so that the thermal imager detected the path of a target in the *xy*-plane and the radar detected the altitude of the target in the *z* direction. The combined system of a vertically pointed radar and thermal imager (VERTRAD/TI) was used to collect data at night at Pendleton, South Carolina and Wallops Island, Virginia and daylight data collection was carried out at McFadden National Wildlife Refuge, Texas. They point out that the orientation of the system need not be vertical and that once the beam is collimated, the radar and thermal imager could be directed at any angle between horizontal and vertical, making it suitable for many different applications. The data was analyzed using a video peak store (VPS) to display the digital videotapes, which apparently works well if a few precautions are taken. The VPS imagery allowed the identification of bird tracks, which were easily distinguished from bat and insect tracks. Furthermore, the wing beat patterns of flying birds clearly produced modulations in the VPS tracks generated from the video clips of the thermal imager and produced pulsations of echoes in the radar beam, suggesting that in the future these patterns might be useful in bird identification. The photographs of their scans are excellent and are shown in Figure 2 in their paper and are well worth consulting to see the high-contrast imagery that can be achieved when using the cold sky as a thermal background.

A review of different remote technologies that can be used to study bird behavior in relation to bird-wind turbine collisions at offshore wind farms was conducted by Desholm et al. (2006). The hazards posed for several species of resident and migratory birds by the construction and placement of offshore wind farms in the waters of Europe are summarized under three broad headings: (1) displacement and flight avoidance responses, (2) habitat loss and/or modification, and (3) collision risk. All of these effects are serious and can represent significant losses in the long term when consideration is given to the impact of the construction of many offshore wind farms along the length of a migratory bird species' corridor (Fox et al., 2006). The amount of data required to make assessments of migratory birds for both pre- and postconstruction evaluations is prohibitive for human observers. As a result, existing remote technologies such as radar systems and infrared camera systems (see Chapter 3) were evaluated to assess their degree of usefulness for studying bird behavior in relation to offshore wind energy facilities. Their Figure 3 is an excellent photograph of a thermal image, showing a flock of common eiders (*Somateria mollissima*) passing through the FOV of their thermal imager at a distance of 70 m. They indicate that they used the same imager as that used by Liechti et al. (1995), which was an Inframetrics IRTV-445L with an NETD = 0.1°C and an IFOV = 2 mrad.

They suggest several ways to implement collision monitoring using the infrared cameras, and depending on the objectives the planned program may call for several to a large number of thermal cameras for measuring the daily collision frequency. They further suggest that, in the event that the direct measurement of avian collisions is not feasible, an indirect approach of modeling the avian risk of collision can be applied and the infrared monitoring system can contribute with important data to these models. Specifically, it could provide estimates for the following parameters: (1) near turbine blade avoidance behavior; (2) flight altitude; (3) flock size (especially at night); and (4) species recognition (especially at night).

Their advice and subsequent discussion of designing a thermal monitoring program (p. 82) contains information that is essential to accomplishing the goal of setting up thermal cameras to acquire imagery that can be used to satisfy the needs of a remote 24-h monitoring program. Understanding the limitations of a specific thermal imaging camera and how best to optimize the camera for a specific task is the most important step in this process (see Chapter 8).

The environmental consequences of terrestrial and marine wind farms on nocturnally active birds and bats are a cause of great concern as the wind energy industry continues its expansion around the world. Wind energy facilities have been demonstrated to kill birds and bats and there is evidence that wind energy development can also result in the loss of habitat for some species. To the extent that we understand how, when, and where wind energy development most adversely affects organisms and their habitat, it will be possible to mitigate future impacts through careful site selection decisions (NRC, 2007). This report by the National Research Council outlines the myriad of problems facing biologists regarding the determination of the causes and specific information about actual collision rates (fatalities) and how the presence of industrial-sized wind energy facilities impact the behavioral patterns of nocturnally active birds and bats. Note that what is mentioned here regarding the effects of wind energy development on ecosystem structure and functioning, through direct effects of turbines on organisms and indirectly on landscapes through alteration and displacement, is a very complex problem. The ecological influences can vary with spatial and temporal scale, location, season, weather, ecosystem type (landscape or at sea), species, and other factors. This report reviews the documented and potential influences of wind energy development on ecosystem structure and functioning, focusing on scales of relevance to site selection decisions and on influences on birds, bats, and other vertebrates.

Kuntz et al. (2007) prepared a comprehensive (246 references) guidance document for assessing impacts of wind energy development on nocturnally active birds and bats. This guidance document considers the methods and metrics for assessing both the direct and indirect impacts of wind energy facilities. The direct impacts of wind energy facilities refer to fatalities resulting from night-flying birds and bats being killed directly by collisions with wind turbine rotors and monopoles. Indirect impacts refer to disruptions of foraging behavior, breeding activities, and migratory patterns resulting from alterations in landscapes used by nocturnally active birds and bats. They reemphasize that direct and indirect impacts on birds and bats can contribute to increased mortality, alterations in the availability of food, roost and nest resources, increased risk of predation, and potentially altered demographics, genetic structure, and population viability.

They suggest that a combination of techniques, tools, and methodologies might be necessary to unambiguously assess the impact of wind turbines on the temporal and spatial variations in natural populations of nocturnally active birds and bats. Each and any device, method, or protocol has its own strengths, limitations, and associated biases that must be completely understood by the field biologist charged with conducting the field studies. The techniques and equipment presented by Kuntz et al. (2007) used for visual monitoring of nocturnal activity such as moonwatching, ceilometry, night-vision goggles (image intensifiers), and thermal infrared imaging have been covered in other sections of this book.

The size, location, number of turbines, and their spatial distribution within a particular wind energy facility is highly variable, which makes nocturnal observations a very difficult undertaking to determine where, how, when, and most importantly, why nocturnally active birds and bats come into contact with wind turbines. To answer these questions it is necessary to observe the behavior of bats and nocturnally active birds in the vicinity of these structures, which requires careful observations using appropriate methodologies and protocols to assess the nocturnal and seasonal timing and flight behavior of birds and bats in the vicinity of proposed and operational wind turbines.

When we normally think of ecological research we almost always think of aquatic animals, avian species, and mammals and their relationships to land masses and bodies of water. For avian species the need to find nesting sites, protection from predators, and ample food supplies occupy a large part of the research efforts devoted to understanding their ecology. Much of this research is undertaken by field biologists who rely primarily on visual techniques to gather data and make observations that will add to the knowledge of avian species. A review by Hristov et al. (2008) shows that thermal infrared imaging represents the most promising technology that is available to field biologists for making behavioral observations, studying animal energetics (thermoregulation), and for censusing flocks of migrating birds or large colonies of bats. Thermal imaging provides field researchers with a tool that allows them to make these observations and collect data throughout the diurnal cycle since it is capable of imaging noninvasively in the dark at high temporal and spatial resolution. Hristov et al. (2008) point out that the night sky remains a largely unexplored frontier for biologists studying the behavior and physiology of free-ranging, nocturnal organisms and that conventional imaging tools and techniques are not sufficient for research in aeroecology. They selected three examples from their past work to highlight the use of thermal imaging in the study of aeroecology. They point out that the examples selected (behavioral observations, thermographic analysis of animal energetics, and censusing large colonies of bats) represent only a small subsample of possible applications that were chosen to illustrate the diversity, power, and potential of thermal imaging.

This excellent review, while primarily focused on examples of bat aeroecology, extends across a broad spectrum of possible nocturnal research efforts to understand the behavior and physiology for many free-ranging, nocturnal species. It is hoped that the advantages of using thermal imaging as a tool in such research will lead to its broader use in ecological research and stimulate new approaches into the biology of birds, bats, and arthropods in the aerosphere. See above for several efforts involving the integration of radars and thermal imagers for the nocturnal counting and observation of migratory birds. New advances in thermal imaging cameras (cost, weight, sensitivity, and spatial resolution) have provided field researchers (as demonstrated in this paper) with a very powerful tool to incorporate into their studies of aeroecology.

An excellent comparison of thermal imaging and image intensifier performance for improving the efficiency and accuracy of nocturnal bird surveys is given by Lazarevic (2009). This thesis explores some practical and theoretical approaches that can improve the accuracy, confidence, and efficiency of nocturnal bird surveillance. As image intensifiers and thermal imagers have operational differences, each device has associated strengths and limitations. Empirical work established that image intensifiers are best used for species identification of birds against the ground or vegetation backgrounds when lighting conditions permit. Thermal imagers perform best in detection tasks and monitoring bird airspace usage.

Bats

Introduction

Censusing and behavioral studies of bats pose a number of unique problems to the field biologist. Their nocturnal behavior coupled with their propensity to gather together in large numbers in small inaccessible spaces makes the observation of bats and the collection of even the most basic data regarding their ecology difficult. Making population estimates is particularly problematic since not all bats in a given roost will issue before it is too dark to see them and their density at the time of peak issuance can be too large to count. In addition, there may be more than one species present and there may be more than one entrance/exit for cave-dwelling species. Tuttle (1979) noted that population estimates of the gray bat (*Myotis grisescens*) were difficult to obtain for a combination of these reasons and because their numbers are influenced by seasonal movements of bat colonies. These significant challenges have limited the exploration of the night sky and the capability provided by radar, night-vision goggles, flash cameras, and high-power searchlights is inadequate to get the detail and resolution needed to estimate and study nocturnal avian and bat populations. When a thermal imaging capability was added to the list of tools to aid in night counts and observations of migrating birds and bats modest improvements were noted in the detection and enumeration of these species.

Reviews (Bat Surveys and Observations)

Kirkwood and Cartwright (1991) recognized that many observation problems could be overcome if a thermal imager could be used to observe bats in their natural environment. They established the groundwork for using a thermal imager for bat studies by observing big brown bats (*Eptesicus fuscus*) in a laboratory setting to establish the noninvasive nature of thermal imaging. They then studied two different maternal colonies of big brown bats: one roosting in a large hay barn and another in an attic of a residence. At these two sites the big brown bats were located and observed in natural surroundings. They concluded from their observations that infrared imaging has the potential to count emerging bats, to count or estimate the size of nesting colonies or area occupied *in situ*, to locate bats roosting in inaccessible sites, and to record behavioral patterns of individual bats in the roost. This was accomplished in a nonintrusive manner and they suggested that similar treatment could be used to document the behavior of other bat species, including the Indiana bat (*Myotis sodalis*), which is currently on federal and state endangered species lists (Kirkwood and Cartwright, 1991, p. 371).

The performance of a night-vision device (image intensifier) and an infrared thermal imager were compared under low-light conditions for making behavioral observations of big brown bats (Kirkwood and Cartwright, 1993). The night-vision system was mated to a camcorder to facilitate the collection of imagery. This system could be handheld or mounted on a tripod. The scanning infrared imager was a cryogenically cooled Inframetrics® 525 LWIR imager with a modest spatial resolution of 2 mrad, FOV (14° × 18°), and an NETD = 0.1°C. The unit was rather bulky/heavy, requiring it to be transported on a cart. This made it difficult to transport into the attic of a church where the bats were roosting. Both systems outperformed visual observation and images recorded with the camcorder. The thermal imaging system allowed detection of bats that were located in inaccessible crevices, which the night-vision system could not detect. One bat detected was in torpor and it was purposely disturbed. As the bat's metabolic rate increased they estimated that the thermal contrast ΔT of the bat increased from about 0.5°C to about 4°C and the bat appeared white (warm) with respect to its background. At the time of these studies the cost of the cumbersome thermal imaging system used was ~$40,000.

We point out that a handheld thermal imager serves as the replacement for the cryogenically cooled Inframetrics 525 LWIR imager that has a thermal/visual/fused image capability, uncooled FPA, and a thermal sensitivity or NETD = 0.04°C at a cost of ~$10,000. This device would have provided much better thermal resolution and would have been easily transported into tight roosting areas, making behavioral studies such as mother–young interactions quite manageable.

Estimating the size of bat populations during nocturnal emergence from caves relies on the technique of visual counting, which is subjective and depends to a large degree on the experience of the observer. A comparison of visual counting methods and counts obtained from digital analysis of thermal imagery of emerging gray bats (*M. grisescens*) was carried out by Sabol and Hudson (1995). The gray bat is federally listed (1976) as endangered. For the visual count an experienced observer was positioned inside the cave opening and counted bats crossing the field-of-view. They used a skip-minute method in which the observer would count bats for 1 min and then rest for 1 min, repeating this cycle to completion and resulting in a count of one-half of the bats. When the emergence ratio was high the bats could not be counted and were estimated. Three population estimates were made at two different limestone karst caves (two at one site and one at the other).

An Agema model 782 LWIR scanning thermal imager with a $20 \times 20°$ FOV and a 1.9 mrad spatial resolution was used to collect imagery on videotape, which was subjected to digital image processing routines to extract bat numbers from the imagery. For this method of counting to work it is critical that all bats pass through the field-of-view of the imager as they are emerging from the cave. It is also a requirement that the bats fly a unidirectional path across the field-of-view so that the digital processing of the imagery will be unambiguous. The other conditions that need to be met deal with the thermal contrast and background clutter. Good thermal contrast will be obtained if the background is a cool one like the night sky, cave roof, or some other backdrop that has come to thermal equilibrium. Background clutter will be minimized if the imagery is recorded against a sky background or the cave roof if the FOV of the system will allow it and if there is room enough in the cave opening to position the camera. The thermal imagery captured in these experiments was affected by thermal contrast problems associated with warm vegetation at the mouth of the cave and emerging bats scattering in all directions (loss of unidirectional flight) while still in the FOV of the imaging system. These problems made it necessary to reject the data collected from two of the thermal counts. The remaining set of imagery was subjected to digital image processing and gave a bat count of 46,950. The visual estimate of emergence from the same cave was 49,000 bats.

Kuntz et al. (2007) prepared a comprehensive guidance document for assessing impacts of wind energy development on nocturnally active birds and bats. This document considers the methods and metrics for assessing both the direct and indirect impacts of wind energy facilities. The direct impacts of wind energy facilities refer to fatalities resulting from night-flying birds and bats being killed directly by collisions with wind turbine rotors and monopoles. Indirect impacts refer to disruptions of foraging behavior, breeding activities, and migratory patterns resulting from alterations in landscapes used by nocturnally active birds and bats. They reemphasize that direct and indirect impacts on birds and bats can contribute to increased mortality, alterations in the availability of food, roost and nest resources, increased risk of predation, and potentially altered demographics, genetic structure, and population viability.

Betke et al. (2008) report on census results taken of Brazilian free-tailed bat (*Tadarida brasiliensis*) colonies at eight caves conducted between 2000 and 2006; six of those colonies are considered to be among the 11 largest colonies of this species in North America. A 1957 census based on visual observation of cave emergence flights (which are subject to potentially high errors) estimated that 54 million bats made up the population at that time. The present work used several Merlin MID® MWIR thermal imagers to record emergence rates by placing a single camera pointing perpendicular to the direction of the emerging bats at each cave entrance so that during emergence the bats flew across the FOV of the imagers and allowed for a complete emergence record, regardless of the number of cave entrances used by the emerging bats. They describe a detailed image analysis method and compared the bats-per-frame estimates with estimates made by five independent human volunteers with experience in image analysis in order to evaluate the accuracy of the detection component of the census. Over all of the sample frames, the observers identified an average of 4181 bats and the detection algorithm identified 4215 bats, a difference of $<1\%$. In all they observed a decline of bats from 54 million (1957 estimates) to 4 million and suggested that the observed decrease could be explained by a population decline, a population shift, and/or an overestimate of the population in 1957.

They conclude that it is possible that much of the order of magnitude difference between current and historic census estimates reflects a population decline; however, they note several factors affecting the emergence of bats that are observed with thermal imaging and not with the conventional counting methods. The length of the single or multiepisode emergence period increased nonlinearly with colony size. The peak of the first episode occurred during the first quarter of the emergence period and it always occurred before sunset. This is not generally the case for following emergence episodes so they note that if an estimate of the flow rate of bats is made on the presunset emergence episode and based on this estimate, the extrapolation of the number of bats emerging in the dark could lead to invalid generalizations. By using thermal infrared imagery inadequate sampling can be avoided and the complete record of a colony emergence can be analyzed. They conclude that the historic census estimates obtained without this technology may have been too high, and it is quite possible that a combination of both overestimation and population decline occurred.

Betke et al. (2008) directly benefited from the use of thermal imaging because the complete emergence cycle was monitored from start to finish without having to estimate bat emergence after sunset based on estimated flow rates from the presunset counts. These techniques are also important because they are amenable to other species of bats as well as species of nocturnal birds.

The nightly flight activity of bats near operating wind turbines was documented by Horn et al. (2008) from August 2 to 27, 2004, at the Mountaineer Wind Energy Center in the mid-Atlantic highlands near Thomas, West Virginia. The facility was located on a mountain ridge and consisted of 42 wind turbines arranged along the ridge on a cleared access road. Each turbine was 106 m tall and had a 72 m diameter (4071 m^2) area that was swept by the rotating turbine blades. 12 of these turbines were fitted with both steady and strobe aviation obstruction lighting. Operation speeds of the turbines varied from 3 m/s to 18 m/s and rotor blades could turn at up to 17 revolutions/min.

They recorded 9-h sessions of thermal infrared video nightly using three FLIR Systems S60® uncooled microbolometer cameras with 320 × 240 FPAs operating in the LWIR spectral range. Each camera had a 24° FOV and was mounted 0.5 m apart and 30 m from the turbine base such that the plane created by the sweep of the rotor blades was completely covered by the three cameras. The 24° FOV was centered on the rotor blades as it passed through the area covered by each camera so that collisions of bats with the turbine blades and their flight behavior prior to and after blade passing could be recorded. Figure 6 in this paper shows a photograph of a time-lapsed series of thermal images depicting a bat being struck with a turbine blade.

They suggest that there are three hypotheses that may account for the discovery of injured, dead, or moribund bats on the ground beneath and near turbines: (1) flying bats may randomly come into contact with rotating blades, (2) bats may be attracted to wind turbine structures, leading to an increased potential for contact, and (3) factors resulting from the construction (turbine placements) and landscape alterations during construction, which can change behavioral activities such as roosting and foraging. Further complexity may arise from seasonal weather patterns that produce an increase of insect activity, which in turn may attract large numbers of migrating bats. This could be an explanation for increases in bat fatalities at certain times of the year. The original study was designed to test the effect of slowing or stopping the blade rotation on bat behavior around turbines and on the number of fatalities but was not conducted because the operators of the wind turbine facility would not permit them to experimentally stop rotation by feathering the blades at predetermined times.

The results of this study indicate that bats (1) approached both rotating and nonrotating blades, (2) followed or were trapped in blade tip vortices, (3) investigated the various parts of the turbine with repeated flybys, and (4) were struck directly by rotating blades. Furthermore, bats may be at higher risk of fatality on nights with low wind speeds or when the rotational speed of the blades is lower.

The description of the viewing geometry for the turbines in the experimental setup is a little confusing. The cameras are viewing the sweep area of the plane of blade rotation in a direction more parallel as opposed to a perpendicular direction. The viewing angle relative to the plane of rotation is ~23.25° based on the camera placement, as shown in their Figure 1. The photographs shown in Figures 2 and 6 appear to be looking parallel to the plane swept by the rotor blades as well. This does not affect the results of this important work. Understanding the behavior of bats in the vicinity of wind turbines is a complex issue with many interconnected variables that need to be sorted out. Perhaps the extent of this complexity (Kuntz et al., 2007) can be unraveled (i.e., the basic questions about where, when, how, and why wind turbines kill bats) and mitigated to some extent by the use of thermal imaging cameras to make direct observations of night-flying birds and bats.

A census of the endangered Mexican long-nosed bat (*Leptonycteris nivalis*) in Emory Cave, Big Bend National Park, Texas using thermal imagery was conducted by Ammerman et al. (2009). A FLIR ThermoCAM S65® LWIR thermal imager with a spatial resolution of 1.3 mrad, a FOV of (20 × 18°), and an NETD = 0.08°C was used to develop and evaluate new monitoring methods and to establish and monitor colony size. Past efforts of counting roosting bats/m^2 and then estimating the portion of the roost that is covered by bats can be biased and are disruptive to the roosting colony (Kuntz, 2001). The bats in Emory Cave were also found to be dispersed in passageways of the cave that were inaccessible to observation. Collecting visual data to estimate colony size as they emerge at dark is also difficult because of low-light levels, dense vegetation, and the presence of large numbers of individuals circling the cave opening prior to and during nightly emergence. Another particularly difficult situation for censusing Emory Cave arises because two other species of bats, *Myotis thysanodes* and *Corynorhinus townsendii,* also roost in the cave, albeit in lesser numbers than *L. nivalis*.

The thermal imager was tripod-mounted and positioned inside the cave opening with a perpendicular view of the emerging bats. Data collected at 60 frames/s was recorded and saved to the hard drive of a computer and each emergence was played back so the bats could be manually counted. While they could not distinguish between *M. thysanodes* and *C. townsendii*, they could separate *L. nivalis* from the two by the differences in the thermal signatures associated with the body shape and wing structure of the flying bats. Ammerman et al. (2009) point out that thermal imaging provides a high-resolution, permanent record of the emergence and results in less human error. They find it is less disturbing to bats than entering a cave to count them or from using supplemental lighting during emergence to facilitate filming. Their data suggests that thermal infrared imaging provides a more reliable and accurate estimate of the number of *L. nivalis* at Emory Cave than has previously been possible using extrapolation from the surface area of the cave ceiling covered by bats. The use of thermal imaging provided a way to isolate the Mexican long-nosed bats from other bats present during the census and allowed a count that was probably independent of occupancy preference within the cavern, thereby eliminating biases associated with not counting bats occupying the deeper chamber portions of the cavern.

A major maternity colony of Brazilian free-tailed bats (*T. brasiliensis*) is present from March through November at Carlsbad Cavern, New Mexico. Their behavior and population dynamics have been studied since 1928. Hristov et al. (2010) studied the variation in colony size of this species using Merlin MID MWIR thermal imagers to record emergence rates throughout the occupancy period. The size of the colony was estimated several times monthly from March to October and ranged from 67,602 to 793,838 bats. The imaging cameras were equipped with a high-resolution InSb (320 × 256) FPA having an NETD = 0.025°C. One to three cameras were used, depending on the direction and pattern of emerging bats observed on the previous nights. They were placed at strategic locations near the top of the sinkhole entrance where the emerging bats formed a tight unidirectional column. They were aimed so that they were viewing the emerging spiral of bats perpendicular to their flight direction and after the emerging column became less organized but still within the FOV of the cameras. The collected imagery was subjected to computer vision algorithms to automatically count individuals as they emerged from the cavern by using a two-step process. Bats were first detected and then tracked from frame to frame and those that were successfully tracked beyond a preset threshold value were recorded and added to the total count of emerging bats (see Betke et al., 2007, 2008 for a discussion of the image analysis methods and the effects of clutter in the background when tracking large variable numbers of objects). This work demonstrated that the size of the bat colony at Carlsbad Cavern is highly variable on a seasonal and daily basis and the factors that influence the number of bats and their emergence behavior are not entirely clear. The frequency and accuracy of their measurements indicate that these fluctuations are not a by-product of uncertain population estimates and sampling bias but rather a function of the complex colony dynamics of *T. brasiliensis*.

The combination of thermal infrared imaging and computer vision analysis provides an effective method for estimating colony size and emergence behavior of Brazilian free-tailed bats. Reliable estimates of colony size are critical for evaluating the natural and anthropogenic forces that can affect current and future population trends for this and other bat species.

From this work we see that placement of the thermal imagers is critical for several reasons: (1) to obtain good imagery of emerging bats, (2) to record and count individual bats, (3) to avoid double counting by being too close to the entrance where the vortex of spiraling bats could be counted multiple times, and (4) to have a good background to keep clutter to a minimum (if it is possible the night sky would be preferred as a background for the imagery for a particular location under study).

There are several good review references that provide a description of various techniques (where several different methods, techniques, and observation tools are used in conjunction) for observing the nocturnal activity of birds and/or bats that have been successful for a number of species. Some of these methods, techniques, and observation tools are discussed in this section and others are discussed in Section "Literature Reviews" of this chapter.

Nests (Dens, Tree Cavities, Lairs, Burrows)

Introduction

Occupied nests, lairs, and den sites such as tree cavities and hollow logs present thermal signatures that are easily detected with modern thermal imagers operating in the MWIR or LWIR wavelength bands. Depending on the thermal inertia or heat capacity of the nest or den materials, recently vacated nests, dens, and tree cavities are also easily detected. There are, however, certain measures that must be taken to ensure that the imagery collected in the field will allow for this level of data extraction. Thermal imagers only detect radiation and all objects radiate heat, regardless of their temperature. In order to create a useful thermal image from the radiation in a scene (i.e., can we observe objects of interest in the scene) there must be a difference in the amount of radiation emanating from objects of interest and their backgrounds.

The trick is to provide a window of opportunity for data acquisition that will allow us to identify the objects of interest (e.g., tree cavities) in a scene. In many cases the objects of interest may not be producing the brightest signatures in the scene. The number of thermal signatures in the scene that are not of interest (any hot object, such as bare soil, rocks, deadfall, standing water, etc., that had been previously warmed by solar radiation) must be eliminated prior to data collection. Their presence only serves to confound the interpretation of the infrared imagery being collected. The place and time of the data acquisition, the atmospheric conditions during the 24-h period preceding data acquisition, and the prevailing atmospheric conditions during data acquisition must be collectively optimized to provide the necessary conditions for successful detections. In some of the examples provided below these problems were encountered and if some form of mitigation was not implemented then the results were generally poor.

Two examples are provided to illustrate the type of imagery that is collected when occupied dens, nests, or tree cavities are viewed with a thermal imager. Once a tree is near thermal equilibrium with its surroundings there is still considerable heat being radiated from the tree but the background cover (forest floor, shrubs, vegetation, and exposed soil) is also radiating at nearly the same level as the tree so a condition of quasi equilibrium exists. If each individual tree has reached a near-uniform temperature throughout the sampling area then tree cavities will be very hard to observe when they are unoccupied. However, by achieving this state of quasi-thermal equilibrium, an ideal situation for detecting occupied tree cavities is available since there are only two radiating bodies to compare. In this situation the two bodies radiating the energy to form the thermal image consist of the uniform background of the extended forest and the embedded bright thermal signatures associated with occupied tree cavities.

After an extended period of zero thermal loading (absence of solar radiation) the trees (both dead and alive) are at their coolest and most uniform temperature. The desired temperature uniformity applies to both individual trees (nearly uniform core and exterior temperatures) and to the forest and its background (forest floor, deadfall, and under story). If the period of no solar loading is long then the state of quasi-thermal equilibrium approaches the ideal. In general, the exterior surface of the individual tree cools first through convection, radiation, and possibly evaporative cooling (if the surface is damp). The interior of the tree will still be warmer than the exterior since the interior of the tree loses heat only through conduction and wood is a relatively good thermal insulator. The rate of heat transfer from the interior to the exterior surface is therefore low and hence the longer the period of no thermal loading the better. A period of rain followed by cloudy or overcast skies during drying will generally provide the best atmospheric conditions for achieving quasi-thermal equilibrium.

FIGURE 10.6 **These two images (one formed with visible light and the other with thermal radiation) of the same tree cavity are provided for comparison.** Both images were taken during daylight hours within minutes of one another. (A) A frame from a video taken with an 8 mm Sony Handycam®. (B) A frame from a video recorded on 8 mm film taken with a 3–5 μm InfraCam thermal imager. This cavity was occupied at the time by a family of raccoons.

Examples of Nest Imagery

Keep in mind that a thermal imager only responds to heat in the form of radiation and only radiation is capable of inducing a response in the detectors comprising the FPA. Tree cavities radiate in all directions but we are only detecting the radiation that exits the cavity and is captured by the collection optics of the thermal imager. In Figure 10.6 we compare two images (one formed with visible light and the other with thermal radiation) of the same tree cavity. Both images were taken during daylight hours within minutes of one another. This deep cavity was occupied at the time by a family of raccoons (*Procyon lotor*). The thermal image is formed with only a small fraction of the radiation emitted from the cavity and an even smaller amount of the total radiation originating in the cavity. Most of the radiation is directed back into the interior of the tree and is reabsorbed and subsequently re-emitted. This absorption and re-emission contributes to the slowness of core cooling. Rarely is the radiation emanating from the animal within the cavity directly observed from the cavity. Typically, we only see a thermal signature of the cavity after undergoing a very complex redistribution of heat through convective, conductive, and radiative processes within the cavity, which eventually leads to a radiative component emanating from the cavity opening.

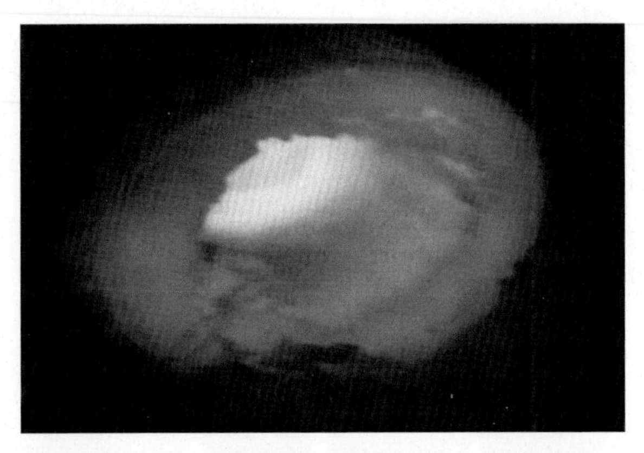

FIGURE 10.7 **This photograph shows a thermal image of a large hollow log lying on a forest floor, which served as a den for a striped skunk.** This image was collected at night using a 3–5 mm imager.

Another example of detecting an animal den is given below. Figure 10.7shows the thermal image of a "glowing" den (consisting of a large hollow log lying on the forest floor) that was occupied at the time by a striped skunk (*Mephitis mephitis*). These images were collected during a cool fall evening in coastal Virginia using a handheld MWIR imager.

After waiting a few minutes the resident skunk made an appearance, as shown in Figure 10.8. The signatures shown in these two examples are typical of dens or tree cavities occupied by fairly large mammals.

Other examples are provided in the following chapter where we take up the task of discussing ways to increase the detectability for a variety of field situations.

Reviews (Nests, Dens, Tree Cavities, Lairs, and Burrows)

Wilson and Delahay (2001) discuss the problems associated with counting conspicuous visible structures constructed by some carnivores. They caution

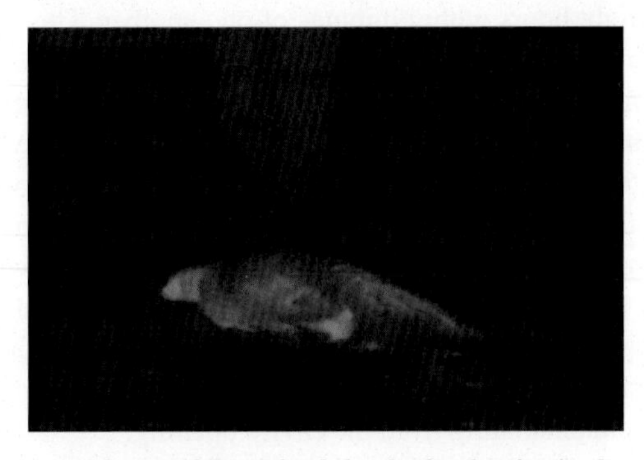

FIGURE 10.8 **A thermal image of the striped skunk after leaving its den.**

that the use of den sites to estimate carnivore numbers relies upon accurate identification of the resident species. For example, they point out that in parts of Europe red foxes and raccoon dogs will opportunistically inhabit disused badger dens, which are generally characterized by the size and shape of the entrance holes and by the presence of large spoil heaps. Probing these structures with thermal imagers for the purpose of estimating badger populations must be done with care because unless there is a way to tell what animal is occupying the dens then an imager will only yield information on occupancy and not occupants. A number of field studies and experiments where thermal imagers were used to locate and monitor animal dens, nests, and lairs are discussed below.

A handheld thermal imager was used by Hirons and Lindsley (1986) to locate nesting woodcock (*Scolopax rusticola*) on a woodland floor at Whitwell Wood, a 126 ha mainly deciduous woodland in Derbyshire, UK. Radiotagged woodcock were detected on the ground at ranges of up to 30 m. Both day and night surveys were conducted using transects 20 m apart. Arc scans of 180° of the forest floor were made every 20 paces. No woodcock or other ground-living animals were detected under sunny conditions but at night and under overcast conditions woodcock and other animals (pheasants and hares) were detected. Since many of the woodcock left the forest at night, it was estimated that the detection rate might have been 25 times higher at night than during the day. Here the significant observation was that no animals were detected in the forest under sunny conditions yet that was the time when most of the woodcock were in the forested habitat. This points out, once again, the deleterious effects of sunlight on thermal imagery.

Based on the success of their night surveys Hirons and Lindsley (1986) suggested that thermal imaging would provide better estimates of nest densities than hitherto and would perhaps also allow calibration of other nest-finding methods. Unfortunately, there was no information or specifications provided regarding the thermal imager used in this study. It was also unclear as to what time of the day the surveys were made so it is difficult to provide information that might improve upon their detection capability. Perhaps starting their surveys just after the woodcock return to their nests in the morning on an overcast day would provide the best opportunity for detection.

Kingsley et al. (1990) used an LWIR thermal imager to locate the lairs of ringed seal (*Phoca hispida*) under snow. Lairs are dug under winter snowdrifts covering breathing holes in the unbroken arctic sea ice by ringed seal for use as haul-out and birthing areas. Heat is lost from lairs by convection and radiation through the snow roof and there is a direct correlation between detection rates and roof thickness. Other factors improving the detection of ringed seal lairs were shown to be low wind speeds, low temperature, and predawn flights or cloud cover. Interestingly, the temperature inside a lair is stabilized by the heat input from the unfrozen water in the breathing hole and varies less than the outside temperatures, so the heat loss is greater when the ambient temperatures are lower. They used a FLIR Model 1000A® thermal imager with an HgCdTe detector operating in the LWIR (8–14 μm) with an FOV of 28° × 17°. The imager was mounted on a helicopter flown at elevations of 45–189 m agl at 20 knots and the imagery recorded to VHS videotape. They studied the imagery acquired from flying over known lairs and recorded signatures, which were graded from 0 to 5 depending upon the brightness relative to an otherwise monochrome snowdrift. The thickness and compression of the snow forming the lair roof were correlated with the signature brightness. A brighter signature was obtained when thinner and more compact roofs covered the lairs.

Once the operators were able to positively identify lairs by way of ground truthing they became familiar with what to look for in the imagery recorded. The apparent temperature difference available for signature formation was affected by a number of environmental factors, which place strict conditions on achieving a uniformly radiating background. That is, the lairs were more easily detected under cloudy skies with light or calm winds because these conditions provided uniformity and reduced the heat loss due to convection. The insulation provided by fluffy and thicker roofs reduced the apparent temperature differences achievable to values that were beyond the sensitivity of the imager. The exact specifications of the imager used in these studies were not provided but they were probably similar to those used by others in the same time frame, which would make the NETD near 0.5°C. By pushing the imager to its limit Kingsley et al. (1990) made this important observation: "Weather conditions may prevent observations of many kinds, or make them difficult or unreliable, and can also affect the behavior of subject species. Survey methods also have time constraints based on daily life cycles of behavior of the subject species, available daylight, daily wind patterns, etc. This survey technique, therefore, is not unusual in having specific requirements for ambient conditions, but it is unusual in that they are so strict".

Boonstra et al. (1994) tested three types of infrared imagers: a pyroelectric imager, an older model LWIR scanning imager that was cryogenically cooled to liquid nitrogen temperatures, and a thermoelectrically cooled MWIR imager – the Thermovision 210® ($\Delta T = 0.1$°C) – that was manufactured and commercially available from Agema Infrared Systems. The pyroelectric device was found to be unsuitable for fieldwork since it operates as a capacitive bolometer and required constant motion of the camera to detect temperature differences. They discontinued using the LWIR scanning imager because it was heavy and awkward to use in the field. As a result they conducted a series of ground-based field experiments with the handheld MWIR imager to survey nests, burrows, and free-ranging small mammals. Animals as small as meadow jumping mice (*Zapus hudsonius*) were located both in and out of their nests at ranges limited to 2 m by 30–60 cm of high grass. In a survey of Arctic ground squirrel (*Spermophilus parryii*) burrows they found a strong correlation between burrow occupancy and the observed thermal contrast presented by the burrow.

They were unable to determine if red squirrel (*Tamiasciurus hudsonicus*) nests were occupied from survey experiments, even when they had prior knowledge that the nest was indeed occupied at the time of the survey. In two instances they detected a faint glow from the entrance of nests after they saw a squirrel depart the nest. Since they easily imaged all red squirrels not in nests and since their fur does not provide significant insulation (Hart, 1956) it was suggested that the nests provide such superior insulation that it minimizes heat loss to the environment. Free-ranging snowshoe hare (*Lepus americanus*) were easily detected "appearing as glowing silhouettes" when viewed from 20 m away in dense undergrowth but were less distinct from a distance of 40 m. They observed that a direct line of sight was necessary as dense undergrowth could block the image yet as long as part of the body could be detected by the imager they perceived a glow on the observation screen. After these series of observations they pointed out the following limitations of thermal imaging in the field: (1) it can be used optimally only during certain times of the day or night and/or under certain weather conditions, (2) it may be difficult to detect some animals or their active nests because of their high insulative properties, (3) random objects in the scene may be emitting more radiation than the animals of interest, which can confound the imagery, and (4) obscuration of thermal signatures due to dense vegetative cover can occur.

Of these observations, (1) and (3) can easily be mitigated by simply waiting until the background is uniform. (4) can be mitigated in many cases by scanning the scene at differing angles (panning with the camera), and (2) surveying occupied nests after dark may enhance the detectivities, at least in warmer climates and/or warmer weather. Many animals are so well-insulated that thermal imaging is taxed to detect them under many situations (e.g., see Figure 4.3).

An aside: The MWIR Agema Thermovision 210 thermal imager was made famous by the US Supreme Court case KYLLO V. UNITED STATES (99-8508) 533 U.S. 27 (2001). Police cannot use a thermal imager to explore the details of your home (albeit a nest) without first getting a search warrant.

Boonstra et al. (1995) reported on the utility of using infrared thermal imaging devices to detect and census birds in the field. They collected data with a bolometric MWIR Thermovision 210 thermal imager ($\Delta T = 0.1°C$ @ 30°C) that is available from Agema Infrared Systems. They reported on using this imager at ~5°C to survey individuals and/or the nests of a variety of species. The thermal imaging was successful in determining activity at all four nests of cavity-nesting species. They found that the imagery would easily show the entrance of active nest sites to be glowing early in the morning (0400–0700 h) when ambient (background) temperatures were still cool (air temperature ~10°C). In particular, the three nesting sites (all in dead aspen) occupied by bufflehead (*Bucephala albeola*), northern flickers (*Colaptes auratus*), and pileated woodpeckers (*Dryocopus pileatus*) and the three nest boxes occupied by barrows goldeneye (*Bucephala islandia*) all showed distinct thermal signatures associated with the nest entrance. Additionally, they were able to image the nests during the afternoon hours (air temperatures 18–25°C) with the exception of the pileated woodpecker nest because it was an east-facing cavity and the external surface of the tree had warmed to the cavity temperature so there was insufficient thermal contrast to form images. They reported that they experienced difficulty with residual hot spots in the background when attempting to find nesting cavities without prior knowledge of their location even during the coolest part of the night (1200–0300 h), and that these hot spots were sufficient to mask the location of any active nests.

They also attempted to locate the nests of ground nesting waterfowl including green-winged teal (*Anas crecca*) and mallards (*Anas platyrhynchos*). Unsuccessful attempts were made to collect thermal images of these nests when they contained clutches of eggs during the incubation stage and after the females were flushed from the nests until they were very close to the nests containing eggs (approximately 1 m and standing directly over the nest). These experiments were also hampered by residual hot spots. They were also unsuccessful in obtaining imagery of mallards, green-winged teal, or goldeneye while the birds were resting on water. They attributed the inability to record thermal images of these waterfowl to the superior insulative properties of their feathers, which is probably correct, although there should have been signatures of their heads unless they were under wing at the time. They visually located a geat-horned owl (*Bubo virginianus*) near its nest and were unable to acquire an image of the bird while it was at rest. They did observe a signature from the bird when it was in flight (i.e., hot spots beneath its wings, which is typical of most birds in flight).

When they conducted experiments on cloudy days or early in the morning before sunrise, they were able to easily detect resting or foraging arctic tundra birds such as Lapland longspurs (*Calcarius lapponicus*) and pectorial sandpipers (*Erolia melanotos*). From these experiments Boonstra et al. (1995) concluded that FIR imaging will have limited utility in assessing avian populations but could serve as a useful tool for monitoring the activity of nests at known locations. They point out a number of limitations that FIR thermal sensor systems must overcome to be field worthy in locating and surveying avian populations and their habitat. Fortunately, these limitations can be overcome since they stem from incorrect methodology rather than inadequate thermal imagers.

They experienced difficulty with residual hot spots in the background when attempting to find nesting cavities without prior knowledge of their location. These hot spots are due to objects in the background that are reradiating previously absorbed solar radiation. When the magnitude of ΔT associated with nest cavities is similar to those associated with other objects in the background that are not of interest then the imagery is "cluttered" with false signatures and sorting out the nesting cavities from the collected imagery becomes difficult. In these situations it is necessary to let the background reach thermal equilibrium prior to data collection. Some survey situations may require several days of overcast skies to obtain suitable background uniformity, which may have been the situation in these particular surveys based on their comments of hot spots hampering the imagery at 1200–0300 h. The overall sensitivity of their imager does not seem to be adequate for the tasks attempted based on the signatures contained in their Figure 1, which shows Canada geese (*B. canadensis*) at a relatively close range.

They seem to have encountered the same difficulties that they did during their earlier work using the same imager (Boonstra et al., 1994) but did not make corrections for the deficiencies they pointed out in their earlier work. As should be expected the same problems persisted during this effort as well.

Several experiments were conducted by Hubbs et al. (2000) to determine the feasibility of estimating the abundance of semifossorial mammals. They compared the technique of using powder tracking boards placed at the entrance of Arctic ground squirrel (*S. parryii*) burrows to that of detecting the thermal signatures of the entrances of occupied burrows. They used an Agema MWIR Thermovision 210 imager to monitor the entrances of all the burrows found on eight live-trapped sites ranging in area from 1.6 ha to 7.3 ha. They considered burrows occupied if they exhibited a white thermal image against a dark background as a result of the heat generated by the resident squirrel. The results of these experiments were compared to densities obtained via live-trapping of animals and both the techniques tested showed strong associations with squirrel abundance. There was no evidence found of overestimating or underestimating population abundance relative to live-trapped densities. They concluded that both techniques are simple, rapid, and efficient and are useful methods for predicting densities of burrowing mammals, especially for sedentary species in which a single animal occupies a burrow system with only a few entrances (e.g., eastern chipmunk, *Tamias striatus*). These experiments point out that thermal imaging is capable of rapidly determining populations of burrowing mammals by simply counting the number of active burrows without intrusion of any kind.

As with all applications, when using thermal imaging to detect animals or animal presence care must be taken to optimize the thermal contrast between the animal, or in this case the burrow opening, and the background used to create the apparent temperature difference detected by the imager. When seeking animals there are levels of discrimination that let the observer determine detection, recognition, and identification of species, depending on the image quality (see Chapter 9). When examining burrows, dens, or nests there is no definitive discrimination possible without additional information regarding the occupants. One of the biggest problems encountered is that field researchers do not let the background temperature come to equilibrium before they begin their search of burrows, nests, dens, cavities, and lairs and usually the apparent temperature differences between these structures and their backgrounds is small but easily detected if care is taken. The uniformity of the background temperature (no hot spots or background clutter, e.g., see Section "Background Clutter" of Chapter 11) is far more important than the actual background temperature. Even if the imagery of a burrow is good and positively identified as being occupied there is still work to be done to identify the occupants.

Careful field studies devoted to detecting ground nesting birds in dense and tall grass were successfully carried out by Galligan et al. (2003) using an LWIR handheld FLIR ThermaCAM PM 575® thermal imager. During these qualitative experiments they used excellent field methodology and obtained images of the highly cryptic nests of grasshopper sparrow (*Ammodramus savannarum*) and Henslow's sparrow (*Ammodramus henslowii*). They clearly demonstrated that by understanding the basic physical principles of thermal imaging and by using the thermal imager accordingly they could locate nest sites with minimum disturbance to surrounding vegetation.

They correctly hypothesized that because eggs and early hatchlings are not insulated they would present an exposed emitting surface of 30–37°C as compared to a somewhat cooler and yet uniform background in the early morning hours, which would lead to good conditions for acquiring thermal imagery. They found that even though the eggs and hatchlings would probably be screened by thick vegetation they should still be visible through openings in the vegetation when viewed from above at close range or at an advantageous look angle. Furthermore, the thermal imager allowed them to locate nests more efficiently in tall vegetation. The average vegetation height surrounding nests located with the imager was 72.3 cm compared to 48 cm for those found without the imager. They point to the advantages of early morning searches for nests since the background grasses cool quickly overnight and presented a nearly isothermal background until shortly after sunrise. They noted that the thermal imager could be used for longer periods of time after sunrise on overcast days and that the ambient temperature appeared to have less of an effect on the ability of the device to detect nests than the actual heating of the ground by direct solar radiation.

Useful ranges for detecting new nests were approximately 2 m or less due to the obstruction of the visual path by vegetation and for known nests with a visual path not greatly obstructed ranges of 3–5 m could be expected. The thermal imager was not useful in locating nests prior to incubation but the imager was able to detect an incubating adult bird that did not immediately flush from the nest. Comparative images (visible and thermal) are provided in this paper (Figure 1), clearly showing the advantages provided by thermal images in reducing visibility bias. In addition, the use of the imager reduced the time spent around suspected nests and the disturbance to vegetation inherent in actually pinpointing them. Overall they feel that the device proved to be most effective as an aid in the detection of *Ammodramus* nests during incubation.

This work is instructive since thermal imaging is used to seek cryptic animals or objects in the daytime and allows one to gain an appreciation of the interplay of the thermal properties between the nests and their backgrounds and how they are regulated (intentionally or unintentionally) to produce usable thermal images. Galligan et al. (2003) noted that grasses have very little thermal inertia. Grasses tend to track solar radiation but also tend to cool quickly even under the influence of passing clouds. As a result they can present a nearly uniform radiating background to compare against the radiation emitted by the eggs and hatchlings. At some point solar loading warms the background to the point where the thermal contrast between the two is beyond the sensitivity of the thermal imager. At this point in the diurnal cycle the imager has lost its usefulness but nevertheless it is an invaluable asset that allowed them to not only detect more *Ammodramus* sparrow nests but also to quickly develop an enhanced nest-site gestalt after using the imager. This work provides grassland bird researchers with a useful tool for locating and monitoring nesting sparrows.

To mitigate the potential disruption of polar bear denning activity (giving birth in snow dens during midwinter) from ongoing and planned petroleum activities, such as exploration for new oil reserves that utilize seismic surveys, Amstrup et al. (2004) used FLIR to detect heat escaping from dens. This enabled them to be located and mapped for future reference and to provide guidance for planning industrial activities that will avoid denning polar bears. They flew transects over known dens of radiocollared female polar bears using a Bell 212 helicopter fitted with a FLIR Safire II AN/AAQ-22® thermal imager. The AN/AAQ-22 is an LWIR imager with an NEDT = 0.1°C. The survey consisted of imaging 23 dens on 67 occasions (one to seven times each). The weather conditions were recorded for each survey flight including ambient temperature, wind speed, wind direction, and visibility as well as percentage of cloud cover, cloud ceiling elevation, relative humidity, and dew point. Airborne moisture was recorded if the presence of any of the following was noted to occur singly or in combination at the site of each transect: blowing or falling snow, airborne ice crystals, or fog. If the sun was above the horizon and shining on the snow surface then it was recorded as present even when cloud cover prevailed.

During the study they located 19 polar bear maternal dens on land by radio telemetry. They were unable to obtain thermal images at four of the known den sites due to a combination of malfunctioning equipment and bad weather conditions. They did, however, identify the locations of 12 previously unknown probable polar bear den sites. Dens and targets presumed to be dens both appear as small bright spots (hot) with blurred boundaries that are contrasted against a dark (cold) band of drifted snow. Subsequent spring and summer surveys confirmed that three of the previously unknown sites were in fact not polar bear dens but were due to buried objects that presumably had sufficient thermal inertia to give off enough radiation to be taken as a possible den site. One other site could not be revisited for confirmation and therefore could not be verified as an occupied den.

The conditions required for accurate surveys of occupied polar bear dens are fairly strict with the most severe being zero probability of detecting an occupied polar bear den on FLIR video that was recorded in sunlight. The mean wind speed and the mean spread between temperature and dew point were also significant factors in determining detection and nondetection events. Kingsley et al. (1990) also noted that strict survey conditions must be met to detect undersnow lairs of ringed seal (*P. hispida*). These lairs are dug under winter snowdrifts covering breathing holes in the unbroken arctic sea ice by ringed seals. In the work by Amstrup et al., they concluded that modern FLIR devices may not be satisfactory for detecting dens of either seals or polar bears on ice substrates but they found FLIR imaging to be effective in detecting polar bears denning on land. They argue that bears denning on land are surrounded by colder and more uniform substrates than seals or bears occupying subnivian lairs at sea. Furthermore, they feel that the thermal contrast between polar bears and the more uniform backdrop of land coupled with more sensitive imagers used in this work allowed them to detect polar bears in land dens in early winter. Such work can supply enough lead time to make wise decisions regarding new construction projects relating to seismic surveys by the petroleum industry.

Controlled experiments using handheld thermal imagers to examine three artificial polar bear dens under varied conditions of solar radiation, wind speed, and den wall thickness to determine the efficacy of the imagers in detecting dens were undertaken by Robinson et al. (2014). The purpose of this study was to model the variables that influence the ability of handheld FLIR imagers to detect dens and to identify the optimal conditions for conducting polar bear den surveys with handheld FLIR imagers. In earlier work Amstrup et al. (2004) showed that handheld imagers could detect very small changes (0.01°C) in the apparent temperature differences radiating from the surface of a snow bank as a result of denning polar bears. They note that the threshold for positive den identification using handheld thermal imagers depended on a number of factors, but solar loading, prevailing atmospheric conditions, and imager selection were at the top of the list.

The present study area was located west of Prudhoe Bay, Arkansas in the Prudhoe Bay oil field where three artificial polar bear dens with initial wall thicknesses of 25, 50, and 75 cm were excavated in a south-facing snowdrift ~60 m long and 4 m high. A small unheated structure ~20 m from the artificial dens provided the source of electricity for den heaters. Three meters of spacing were used between the three dens. They were constructed with a plywood lid covering the den entrance, which was backfilled with snow. Ceramic heaters (200 W) were placed in each den and measuring sticks to record wall thickness were embedded in the dens. The dens were allowed to thermally stabilize for a period of 2 weeks before measurements began.

A FLIR ThermaCAM P65HS® LWIR thermal imager with a 72 mm IR lens and an uncooled microbolometer 320 × 240 FPA was used to sample the dens one or more times daily. The sampling was scheduled at stratified intervals to account for variations in solar loading effects and images were taken from three different distances: 60, 100, and 200 m. Solar radiation and wind speed were the most important factors influencing the artificial polar bear den detection using the thermal imager. Excellent imagery showing the effects of solar loading and wind are shown in Figure 8 on page 741 of their paper. Solar loading was the most influential factor as 94% of dens were not detected if solar radiation levels exceeded 100 W/m^2. Regardless of all other variables 96% of the dens sampled at night were detectable at some level.

The present study agrees with the fieldwork of Amstrup et al. (2004) with regard to the influences of solar radiation, airborne moisture, and temperature–dew-point spread as important parameters in polar bear den detection using aircraft-based thermal imaging techniques. The present work did not include airborne moisture as a metric but airborne moisture may have been influential through the effects of wind-driven snow close to the ground when using handheld thermal imagers to sample the dens. Robinson et al. (2014) suggest that managers tasked with locating polar bear dens using handheld thermal imagers do so between dusk and dawn when the wind speeds are slower than 10 km/h, and they feel that their findings probably apply to aerial thermal imaging as well.

Invertebrates (Arthropoda)

Introduction

In a recent review of literature on electronic, remote sensing, and computer-based techniques for observing and monitoring insect movement in the field and laboratory, Reynolds and Riley (2002) reported that thermal imaging cameras are used commercially as a noninvasive technique for detecting active termite (*Isoptera*) infestations in buildings. Nesting insects (bees, wasps, hornets, termites, and others) are cold-blooded animals and are generally in thermal equilibrium with the ambient temperature. However, nesting insects (e.g., honeybee, *Apis mellifera*) maintain a stable core temperature within the colony during cold periods through a combination of cluster contraction (to reduce the surface area of the cluster) and increased activity (shivering or vibrating) (Root, 1972, p. 622). While the heat generated by a single honey bee is negligible, the combination of large numbers of active bees confined in a relatively small volume is sufficient to maintain an inner core temperature of ~57°F (14°C) and a core shell temperature of ~43°F (6°C) throughout the winter months, and this is accomplished with only a 15–18% mortality rate for the colony.

Even though the heat generated by a single insect is small, individual insects can be counted with a thermal imager under the right conditions. We (Havens and Sharp, unpublished) observed wasps (*Vespidae*) leaving and entering a tree cavity perhaps 10 m above ground on a warm summer night in coastal Virginia. The cavity appeared bright white when viewed with a 3–5 μm thermal imager (much warmer than the surrounding branch, which was still visible with regard to the cold night sky). Exiting wasps presented a bright signature and returning wasps presented a very dark (nearly black) signature when compared to the brightness of the cavity. These observations suggest that nesting insects might be found using IR imagers by searching for tree cavities at night and under the right

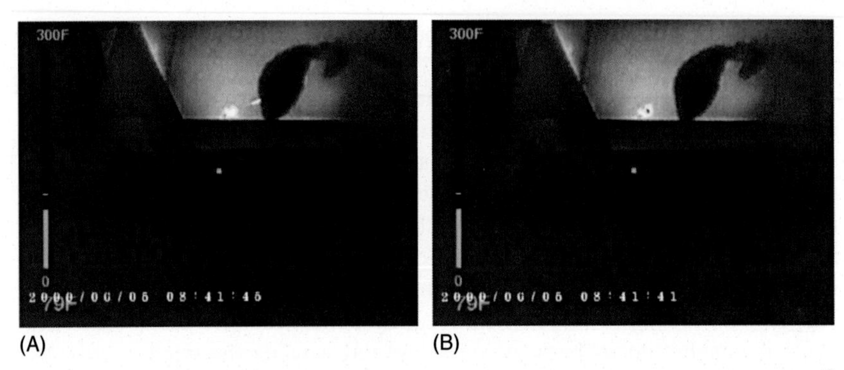

FIGURE 10.9 These are thermal images of a bee hive taken with a 7.5–13.5 µm FLIR® Quark 2 uncooled microbolometer with a NEDT of <50 mK at *f*/1.0. (A) A frame taken as a bee leaves the hot interior of the hive (white dot next to leaf). (B) A frame taken as a bee returns to the hive entrance and because the bee is now cooler than the hive interior it appears dark.

conditions (e.g., using the cold sky as a background). The rate of entry and exit of the cavity during peak activity might be used to estimate hive populations. For example, Figure 10.9 shows honeybees exiting and entering a hive. As the bee exits the hive it can be observed as a white, relatively hot, object (due to the heat generated within the hive) compared with the outside background temperature. Upon returning to the hive, the bee has cooled and appears dark relative to the hive interior.

Reviews (Invertebrate Observations)

Ono et al. (1995) report an unusual thermal defense carried out by honeybees against mass attacks by hornets. The giant hornet *Vespa mandarinia japonica* is the only hornet species known to have evolved en masse predation of other social bees and wasps. A hornet initiates the foraging activity by marking areas (secreting a pheromone) near a selected bee colony. This attracts hornet nestmates to the site where they congregate and attack the marked site en masse. The Japanese giant hornet is a major predator of social bees and wasps. Unlike the Japanese honeybee, colonies of introduced European honeybee (*A. mellifera*) are quickly destroyed by mass attacks of hornets (an entire colony of 30,000 bees can be destroyed in 3 h by 20–30 hornets). The Japanese honeybee (*Apis cerana japonica*) can detect the hornet marking pheromone and responds by increasing the number of defenders at the nest entrance. When an invading hornet is captured by a defending bee, more than 500 other bees quickly engulf the hornet in a ball that contains isoamyl acetate (a communication pheromone of the Japanese honeybee). Thermography (images captured by an Avio, TVS-8100® imager) showed that the ball temperature is very high (~47°C, which is lethal to the hornet but not to the bees). It is thought that these interactions between *V. mandarinia japonica* and *A. cerana japonica* are specifically coevolved.

Cena and Clark (1972) used a thermal imager to study the effects of solar radiation on the temperature of working honeybees (black bees *Apis mellifica mellifica* and Italian bees *Apis mellifica ligustica*). They recorded thermal imagery with an unspecified Aga Thermovision infrared imaging camera of bees returning to, and exiting from, south-facing hives. The camera was placed 1 m from the entrance of two hives: one inhabited by a colony of "black" opaque bees of North European stock and the other by lighter colored, partly translucent bees that were a cross between local and imported Italian bees. The temperatures of the thorax and abdomen were recorded as a function of solar radiation. Emerging bees' thorax and abdomens were the same temperature when the bees exited the hive, regardless of the amount of solar loading or the ambient temperature. The temperatures of 23–24°C were well below the usual core temperature of the hive in summer and did not change in the prevailing weather, that is, for air temperatures from 15°C to 18.3°C and solar radiation from 60 W/m^2 to 710 W/m^2. On the other hand, bees returning to the hives were always warmer than the air, and for bees that appeared to be returning from foraging flights, the thorax temperature was higher than that of the abdomen by up to 10 °C. There was a positive correlation between the solar irradiance and excess temperatures. The thorax temperature of bees arriving in close succession were as much as 6°C hotter in a period of sunshine than when a cloud obscured the sun. The highest thorax temperature was 38°C when the air near the hive was at 18°C and solar radiation was 710 W/m^2. The hottest returning dark bees were warmer than the hottest light bees. They concluded that their results showing variations in excess temperature with solar irradiance under steady conditions were consistent with earlier work by Digby (1955) on painted (black and white) insects.

Care should be taken when collecting thermal images of insects or other invertebrates when they are in direct sunlight, especially if the goal is to determine the body temperature of the invertebrate. Not knowing the wavelength band of the imager used in this study makes it difficult to draw a definitive conclusion regarding their results. An easy way to check results would have been to shade the front of the hives with an umbrella or other sun block that could be utilized instantaneously to provide shade. If the temperature of the bee drops instantaneously according to the thermal imager then the imager is recording a combination of the insect's body heat and a reflected component of solar radiation.

By way of example, we detected and observed orb weaving spiders (*Argiope aurantia*) in a coastal marsh in central Virginia with a 3–5 μm thermal imager for a period of several days. Our goal was to try to determine if the web orientation was dependent on the dominant direction of solar radiation, on polarizing effects the web might display to confuse foraging prey (grasshoppers), or a combination of the two. We systematically altered the direction of several webs and observed that they were returned to their original orientation by the spider overnight. During the course of this work we also shielded the occupied webs from direct sunlight and observed an instantaneous drop in the infrared signal from the spider that was directly correlated with the blocking of the solar radiation. We also observed a much weaker component of the infrared spider signature that slowly decayed away in a few minutes or less while the shade remained in place. After a few minutes the spider could not be detected by the thermal imager while in the shade, which indicates that much of the brightness associated with the thermal image of A. *aurantia* was due to a reflected component of solar radiation rather than the reradiation of previously absorbed solar radiation. Furthermore, by slightly changing the viewing angle of the camera we were able to get imagery of A. *aurantia* as a dark signature against the warmer background of saltmarsh cordgrass (*Spartina alterniflora*). In this situation the apparent temperature difference between the A. *aurantia* and the *S. alterniflora* became reversed since the removal of the reflected component left the background grass brighter than the spider.

A robust noninvasive assessment method for determining the internal population of honeybee colonies was investigated by Shaw et al. (2011) using an LWIR thermal imager. An uncooled microbolometer detector array of 324 × 256 pixels with an FOV of 86° × 67° was used at a distance of 3–4 m from the hives to capture approximately 40 pixels of imagery across the width of the hives. After the imager was stabilized, they would record thermal images of dozens of hives multiple times in differing weather conditions throughout both day and night. The best results were obtained from measurements made just before sunrise with clear skies and calm air. During the day, direct solar heating caused the thermal signature emanating from the hive interior to become difficult or impossible to measure. They made the determination that a quantitative assessment of the influence of nearby objects, trees, and sky conditions would require additional controlled experiments. They also suggested that more work will be needed to investigate the feasibility of using these methods for monitoring wintering beehives.

In addition to the direct benefits of reductions on time and labor (costs) required for monitoring hive populations, which are essential for determining the condition of bee colonies for the tasks of agricultural pollination and honey production, it is also probable that early signs of hive stress due to robbing or diseases may be indicated from this inspection technique. Their reference to conducting the hive monitoring just before sunrise under clear skies points to the benefits derived from letting the hives cool overnight without cloud cover so that they reach a more uniform temperature after the previous day of solar loading. This is sound practice to reduce the possibilities of background clutter, which can confound the recorded imagery.

Thermal imaging was used by Bonoan et al. (2014) to monitor the heat distribution patterns in honeybee hives that contained small populations of adult bees (1000–2500) and brood. The brood nest is sensitive to temperature changes in the hive and must be kept between 32°C and 35°C, although adult bees can tolerate temperatures as high as 50°C. Eight hives were heated for 15 min with a heat lamp to place stress on the resident bees and their behavior was monitored via the thermal imagery. It is well-known that worker bees control the temperature of the hive and a number of techniques are used to accomplish this task. For example, the bees will create heat by balling together and rapidly contracting their thoracic muscles (shivering) if the hive temperature drops below an unacceptable level. On the other hand, when the temperature becomes too hot, the workers use convective cooling by fanning the comb and the hive entrance to expel the heat from the hive. They will also spread fluid to induce evaporative cooling, and when the heat stress is localized they use conductive cooling processes to absorb heat by pressing themselves against the brood nest wall (a behavior known as heat-shielding). This study has shown that the worker bees absorb heat by conduction and then transfer the heat from the brood nest to the peripheral regions of the hive, away from the brood, to dissipate through conduction and convection, thereby redistributing the heat within the hive. The imagery shows that temperatures increased peripheral to the heated regions of the hives as the brood nest began to cool. The thermal images clearly showed that the bees had physically moved the absorbed heat in their bodies to previously cooler areas of the hive.

This work demonstrates that thermal imaging can reveal valuable information regarding the redistribution of heat within a working bee colony without disrupting normal hive activity. It would be very difficult to obtain this type of behavioral activity in any other way. This thermoregulatory activity taking place within the colony supports the theoretical construct of a beehive as a superorganism, reminiscent of bioheat transfer via the cardiovascular system of mammals.

CONCLUDING REMARKS

It should be pointed out that most of the early experiments using thermal imagers for animal surveys produced higher counts than so-called "conventional censusing methods," which are notorious for undercounting. In most of these cases the validity of thermal imaging as a method was questioned as opposed to the other way around (Graves et al., 1972; Naugle et al., 1996). We have determined, after reviewing the literature from the time of the very first attempts at animal censusing with thermal imagers in the wild (Croon et al., 1968) up to the present, that very good data can be extracted from the imagery collected by both MWIR and LWIR thermal imagers. Excellent imagery has been gathered under some of the most extreme conditions, both from a data collection point of view and from a weather point of view. For example, successful counts have been recorded flying at 300 knots in a phantom jet over the Platte River to determine the population of roosting sandhill cranes (*G. canadensis*) (Sidle et al., 1993). Successful counts of bats have also been obtained with a tripod-mounted handheld IR imager in an attic (Kirkwood and Cartwright, 1993). Recording thermal images of animals at ambient temperatures from 37°C to −28°C and at ranges from 2 m to 10 km have all been successfully demonstrated. Studies to locatc animals, record behavioral patterns, and count individual animals or large populations of animals within minutes (Hristov et al., 2008) have all been demonstrated in the wild. Animals (insects, birds, and mammals) have been located, studied, and counted in habitats ranging from dense forest to open ranges and open water, and they have been found and studied in lairs, dens, and nests.

As we mentioned earlier, the primary difficulty with inventorying and monitoring animals is not with the sampling methodology (i.e., distance sampling, aerial transect sampling, quadrat sampling, etc.) or with the statistical methods being used on the collected data but rather lies with the detectability that can be achieved with any particular data-gathering effort. Detection is the single most important function to master in any survey technique. No matter how elegant the statistical methods, the results of a technique that lacks ~100% detection are always inferior. In most of the situations reviewed, where a thermal imager was used as a survey tool and compared with other data collection methods it always produced higher counts, even though the thermal imager was used improperly or under less than ideal conditions. For this reason we examined a considerable body of literature and pointed out in detail the errors and problems encountered by these misuses. Some of the most serious errors have been promulgated for a number of years and the literature is rife with misleading results. Since we are concerned with acquiring the highest detectability possible when surveying or monitoring animal populations it is important to examine past attempts that compared visual surveys with thermal imaging surveys of closed populations. To this end, it is a requirement that each counting technique being compared must be used at its fullest capacity and that the sampled population be confined spatially and the number of animals known. It is important to note that a number of experiments have been conducted that demonstrate that thermal imagers are capable of ~100% detectability when used in aerial censuses of closed populations.

Chapter 11

Using Thermal Imagers for Animal Ecology

Chapter Outline

Thermal Imaging Techniques to Survey and Monitor Animals in the Wild: A Methodology
http://dx.doi.org/10.1016/B978-0-12-803384-5.00011-7

INTRODUCTION

Consistent improvements in the performance of infrared thermal imaging cameras have resulted in widespread use, and they are being touted as the only way to carry out many observations in the field for locating, counting, studying, and photographing nocturnal species. Their nighttime utility does not prevent them from being the best tool to study animal ecology during the daylight hours as well. This chapter is intended to review and extract the concepts, ideas, and methods introduced in earlier chapters that, when executed properly within the specifications of the thermal imager being used, will result in the maximum possible detectability. If the requirements associated with a particular application are not taxing the imager sensitivity and/or spatial resolution of the thermal imager being used, then the performance and the outcome of the application can be predicted with great certainty. As we have seen in the preceding chapters, often imagers are used in ways that inhibit their performance and as a result inhibit the target detectability as well. It doesn't matter if the surveys are done because it is convenient or that it is a nice day since nice days are not necessarily appropriate for thermal imaging surveys. A number of the reviews in the previous chapter point out instances where the reason for conducting a particular thermal imaging survey at a particular time was because it was convenient; the flying conditions were acceptable, they could be compared directly with visual counts, the snow cover was good, or it was just part of the scheduled survey. In many cases very little or no thought was given to what would be expected from thermal imaging observations or even what was being measured and recorded by the imager.

Thermal imagers were explicitly designed to perform in the dark to provide a visual record of objects, scenes, and events that could not otherwise be observed. The fact that thermal infrared cameras work in the dark does not exclude their use during the daylight hours; however, there are a host of conditions that must be met to ensure success when using thermal cameras during the daylight hours and perhaps this point is lost on many researches when they first attempt to use thermal imagers in the field. It appears that many researchers tend to ignore thermal effects because they are invisible and when thermal changes occur, in either the background or the animal, there is no visual change, but there can be significant deleterious changes in the thermal imagery. We point this out because it is difficult to propose another explanation for the widespread misuse of thermal imagers in the field.

METHODOLOGY

Imager Specifications and Use

In earlier chapters we described the essential performance parameters of thermal imagers and, while they can vary greatly from one imager to another, the performance of a particular imager will be dictated by its specified parameters if the methodology for using the imager is sound. Perhaps, in most cases, the best guide would be to use basic common sense when using thermal infrared imagers in the field. This of course implies that users have a good understanding of the general physics governing the operation of thermal imagers, including limitations that might be imposed by the prevailing conditions in the field. The information contained in Chapters 4, 5, and 6 provides the basics for understanding what a thermal imager actually detects. Chapter 7 describes how a thermal camera is capable of detecting animals in the field and what parameters in the design of the camera are responsible for achieving high sensitivity and spatial resolution. Chapter 8 provides guidance in selecting an imager once a particular use has been identified and what circumstances will likely be encountered in the field during data collection efforts. The thermal imager selected will have many features that allow the user to tailor the data collection to acquire the quality and quantity of images needed to satisfy the survey or study. For example, if detection is all that is required then the data collection can be reduced to acquiring images that merely contain signatures of a certain size. When more detailed imagery is required then adjustments to the cameras must be made, and the conditions in the field must be optimized. For the most part this will be the subject matter of the present chapter.

Generally we cannot change the amount of radiation emanating from objects in the field so we must become patiently opportunistic and seek a "window of opportunity" that will maximize the data collection effort. Establishing a suitable window of opportunity for any given field experiment, survey, or census is the single most important thing that we can do to ensure success. In most cases the local weather and atmospheric conditions determine the time the window is open, but it must also coincide with the presence of the species being interrogated with the imager.

Training

Using a thermal imager in the field requires some training to become familiar with the imager. In Chapter 8 it was recommended that the operator become thoroughly familiar with the use of the imager before taking it into the field. This is just common sense and will avoid wasted effort in the field. In addition to knowing how to collect imagery there is another entirely different type of training necessary. It will become necessary at some point to be able to look at imagery and know how to improve the quality of the imagery to be useful for particular tasks. This means that the thermographer must know what factors or parameters in the field influence the image quality and to what degree they can be changed to improve them. This may seem daunting at first but becomes easier with experience and practice. Formal training in the use of the various new imagers is offered by the manufactures and participating in these courses is advised.

SURVEYS

Introduction (Factors Influencing the Detectivity)

The detectivity or detectability depends on a great number of factors, many of which we have very little control over other than waiting until the conditions (weather, atmospheric) are satisfactory and animals are present. Select an adequate imager for the planned work. Is the sensitivity, resolution (both spatial and thermal), and wavelength range of the selected imager capable of meeting the needs of the study? Will the observations be made nocturnally or during the daylight hours? Are there confounding signatures of objects anticipated in the survey area such as trees, shrubs, bare rocks, bare soil, and deadfall that will reradiate or reflect solar radiation? Will there be other species of similar size mixed with the species of interest within the study area? All of these factors must be considered at the same time and if you forgo any one of them there is the probability that you may distort the results obtained, but they will be correct for the methodology you adopted for the study. In short there is no detection rate for a species unless all of these factors are specified during your data collection effort, and then the detection rate is just for that effort.

If one wants to compare thermal imaging with visible techniques for aerial surveys then researchers using visible techniques must provide some method to account for the number of animals that are missed (not counted) due to problems stemming from visibility bias. These problems include animals missing from the count due to obscuration from vegetation, those lost due to observer inexperience, fatigue, or poor eyesight, lack of proper lighting conditions, and a host of others arising from human disturbance or interference that can induce "fear or shyness" in the animals, causing them to be elusive in addition to being cryptic. Note that this can be done by simply creating a proper thermal background to compare the apparent temperature difference between the animal and the thermal background. The issues then are either resolved or very nearly so. You will not get better results with any other method of surveillance or detection period.

Fortunately, the technology is beginning to see more use and favorable results due to proper implementation of counting techniques. The technology is being recommended by more researchers and field biologists as the preferred technique for estimating animal abundance. These endorsements are for both aerial and ground-based surveys and include mammals, birds, bats, feral species, nests, tree cavities, and marine animals. See, for example, the following references: Adams et al. (1997); Ammerman et al. (2009); Amstrup et al. (2004); Barber et al. (1989); Beaver et al. (2014); Bernatas and Nelson (2004); Betke et al. (2008); Burn et al. (2009); Christiansen et al. (2014); Cilulko et al. (2013); Conn et al. (2013); Desholm et al. (2006); Dymond et al. (2000); Edwards et al. (2004); Focardi et al. (2001); Frank et al. (2003); Gill et al. (1997); Havens and Sharp (1998); Horn et al. (2008); Hristov et al. (2008, 2010); Israel (2011); Kirkwood and Cartwright (1993); Kissell and Nimmo (2011);

Kissell and Tappe (2004); Kuntz (2001); Kuntz et al. (2007); Lavers et al. (2009); Lazarevic (2009); Marini et al. (2009); McCafferty (2007); McCafferty et al. (1998, 2010); Melton et al. (2005); Parker (1972); Sidle et al. (1993); Steen et al., (2012); Wiggers and Beckerman (1993); and Wride and Baker (1977). These references are not meant to be all inclusive but rather to give a representative look at work covering a wide range of applications dealing with different aspects of thermal imaging, different species, and several good reviews.

ANGULAR DEPENDENCIES AND EFFECTS

In this section we treat the angular dependent effects on the detectivity associated with both the scanning techniques used in field studies and the naturally occurring dependencies in the environment. For most fieldwork, whether monitoring or surveying from an aircraft or from ground-based platforms, there will be a viewing angle associated with the object under observation. Key parameters that exhibit dependencies on the viewing angle are the size and spatial distribution (shape) of individual thermal signatures, the spatial extent of the survey area, reflectivity, emissivity, apparent temperature difference, heat transfer (through the reflectivity), and the atmospheric transmission. All of these parameters come into play when data is collected in the field with a thermal imaging camera. Not only do they come into play, but more often than not, they come into play simultaneously. For this reason it is necessary to understand the contributions of each and how to manage them in the field to establish an optimized window of opportunity so that the highest detectivities possible can be obtained.

From a strictly geometrical viewpoint, there is an angular dependence on the observed size and shape of any object and its image whether captured on film with a camcorder in the visible part of the spectrum or with a thermal imager in the infrared part of the spectrum. A typical thermal signature of an animal in the field at any point in time is a representation of the animal in two dimensions, and it can have an infinite array of shapes and intensity distributions. This is because the cross-sectional area of the animal viewed by the camera varies with the viewing angle. When a typical thermal image (signature) is displayed on a monitor for detection purposes we almost always see a nearly uniformly white spatial distribution of the radiation incident on the focal plane of the detector, and the spatial distribution is saturated (no contrast within the borders of the signature). Picture a side view and a frontal view of a horse (*Equus caballus*) under these conditions. It is safe to guess that nearly 100% of viewers would recognize the side view as a horse or horse-like animal of genus *Equus* (such as a zebra, mule, or burro), but considerably less would identify the horse from the frontal view. The inverse of these geometric phenomena (absorption rather than emission) was addressed when detailed calculations of cross-sections for various geometric shapes exposed to direct solar radiation were conducted to determine the effects of angular exposure on solar loading (Monteith and Unsworth, 2008, p. 105ff). Note that the problem of recognition and identification can be significantly mitigated if the thermal signatures have some distinctive intensity distribution (i.e., the signatures are not saturated) that can be associated with a species of interest. For example, if it were a zebra in direct sunlight we would see signatures that show the typical spatial distribution associated with the stripe pattern due to the difference in emissivity associated with dark- and light-colored stripes (McCafferty, 2007).

Consider the thermal images (signatures) depicted in Figure 11.1A–C. These photographs of a cat were taken in coastal Virginia with a 3–5 μm thermal imager during early August when the temperature was ~30°C in the late afternoon. These photos were taken within minutes of one another while the cat was moving through a brushy, lightly vegetated area dominated by mature loblolly pine (*Pinus taeda*). The area had been sheltered from direct solar radiation all day by an overcast sky. All the photos were taken from a tripod-mounted imager situated on an elevated platform about 40 m from the cat. Note the thermal uniformity of the background and that the images are not completely saturated. Also note that without the benefit of observing photographs A and C the signature shown in Figure 11.1B would be difficult to identify as a cat without close observation.

While it may seem trivial to mention this here, managing this single geometrical relationship (the shape of uniform spatial distributions) for remote sensing applications is extremely difficult. It has stimulated considerable and intense research efforts for those seeking to advance commercial and military applications such as robotic vision, spacecraft docking, automatic target recognition, and tracking. Consider, for example, the use of the mathematical operation of two-dimensional cross-correlation for determining the degree of similarity between two optical images comprised of uniform intensity distributions but having differing shape and size. Even for images of modest resolution, such as those produced by television imaging devices or thermal imagers, the calculation of

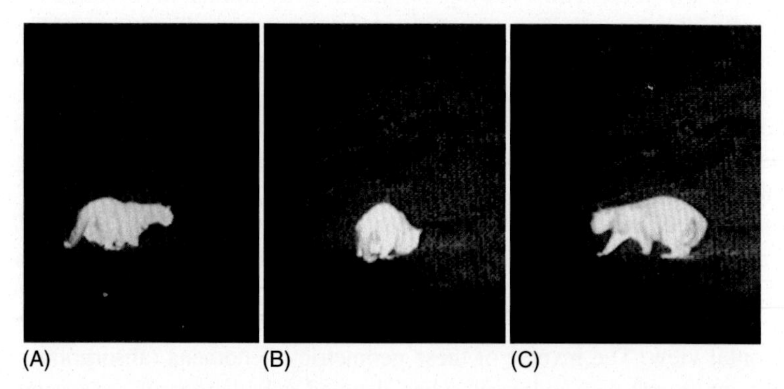

FIGURE 11.1 (A–C) Differences in these three thermal signatures are due to the attitude and position of the cat with respect to the location and viewing angle of the camera. The morphology and the image shapes provide information for identification. Also there are a few noticeable pelage stripes on the cat's tail in photograph (A).

the cross-correlation function is a very extensive exercise in computation. This computational load could be greatly reduced through the development of coherent optical image processors. Optical systems have already been demonstrated that can carry out a number of the preprocessing steps to prepare images for all-optical correlations. For example, the processes of edge enhancement, image cleanup, image storage, image amplification, and image recognition have all been demonstrated (Wood et al., 1995; Sharp et al., 1994). A key component of an optical processor is a spatial light modulator (SLM) that performs the important task of transferring real-world scenes derived from TVs or thermal imagers into scenes impressed on a coherent optical beam (laser). There are many advancements still needed in SLM technology (Efron, 1995) before an all-optical processing capability can become a reality. Anderson et al. (1993) have demonstrated that all-optical correlations can be performed when using a technique called mutually pumped phase conjugation on two-dimensional images of uniform intensity. These experiments were carried out using the most basic of geometrical shapes such as a triangle and a bar pattern. While these results are somewhat encouraging, the construction of an automated all-optical image recognition system for use in the field is very unlikely without significant advancements in a number of critical areas.

Wyatt et al. (1980) made a determination from examining the field data gathered by Marble (1967) and Parker (1972) that thermal contrast (the intensity difference between deer and its background) was insufficient to "identify" to species level. The problem here was that the deer signatures were easily detected (i.e., recorded by the thermal imager) but were imbedded in a background of other thermal signatures of uniform intensity with various shapes or distributions based on the viewing angle. This phenomenon is known as "background clutter" and will be treated in Section "Background Clutter" of this chapter. Wyatt et al. (1980) did not incorporate the spatial distribution of the intensity patterns recorded by the thermal imager into their analysis. Ignoring the spatial distribution of individual thermal signatures and using only the intensity of the signatures gathered with the concomitant presence of strong background clutter proved to be unsuitable for use in an automated data identification system, but this does not mean that thermal contrast, ΔT (actually the apparent temperature difference), is insufficient for detecting, recognizing, and identifying deer in the field.

Fortunately, the human brain can process visual images of a well-known animal, such as a horse or a salamander, in almost any viewing aspect and be able to identify it as such. During this processing the brain can consider the shape, size, color, and perhaps motion (if the images are in video format) to carry out the correlation of an observed image and one stored in our memory. Once this is accomplished, the brain can process the images that the eyes see. This process is nearly instantaneous and nearly 100% correct. In fact, researchers found that individual nerve cells fired when subjects were shown photos of well-known personalities (Quiroga et al., 2005). The same individual nerve cell would fire for many different photos of the same personality, and a different single nerve cell would fire for many different photos of another personality. Follow-up research suggests that relatively few neurons are involved in representing any given person, place, or concept, which makes the brain extremely efficient at storing and recalling information after receiving visual stimulation.

On the other hand, the brain can easily be confounded when attempting to perform correlations of simple plane images of uniform contrast such as a typical thermal signature associated with an animal. When we collect infrared images of animals in the field, the signatures, while easy to detect, can have many varied spatial distributions, depending on the viewing angle (Figure 11.3). These images are subjected to processing by the human brain, and there is some uncertainty because the brain does not have a complete library of two-dimensional images (white spatial distributions on a black or dark background) stored with which to compare all possible input images for validation. When cataloging thermal images collected in the field, a three-level hierarchy (detection, recognition, and identification) comes into play, which will serve to ameliorate some of these deficiencies. This hierarchy is examined in detail in Chapter 9.

Size and Shape of Individual Thermal Signatures

There are several factors that can alter the size and shape of the thermal signatures captured by an infrared imager. Note here that (7.10), relating the smallest observable object X_{min} at range R for a given IFOV, assumes that the target is normal (perpendicular) to the line of sight. The primary factors affecting the size and shape of a thermal signature are range, emissivity, viewing angle, partial vegetative obscuration such as tree canopies, local atmospheric conditions (sunlight, temperature, rainfall, wind), and metabolic activity in the case of homothermic animals. All of these factors will apply to poikilothermic animals as well, excluding that of metabolic activity, in which case the changes to the shape and size of the signatures associated with poikilothermic animals are induced and become noticeable as the animals are warmed or cooled by their environmental surroundings.

We will treat each of these during the course of establishing a working methodology to obtain an optimized window of opportunity for using thermal imaging systems and techniques in the field. A typical signature captured by a thermal imager is displayed on a monitor as a uniformly white spatial distribution (silhouette) of radiation emanating from some object with an apparent temperature that is different from the apparent temperature of the background. The shapes of these signatures are varied and can be intricate; however, since there is usually no variation in the intensity distribution across the signature, very little or no information can be obtained from within the boundary of the white silhouette displayed on the monitor. Intensity uniformity across the spatial distribution in general limits the amount of information that can be extracted from any given signature. For field surveys where counting animals is an objective, bright signatures are more important than those giving information related to the size or shape of the animal. In some instances, however, it is advantageous to reduce the gain and/or the light level of the imager to enhance the intensity variations within the boundaries of the observed signatures. Such information is valuable in determining the thermoregulation of wild animals in their natural environment (McCafferty, 2007; Gade and Moeslund, 2014; Lavers et al., 2009; Cilulko et al., 2013) and ultimately provides assistance in gaining an understanding of the thermodynamic equilibrium conditions for their ecology (Porter and Gates, 1969).

Most thermal signatures of animals will have distinguishing patterns that arise because of the spatial nonuniformities in the radiation distribution. The individual physical characteristics of different species such as hairless areas, eyes, ears, legs and feet of avian species, and coat thickness alter the spatial characteristics of the radiation pattern collected by the thermal camera. See, for example, the thermal signatures shown in the photographs of Figures 9.2 and 9.3B. While conducting aerial surveys these characteristic modulations in the spatial distributions are usually not observed due to the lack of sufficient spatial resolution commensurate with ranges used in aerial surveys.

Spatial Extent of the Survey Area

Depending on the goals of the survey and the species involved, the spatial extent of the survey can range from very small to very large. For example, the survey area for determining the population of a wasp nest would consist of a small area centered around the hive opening; the survey area for a colony of bats might include the area occupied by the opening of a cave; the area for surveying a flock of migrating birds might be the shoreline of an entire lake; the survey necessary for making population estimates for a large herd of ungulates may be thousands of hectares; and the survey area for making estimates of seal or walrus populations could be vast expanses of sea ice. The survey conditions for these examples all use a thermal imager that is directed at some angle to the survey area, and in many cases the angle is critical in obtaining quality imagery.

Survey Geometry

In Section "Spatial Resolution" of Chapter 7, the spatial resolution and the concept of the instantaneous FOV were developed and expressed in (7.10). Recall that $X_{\min} = \theta_{\mathrm{IFOV}}(R)$, where θ_{IFOV} is the instantaneous FOV and determines the size of the smallest object, $X_{\min}$, that can be detected at a range, R. This is true when we are viewing the object at normal incidence, and any differences in the viewing angle from the normal must be included in the analysis of the imagery if size and shape are to be key parameters for signature identification. Here normal incidence refers to the central line of sight of the camera, which we will call R_C. The viewing angles selected for any set of survey conditions should be optimized for obtaining the best thermal signatures of the species of interest observed within the FOV of the thermal imager.

When constructing the survey geometry one of the primary concerns is the determination of a suitable swath width. We will want the swath width to be large enough to cover a sufficient percentage of the survey area to be representative of the animal populations yet small enough to preserve adequate resolution for the resulting ranges selected. Basically, the larger the swath width the lower the resolution will become. Besides the swath width there are many other factors that will come into play when the window of opportunity is constructed for the experiment at hand. For example, the animals must be available for observation, the thermal contrast and the prevailing atmospheric conditions must be optimized, the spatial distribution of animals within the FOV representative of the animal population must be thought out, and a host of other factors and parameters must be considered when selecting how the data is to be captured by the imager.

Simple geometrical relationships will determine the available size and shape of the viewing area based on the FOV, θ_{FOV}, of the thermal imager. Typically, imaging cameras are equipped with optics to permit both a wide and a narrow FOV, which also determines, to a large extent, the spatial resolution along with the detector pitch (see (7.9)) that can be obtained. Since the focal-plane-arrays of thermal imagers are not always square, the horizontal and vertical FOVs are not always equal.

There are four basic geometries that will suffice for most fieldwork, depending on the location of the viewing platform (tower or tree, hilltop, aircraft, ground vehicle, watercraft, etc.). These four geometries are described later in the chapter and could be extended or adapted to other specific geometries as well. By selecting oblique viewing angles with respect to the nadir or zenith we have developed a consistent set of relationships, regardless of how the survey is set up (air-to-ground or ground-to-air).

Air-to-Ground

When establishing a survey geometry there are many considerations (e.g., topography, possibility of abnormal weather patterns, ground cover, time of day, imager capabilities and specifications, aircraft type, flight speed, species ecology, and survey design) that must be taken into account when determining the best swath width for a successful survey. The primary goal is to optimize the detectability with the aim of detecting all animals of interest within the FOV of the detector. The basic idea is to cover a predetermined area of interest that will best represent the species ecology under study. In an aerial survey, as discussed here, the altitude of the aircraft (h), the viewing angle (θ_C relative to the nadir), and the FOV of the imager will determine the available ground swath width. The basic parameters needed to describe survey geometries are shown in Figure 11.2. Here the aircraft is flying with the thermal imaging camera pointed outward from the side of the aircraft and perpendicular to the flight line (direction of flight) toward the ground at an angle θ_C with respect to the nadir.

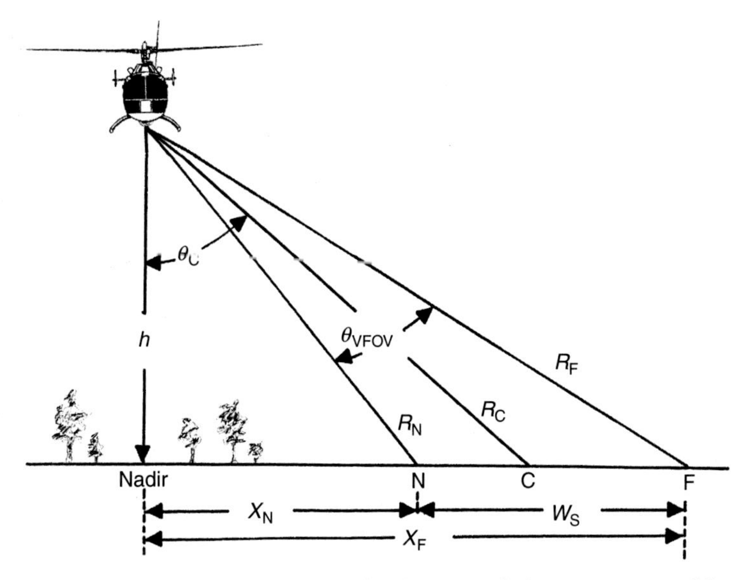

FIGURE 11.2 A sketch of a helicopter showing the geometrical arrangement used for scanning along transects from the side of the aircraft while following a flight line. The altitude, viewing angle with respect to the nadir, and FOV of the imager determine the swath width.

The area on the ground being recorded by the thermal imager is determined by the FOV of the imager. The FOV is usually rectangular in area, so the FOV has a horizontal HFOV and a vertical VFOV component. When used in the geometry, as shown in Figure 11.2, the VFOV determines the extent of the "footprint" dimension in the direction perpendicular to the flight line, and this dimension of the footprint determines the swath width, W_S. The dimension of the footprint parallel to the flight line is determined by the HFOV.

In general, if we were to look at the subtended footprint cast by an array of detectors it would appear distorted because of the oblique viewing angle. If the line of sight were exactly along the nadir then this distortion would be minimized. Most survey conditions will use a scan angle that casts the vertical footprint off to one side of the aircraft so that the footprint of the projected detector angular subtense is distorted, as shown schematically in Figure 11.3 and as would be typical for aerial surveys (see "Detector-Limited Resolution"). In this situation the projected angular subtense of each pixel is a distorted trapezoid or keystone-like.

Figure 11.4 shows a typical delineation of possible transects, which are arranged here in a linear fashion and aligned parallel to the flight direction of the aircraft. Each transect will have a specific group of parameters to determine the arial coverage on the ground. There are three parameters that can be adjusted to establish and tweak the swath width: (1) altitude, (2) viewing angle, and (3) FOV (the vertical FOV for the geometry of Figure 11.2). The widths of

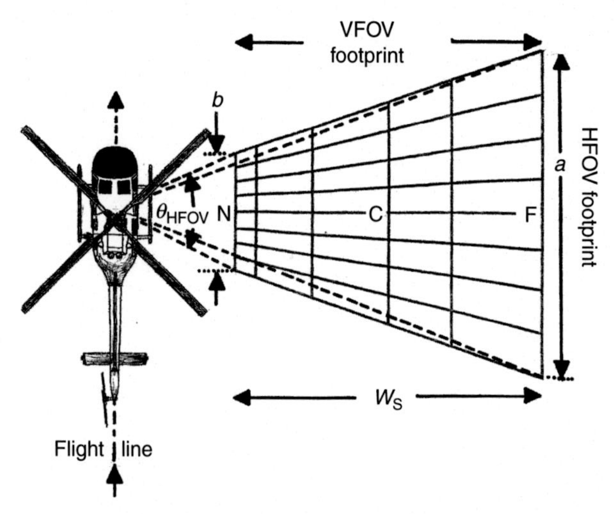

FIGURE 11.3 **The subtended footprint shown as cast on the ground as a result of oblique viewing with respect to the nadir in a direction perpendicular to the flight line.** Equation (7.10) would apply to an object located at R_C, which would be on the center line located in the center of the vertical footprint at C. Note that there is also a horizontal component to the footprint as a result of the HFOV, and the width of the HFOV increases across W_S from N (near) to F (far).

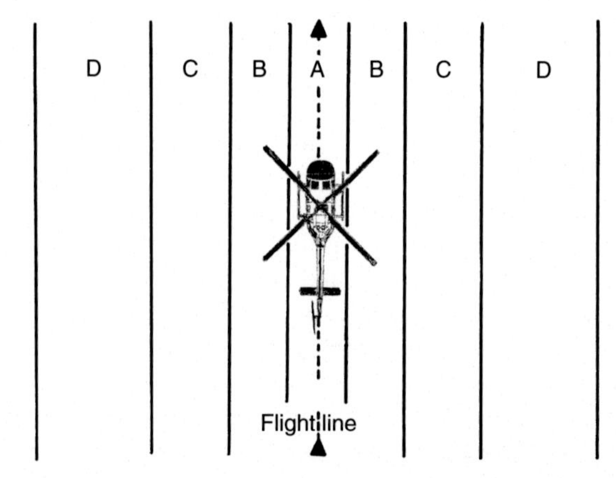

FIGURE 11.4 **A schematic of possible selections for transect widths and positions on the landscape are shown here on both sides of the aircraft where the width of the selected transects are equal to the VFOV footprint, which is established by the viewing angle θ_C and the flight altitude h.**

the different transects (A, B, C, and D) and the distance from the aircraft are adjusted by selecting these parameters to satisfy the goals of the survey. The transect length is determined by the flight length or where the recording of data is terminated. The length of the transect L_T times the swath width W_S is the area surveyed.

For transect (A) the camera is either pointed straight down or at an oblique angle with respect to the nadir pointing in the forward direction along the flight line. The survey geometry where the camera is directed straight down along the nadir has been demonstrated by Kissell and Nimmo (2011) for estimating deer density using distance sampling in conjunction with airborne infrared thermal imagery integrated with GPS and GIS data (for computing distance calculations for individually detected deer). The important point about using an imager and the defined footprint is that the imager will only see animals within the FOV of the camera. This is an important consideration since any animal outside the swath width will not be detected so counting animals outside of the transect width cannot happen as it often does during visual surveys.

Example I

Suppose we are viewing the surface of a marsh from an elevation, h = 300 m, above ground level (agl). The VFOV of the imager is 10°, and the HFOV is 8°. Objects in the marsh that are within the FOV will be at different distances from the camera in accordance with their location within the footprint subtended on the surface of the marsh.

In Figure 11.2 we have chosen an oblique viewing angle $\theta_C = 45°$ with respect to the nadir. As a word of caution, oblique viewing angles greater than 45° should be avoided because of the strong angular dependence of the emissivity at viewing angles greater than 45° and because the size of the images for identical objects will be different across the footprint. The significance of the angular dependence of the emissivity was discussed in Chapter 6. We can now calculate the subtended vertical and horizontal footprints for the geometry depicted in Figure 11.2 by determining the values of W_S or VFOV and the horizontal dimensions of the VFOV at R_N and R_F, which are the ranges of the nearest and farthest objects in the VFOV. At any point in time during a flyover the distance from the nadir to the nearest point in the swath width is given by

$$X_N = h\left[\tan\left(\theta_C - \theta_{VFOV}/2\right)\right].$$

The distance from the nadir to the farthest point in the swath width is given by

$$X_F = h\left[\tan\left(\theta_C + \theta_{VFOV}/2\right)\right].$$

We can now determine the swath width from $W_S = X_F - X_N$ as

$$W_S = h\left[\tan\left(\theta_C + \theta_{VFOV}/2\right) - \tan\left(\theta_C - \theta_{VFOV}/2\right)\right].$$

For the example given earlier with h = 300 m, $\theta_{HFOV} = 8°$, $\theta_{VFOV} = 10°$, and $\theta_C = 45°$ we find the swath width W_S = 106 m. W_S is the length of the vertical footprint of the VFOV and the width of the transect. This trapezoidal-shaped footprint is shown schematically in Figure 11.3 and illustrates the distortion of each pixel in the instantaneous footprint as a result of the oblique viewing geometry. Here, a, is the horizontal width of the vertical footprint at R_F, and b is the width at R_N. The geometrical relationships between θ_{VFOV} and θ_{HFOV} and the footprint of the detector angular subtense can be seen in Figure 11.3. The notation used in Figures 11.2 and 11.3 will be used to define the parameters and relationships needed to derive equations for a number of different survey geometries that will allow us to determine the footprint shape and area covered for the survey.

We can estimate the horizontal width of the vertical footprint at R_N and R_F, which can be thought of as the horizontal footprint in the static mode or when the helicopter is hovering. The values of R_N and R_F can be written as

$$R_N = h\left[\cos\left(\theta_C - \theta_{VFOV}/2\right)\right]^{-1},$$

and

$$R_F = h\left[\cos\left(\theta_C + \theta_{VFOV}/2\right)\right]^{-1}.$$

From the values of R_N and R_F we can determine the values of the width at the farthest and nearest points of the swath width as

$$a = 2R_F\left[\tan\left(\theta_{HFOV}/2\right)\right],$$

and

$$b = 2R_N\left[\tan\left(\theta_{HFOV}/2\right)\right].$$

For our example we find $R_N = 392$ m, $R_F = 467$ m, $a = 65$ m, and $b = 55$ m. The area of our static footprint is $A = W_S(a + b)/2 = 6360\ m^2$.

Basically, the larger the swath width the lower the resolution will be since for a given optical system the spatial resolution is determined by (7.10). Setting a wider swath would require flying at higher altitudes or using more oblique viewing angles. Both of these options would increase the viewing range so that according to (7.10), the minimum size of a detectable object can become quite large. Using large oblique viewing angles may introduce several deleterious consequences from the local atmospheric conditions and the reflection and scatter of unwanted background radiation. At very high altitudes the swath widths become large and the ranges become so long that the resolution of the optical system is no longer adequate to resolve small objects within the swath. This does not prevent using thermal imaging to survey vast areas, however.

We can get a feel for the size of animals that can be detected by continuing with the examination of higher-altitude flights. If for example we were flying at 1000 m the values of the swath width would be 353 m and $R_N = 1305$ m, $R_F = 1556$ m. Suppose that the imager has an IFOV = 0.6 mrad for the VFOV and HFOV given previously. For the 300 m agl flights the values of X_{min} for R_N, R_C, and R_F are 0.235, 0.255, and 0.280 m, respectively. For flights at 1000 m agl the values of X_{min} become 0.783, 0.858, and 0.993 m. If all else is equal for the two flights except for the altitude then in the low-altitude case we should detect animals down to one-fourth of a meter in size, and in the other case animals approaching a meter in size would be the smallest detected. Also note that the size of the images detected will vary across the VFOV by approximately one-fourth of a meter in the high-altitude case. This variation in the minimum detectable size is not as severe as one would think for doing fieldwork.

Note that the extreme cases of $\theta_C \sim 0°$ (the nadir) would result in a rectangular footprint for the case at hand and would be a square for imagers with equal vertical and horizontal FOVs. If $\theta_C \sim 85°$ the camera would be looking very nearly parallel to the surface or at very long distances, depending upon the topography of the landscape or habitat being surveyed. See Chapter 6 and the discussion relating to the loss of contrast in the thermal image due to using large oblique viewing angles, which is the result of the angular dependence of the emissivity on viewing angle. At these large viewing angles the atmospheric conditions close to the surface of the earth become problematic, the emissivity of objects tends to zero, and the reflectivity of objects tends to unity. Except for perhaps very specialized surveys neither of these cases would be a desirable situation for collecting thermal imagery. Normally, transects would be flown over the marsh and the oblique viewing angle would be adjusted to include adequate resolution of the animals within the vertical footprint.

In practice, when the camera is pointed downward along the nadir it is advantageous to slowly (with respect to the aircraft speed) pan the camera along the direction of travel so that the fore/aft viewing angle at any point in time is changing with respect to the direction of travel. As the aircraft advances the camera is slowly panned back and forth in the direction of travel and the flight speed will determine the amount of overlap of each forward/backward scan with the camera. The advantages of this panning technique while collecting imagery will allow for a number of different looks along transects, which helps to "see" under and through tree canopy.

Ground-to-Air

The FOV of the 3–5μm Inframetrics Infracam® thermal imager is (8° × 8°). If one were looking upward directly along the zenith (straight overhead), as in Case I depicted in Figure 11.5, the cross-sectional area (footprint) at any given height would be a square (because the VFOV and HFOV are equal), and the area of the square depends directly on the height.

Case I

If the vertical and horizontal FOVs are equal ($\theta_{HFOV} = 8°$ and $\theta_{VFOV} = 8°$) and the camera is pointed straight up along the zenith (i.e., directly overhead) then the projection of the footprint has area *A*.

$$\text{Area} = \left[2h\left(\tan\theta_{VFOV}/2\right)\right]\left[2h\left(\tan\theta_{HFOV}/2\right)\right] = \left[2h\tan 4°\right]^2$$

where h is the height and $\theta_{VFOV}/2 = 4°$ is the half angle of the vertical and $\theta_{HFOV}/2 = 4°$ is the half angle of the horizontal FOVs. Since both the horizontal and vertical FOVs are the same we can simply square the result of the vertical $2h(\tan 4°)$ or horizontal $2h(\tan 4°)$ projection at height h to get the footprint area. For example, if $h = 300$ m, Area = 1760 m^2. If $\theta_{HFOV} \neq \theta_{VFOV}$ then the cross-sectional area of the footprint is a rectangle.

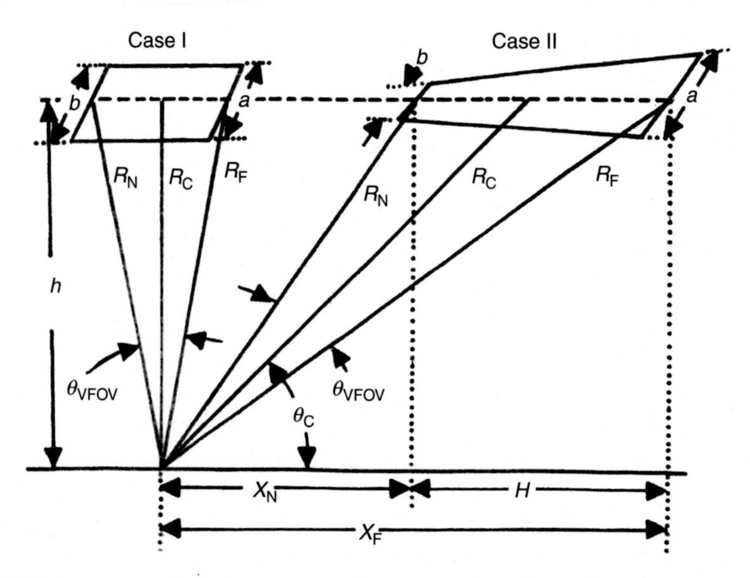

FIGURE 11.5 The geometrical relationships for two different ground-to-air survey conditions that show the shape of the footprints in a plane parallel with the surface of the ground.

Case II

If we are not looking along the nadir then there is an oblique viewing angle (θ_C) with respect to the zenith and the cross-section of the projected area at any given height is a trapezoid, as shown in Case II (see Figure 11.5) depicted later in the chapter. If the same camera is pointed upward with some viewing angle θ_C relative to the zenith, resulting in a vertical height of h, then the cross-sectional area viewed by the camera is a trapezoid whose height H (called the vertical FOV or vertical footprint) is calculated from the expressions given by

$$X_N = h\left[\tan\left(\theta_C - \theta_{VFOV}/2\right)\right]$$

and

$$X_F = h\left[\tan\left(\theta_C + \theta_{VFOV}/2\right)\right].$$

$$H = \left(X_F - X_N\right) = h\left[\tan\left(\theta_C + 4°\right) - \tan\left(\theta_C - 4°\right)\right]$$

Here θ_c is the viewing angle of the camera relative to the zenith. The area of a trapezoid with two parallel sides is given by $A = H(a + b)/2$ where a and b are the parallel sides. The length of side (a) of the trapezoid is just the horizontal projection of the footprint at the farthest range from the camera $R_F = h/\cos(\theta_C + 4°)$, and similarly, side ($b$) is just the horizontal projection of the footprint at the nearest range from the camera $R_N = h/\cos(\theta_C - 4°)$. These projections are computed as $a = 2R_F(\tan 4°)$ and $b = 2R_N(\tan 4°)$.

We calculate the area of the projected trapezoid as

$$A = h/2\left[\tan\left(\theta_C + 4°\right) - \tan\left(\theta_C - 4°\right)\right]\left[2\left(R_F + R_N\right)\left(\tan 4°\right)\right]$$

For example, $h = 300$ m and $\theta_C = 45°$.

Here we calculate the projected area of the footprint $A = 5058\ \text{m}^2$ as seen by the camera at a vertical height of 300 m above the ground when the camera is pointed upward at an angle of 45° relative to the zenith.

Air-to-Air

The geometry for an air-to air survey would be the same as that for air-to-ground except that the footprint would be calculated for a survey area that could lie parallel with the surface of the earth, perpendicular to the surface of the earth, or anything in between. In these situations the viewing angle can be adjusted to accommodate the best survey conditions without losing any of the geometrical simplicity. For example, a survey of birds roosting on the side of a tall vertical cliff (wall) could be surveyed by aircraft (helicopter or UAV), as depicted in Figure 11.6. The same habitat could also be surveyed from the ground, and in both cases the equations above can be used to calculate the vertical footprint subtended on the cliff.

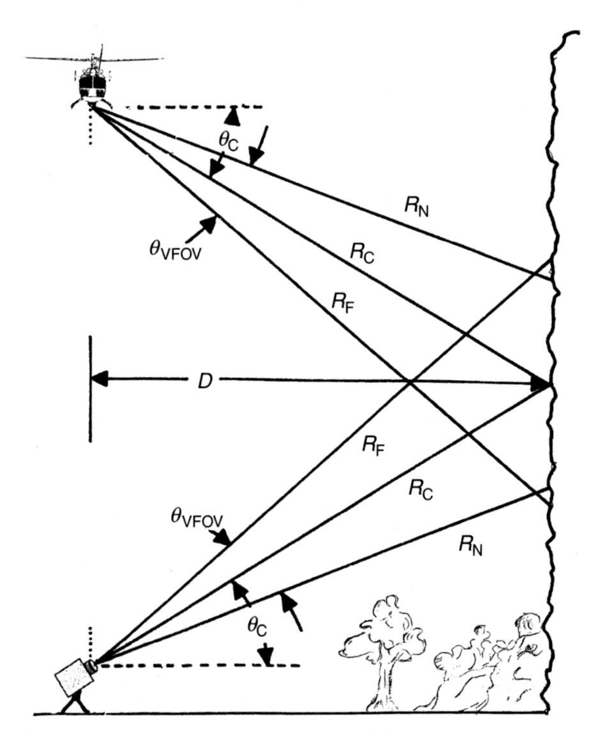

FIGURE 11.6 A schematic showing the geometry for two different viewing platforms surveying the same habitat is depicted. If the distance to the cliff and the oblique viewing angles are the same for these two cases then their vertical footprints are the same relative to the observer. The equations developed in Section "Air-to-Ground" of this chapter for Figure 11.2 would apply to both the air-to-ground and the ground-to-ground for these examples.

Ground-to-Ground

The geometrical possibilities include using imagers as handheld devices while moving along transects by walking, driving a motor vehicle, or perhaps viewing a segment of shoreline from a boat. In these cases the viewing angles are nearly parallel with the surface or ~85°–90° from the zenith. When the ranges required for the fieldwork become large, local atmospheric conditions can dominate the radiative signal detected by the imager due to scattering and absorption (dust, fog, water vapor, aerosols). This is especially true for large viewing angles relative to the zenith. Distance sampling is typically performed on the surface of the planet, either at sea or on the ground. It is important to capture the distance from the observer to the detected animal so that the distance from the transect line can be determined. The GPS position at the observer and at the animal is provided with the new microbolometric uncooled imagers via a built-in range finder. The following example demonstrates the use of a thermal imaging camera with field conditions for pointing angles that are large with respect to the zenith.

Signature Dependence on Viewing Angles

The geometry for the following field observations would use that shown in the sketch in Figure 11.6 for the ground-based platform using a thermal imager either in a handheld mode or tripod mounted to examine elevated habitat at a distance (tree cavities in this instance). For any particular field experiment designed to capture thermal images of occupied tree cavities, we need to consider the processes involved with heat exchange in dens, nests, and cavities for each particular situation and species we are searching for in the field (see Section "Reviews (Nests, Dens, Tree Cavities, Lairs, and Burrows)" of Chapter 10). For example, the federally endangered red-cockaded woodpecker, *Picoides borealis,* drills its nesting cavities in living pine trees, a behavior unique to these woodpeckers. These cavities are used by many generations of woodpeckers and other animals, in some cases for up to 50 years. This unique nest fabrication, which includes cavity protection via sap and resin flow around the entrance from an array of drilled holes, ties the origin of these cavities to *P. borealis.* In a given region of habitat the location of these tree cavities can be cataloged and a survey conducted to test for occupancy when the birds are roosting. If the previously mentioned prerequisites for obtaining a suitable window of opportunity have been accomplished then observing these cavities with a thermal imager would certainly be successful. For example, Boonstra et al. (1995) reported on the utility of using infrared thermal imaging devices to detect and census birds in the field. The thermal imaging was successful in determining activity at four nests of cavity-nesting species. They found that the imagery would show the entrance of active nest sites to be glowing. In particular, the three nesting sites (all in dead aspen) occupied by Bufflehead (*Bucephala albeola*), Northern Flickers (*Colaptes auratus*), and Pileated woodpeckers (*Dryocopus pileatus*) as well as three nest boxes occupied by Barrows Goldeneye (*Bucephala islandia*) all showed distinct thermal signatures associated with the nest entrance, indicating occupancy.

A word of caution is necessary with regard to locating (detecting) nesting sites with thermal imagery. Boonstra et al. (1995) reported that they experienced difficulty with residual hot spots in the background when attempting to find nesting cavities without prior knowledge of their location. These hot spots are due to objects in the background that are reradiating previously absorbed solar radiation. When the magnitude of ΔT associated with nest cavities is similar to those associated with other objects in the background that are not of interest then the imagery is "cluttered" with false signatures, and sorting out the nesting cavities from the collected imagery becomes difficult. In these situations it is necessary to let the background reach thermal equilibrium prior to data collection. As mentioned earlier, some survey situations may require several days of overcast skies to obtain suitable background uniformity. Therefore, it is not recommended to look for tree cavities in the evening unless there has been extensive cloud cover during the daylight hours.

In Virginia, translocation of the red-cockaded woodpecker began in 2000 to supplement the then two remaining clusters in the state. During the breeding season of 2002 Virginia supported only two breeding pairs and two clusters with solitary males. Dramatic habitat management, population monitoring and management, and translocation of birds into the population have been ongoing and are beginning to show promising results (Watts et al., 2006).

In February, 2006, we had an opportunity to conduct a simple experiment with Watts et al. to test the hypothesis of whether or not it was feasible to determine if tree cavities were occupied by *P. borealis* with a handheld thermal imager. During their ongoing work with the population of *P. borealis* at the Piney Grove Reserve in Surry, Virginia, Watts et al. (2006) typically examined the interior of the nest cavities with a peeper camera (see Section "Low Light Level Cameras" of Chapter 3) to determine if the cavity was occupied and to check on the condition of the nest cavity. They marked individual trees that had red-cockaded woodpecker cavities by placing a band of marking tape or paint at the base of the tree. At the time of these observations, not all of the cavities were occupied, so we examined each cavity in the marked trees from the ground with the handheld 3–5 μm thermal imager to determine whether the cavity was occupied. In the days leading up to the experiment Watts et al. determined which cavities were occupied by probing with the peeper camera. We made the thermal observations early in the morning prior to sunrise (thermal equilibrium was established) at Piney Grove Reserve (see Figure 3.2). The equipment package used for these observations and data collection is shown in Figure 11.7 and easily fits into a small backpack. We used a cavity that was known to be occupied to get a baseline infrared image for making a determination of occupancy and then examined the remaining cavities in the cluster of banded trees. We identified every *P. borealis* cavity that was occupied and did not identify any unoccupied cavities as occupied, but we must point out that an imager will only yield information on occupancy and not the occupants. For example, Wilson and Delahay (2001) discuss the problems associated with counting conspicuous

FIGURE 11.7 A photograph of the equipment package used for the observation of red-cockaded woodpecker tree cavities.

visible structures constructed by some carnivores. They caution that the use of den sites to estimate carnivore numbers relies upon accurate identification of the resident species. They point out that in parts of Europe, red foxes and raccoon dogs will opportunistically inhabit disused badger dens, which are generally characterized by the size and shape of the entrance holes and by the presence of large spoil heaps.

The tree cavities drilled by red-cockaded woodpeckers are also subject to being occupied by other species of woodpeckers and flying squirrels (*Glaucomys volans volans*). Probing these structures with thermal imagers for the purpose of estimating populations must be done with care because unless there is a way to tell what animal is occupying the dens and cavities then an imager will only yield information on occupancy and not occupants, as mentioned previously. Figure 11.8 shows a typical 3–5 μm thermal image taken with a tripod-mounted imager after the basic geometry shown in Figure 11.6. To draw conclusions based on the collected imagery of nests in tree cavities, it is necessary to calibrate the camera for comparing the signatures of occupied and unoccupied cavities and to determine the signature decay dynamics after the bird is observed to leave the cavity. For example, the handheld 3–5μm thermal imager was successful in detecting occupied tree cavities, and a bright glowing signature of the tree cavity was observed during the time the cavity was occupied. Once the cavity was vacated there was an immediate drop in the intensity of the cavity signature, which was then observed to continue to diminish on a timescale commensurate with the dynamic processes of heat loss due to conduction, convection, and radiation. The size of the cavity opening was not a factor in affirming occupancy. Cavity openings on the order of 5 cm are typical for those occupied by the red-cockaded woodpecker (*P. borealis*). When the cavity opening drilled by a *P. borealis* is enlarged by other woodpecker species the positive

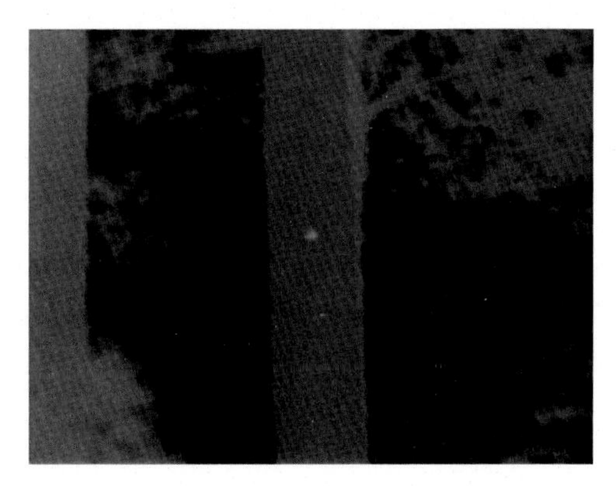

FIGURE 11.8 **A photograph showing the MWIR thermal image of an occupied red-cockaded woodpecker nest cavity located in a mature loblolly pine in the Piney Grove Reserve in Surry, Virginia.**

determination of occupancy is unaffected. The thermal signature of an occupied cavity is markedly distinctive, regardless of the size of the cavity opening.

Unoccupied tree cavities were also observable and exhibited a weak glowing signature commensurate with a long-term (~12 h) heat transfer processes (primarily conduction and convection) driving the inner core of the tree to equilibrium. We examined these cavities within 12 h following a day of uninterrupted direct solar loading. Nonetheless, the signatures of these unoccupied cavities were very weak as compared to those associated with occupied cavities. If the trees were shielded from solar radiation by cloud cover the day preceding the period of cavity observation then the difference in the thermal signatures presented by the occupied and unoccupied cavities would be even greater. In fact, under these conditions, it is unlikely that the unoccupied cavities would even be observable with the thermal imager. It is easy to determine the ratio of the occupied to unoccupied cavities when the survey is conducted in the predawn hours even after a full day of solar loading the previous day.

Most of the pine trees at the Surrey site are loblolly pine (*Pinus taeda*), which are typically 80–100 ft. tall at maturity, with the cavities located in the tree at approximately 50% of the tree height or greater. Sap flow around the cavity is maintained by *P. borealis* to provide protection against predators (snakes). Once the tree dies the birds abandon the tree cavity.

The fact that the tree cavities were generally facing southwest did not influence the results but was helpful in finding the cavity in any specific tree. When the cavity was facing such that it was radiating directly toward the camera the strongest signal was obtained. If the angle of observation deviated from the condition of straight ahead viewing then the intensity of the thermal signature from the cavity decreased proportionally with the observed cross-sectional area presented to the camera. When the angle deviated from the straight ahead view such that no portion of the interior of the cavity was observable then the signal associated with the cavity went to zero (i.e., radiation is directional). If the cavity was located high in the tree (>25 m), as was the case for many of them, then standing off from the tree to accommodate at least a 45° angle of elevation relative to the horizontal was beneficial. In the next example (tree cavity analysis) we show the details of setting up a thermal imager on a tripod to study these cavities that were drilled high in the tree.

Initially, identifying and marking the location of *P. borealis* tree cavities would best be accomplished through visual inspection of suitable habitat by trained field personnel (Hovis, 1997). The characteristic features associated with these cavities (size of opening, tree species, flowing sap, etc.) are positive identifiers that are not amenable to thermal imaging. Monitoring the use of these sites afterward would be accomplished best by utilizing thermal imagery. This would reduce the time field personnel spend monitoring unoccupied cavities and cavities occupied by species other than *P. borealis*. Based upon the openness of the habitat containing the clusters of *P. borealis* at the Surrey site it would be a routine matter for a single field biologist to determine the occupied-to-unoccupied ratio on a daily basis using a handheld thermal imager

Using the 5 cm diameter cavity opening as a standard, a calculation can derive a maximum viewing angle that will still allow us to determine if the cavity is occupied. The angle formed by the line of sight of the camera directed toward the cavity opening and the line perpendicular to the 5 cm plane formed by the cavity opening that allows 80% of the cavity opening to be viewed can be determined as a minimum standard for determining occupancy.

Tree Cavity Analysis

When scanning for tree cavities drilled by *P. borealis* it is suggested that at least 80% of the cross-sectional area of the opening be observable to determine if the cavity is occupied or vacant. Since the cavity is recessed well into the heartwood of the tree and the cavity opening is only ~5 cm in diameter, it is necessary to maintain a suitable angle of incidence relative to the plane formed by the cavity opening. A good rule of thumb would be the angle that allows at least 80% of the cavity opening to be viewed by the thermal imager. This estimate needs to be tested in the field. The radiation emanating from an occupied tree cavity is decidedly directional as a result of the masking effect by tree bark surrounding the cavity opening. Radiation emitted from the interior of the cavity is directed outward through the cavity opening and will only be detected by the thermal imager if the radiation is captured within the FOV of the thermal imaging optics. To ensure that this criterion will be met, we can calculate the angle of incidence required.

The following calculation is based upon the assumption that at least 80% of the cross-sectional area of the tree cavity opening must be captured by the thermal imager to determine occupancy by *P. borealis*. The actual angular dependence will be determined in the field, but the calculation would be the same. When the cross-sectional area of a circle is viewed at an angle the resulting cross-section is an ellipse with the semiminor axis lying in the plane of incidence. When the area of the ellipse is 80% that of the area of the cavity opening, viewed at normal incidence, we would achieve the minimum requirement of 80%.

Here $A = \pi r^2$ and $A' = \pi ab'$ are the areas represented by the cross-sections shown in Figure 11.9, where $b' < a$. To obtain an area A' that is 80% that of area A, $A' = \pi ab'$ and $A = (0.8)\pi a^2$, and therefore b' must be $0.8b$. Tree cavity openings are ~5 cm in diameter or $b = 2.5$ cm. Therefore, $A = 19.62\ \text{cm}^2$ and $A' = 15.69\ \text{cm}^2$. To find the maximum angle acceptable with respect to the normal of the plane of the cavity opening for viewing at least 80% of the cavity opening consider the following in Figure 11.10. Note that $b' = 2.0$ cm, and therefore $\sin\theta = 2/2.5 = 0.8$ or $\theta = 53°$ and $\phi = 37°$.

We want to keep the angle of incidence $\phi < 37°$ with respect to the normal of the cavity opening or $\theta > 53°$. If the tree bark were thin this would allow the camera to see 80% of the radiator. In reality the bark is not thin and the cavity

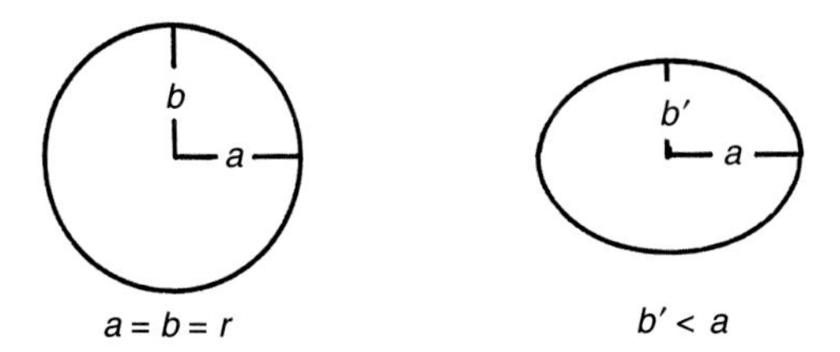

FIGURE 11.9 A sketch showing the cross-sectional area of a circle and the result of viewing the circle at an angle. The resulting cross-sectional area is an ellipse with the semiminor axis lying in the plane of incidence.

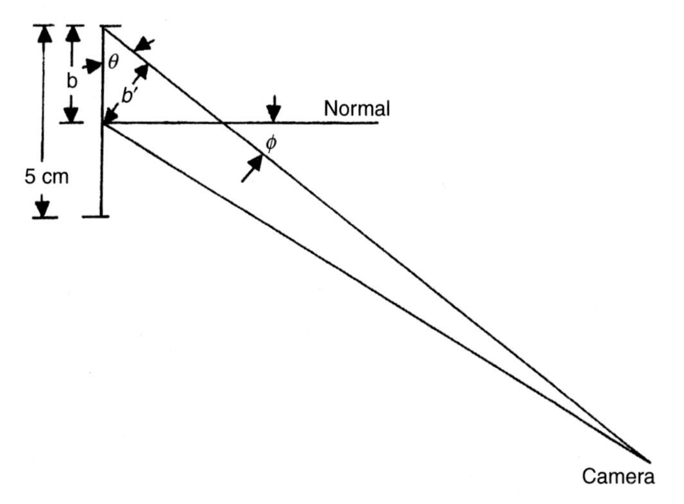

FIGURE 11.10 The viewing geometry used to observe the tree cavities of the red-cockaded woodpecker.

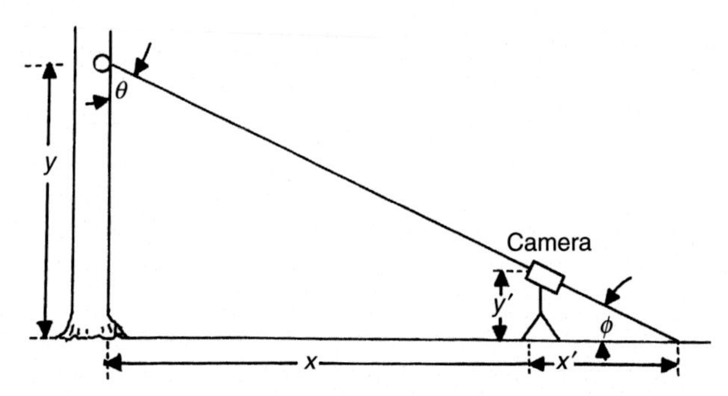

FIGURE 11.11 **Setting the thermal imager to determine the survey dimensions in the field.**

(hence the radiator) is recessed into the core of the tree so the actual cross-section captured by the camera optics is something less. We have determined that 37° is a good estimate by field measurements.

To determine the cavity height use a tripod so that the camera is about 5 ft. above ground level and has the cavity imaged in the center of the FOV. Determine the line of sight from cavity to ground level behind the camera (see Figure 11.11).

In Figure 11.11 $\tan\theta = (x + x')/y = x'/y'$ or $y = y'(x + x')/x'$. As a result the cavity height $y = y'(1 + x/x')$ or $y =$ tripod height $(1 + x/x')$. These parameters y' (tripod height), x (distance from tripod to tree), and x' (distance from tripod to the view line intersection with the ground) are all conveniently measured in the field. Here the angle of incidence $\phi = (90° - \theta)$ and $\theta = \sin^{-1}(x'/y')$. Note that ϕ only depends on θ and for any selected θ the thermal signature is constant with respect to a 360° rotation about the normal of the plane of the cavity opening.

From the field data of February 24, 2006, taken at tree #3 in cluster #3 (Piney Grove, Virginia), we saw that an angle of incidence ϕ near 60° with respect to the normal of the cavity opening was far too large to see a uniform intensity distribution across the cavity opening. A more favorable viewing angle is obtained by looking into the depth of the cavity as much as possible. This was best achieved by standing off from the tree with the imager such that the angle of incidence was approximately 35° or less for the cavity of tree #3 at 26.5 ft. above the ground. When we viewed the cavity at distances closer than 70 ft. we began to see a signature that was not uniformly bright across the diameter of the cavity opening.

The 5 cm diameter tree cavities of the *P. borealis* present a distinct signature associated with the angle of incidence when the angle is large. The radiation emanating from the recessed interior of the cavity is constricted by the cavity opening such that the radiation is predominantly guided in the forward direction. If the cavity is viewed in a near normal direction or at small angles of incidence (commensurate with a viewing distance of ~2.5 times the height of the cavity above ground), we observe a signature that is circular in cross-section and uniformly bright from edge-to-edge across the signature.

If on the other hand we are viewing the cavity from the ground at an angle of incidence that is large ($\phi > 45°$) the lower edge of the cavity masks the radiation emanating from the lower portion of the cavity. This produces a distinct signature that is slightly elliptical in cross-section where the lower half of the minor axis is brighter than the upper half. Additionally, the lower edge of the signature is in sharp contrast while the upper edge of the signature is indistinct and somewhat blurred (see Figure 11.12).

In the morning the birds will leave the tree cavities shortly after first light and gather together in a group before heading out to forage. When the bird moves near the cavity opening, as shown in Figure 11.13, it is possible to observe its movements since the thermal signature of the bird is stronger than that of the cavity interior.

Once the bird exits the cavity the intensity of the cavity signature is immediately reduced (see Figure 11.14). As a result of these observations it is clear that there are three distinct situations that can be identified by comparing the

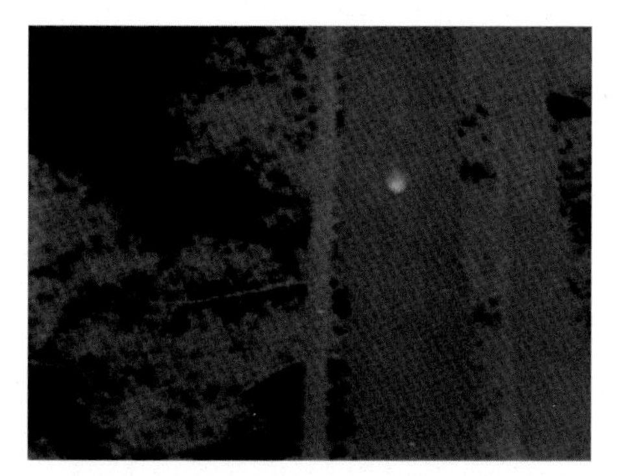

FIGURE 11.12 **A thermal image showing an occupied red-cockaded woodpecker tree cavity that displays the elliptical features in the spatial distribution of the cavity signature.**

FIGURE 11.13 **A red-cockaded woodpecker just as it exits a tree cavity. The signature of the cavity is just above the bird in this image.** It has already lost much of its intensity and is considerably less bright than the signature of the bird.

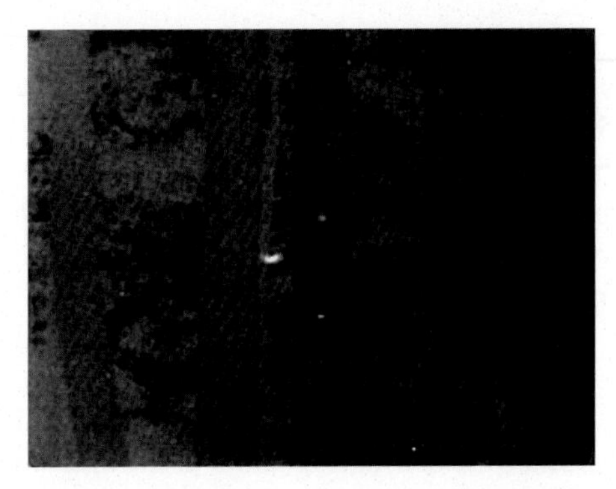

FIGURE 11.14 **The thermal signature of the unoccupied tree cavity is not as bright as the bird flies from the cavity. In this photograph the bird is approximately 10–15 m from the cavity.**

thermal signatures of tree cavities: the cavity is occupied, the cavity has been recently vacated, and the cavity has been unoccupied for an extended period of time (>24 h). A survey of known tree cavities could be conducted in a few hours before dawn to determine the occupancy rate within any given cluster(s).

Thermal Shadows

Thermographers must be aware that shading effects caused by passing clouds can affect the quality of the thermal imagery collected during a survey or census. When imagery is collected in direct sunlight, the animals cast a "shadow" that appears cooler than the rest of the surface around the animals (see Figure 11.15, which illustrates this phenomena). The darker area on the ground to the right of each cow is due to the shading effect of the cow as it blocks the solar radiation arriving at some incident angle from the left in the photograph. It appears as a shadow in the image because the body of the cow is blocking the direct

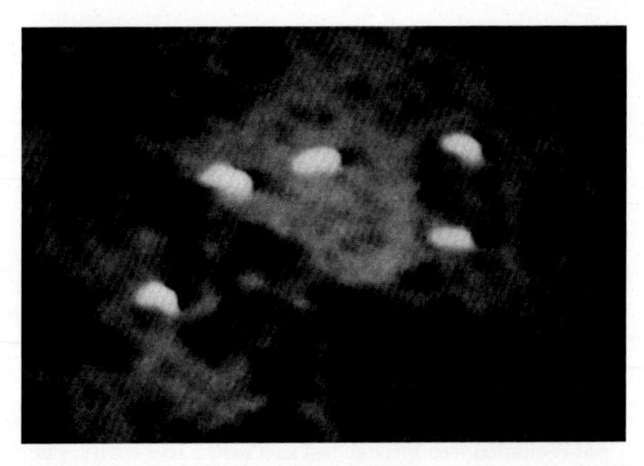

FIGURE 11.15 **A small herd of cows in direct sunlight at approximately 60° from the nadir on a sunny morning.** The sun is to the left in the image.

sunlight next to the animal. This loss of energy in the thermal imagery is due to the fact that solar radiation is not being reflected from the ground in the area of the shadow. It should not be interpreted as a thermal change in the background even though it appears cooler in the image. Here the earth adjacent to each cow is momentarily being shielded by the body of the animal, which only reduces the apparent temperature by preferentially changing the local spatial distribution of the reflected component. See Section "Invertebrates (Arthropoda)" of Chapter 10 for similar observations of reflected solar radiation on the collected imagery of invertebrates. These effects are easily identified by noting the time response associated with the shading effects. The time response for thermal shading (loss of reflectivity) and thermal cooling due to shading are vastly different, that is, the loss of reflectivity (an optical effect) is instantaneous, and losses due to thermal cooling occur on the order of thermal processes. The phenomena of shading can provide the field scientist with a valuable opportunity to use passing cloud cover as a solar shield to reduce unwanted solar radiation reflected from the background so the contrast between the signatures of the animals and their backgrounds can be enhanced.

Atmospheric Transmission

The atmospheric transmission changes with the viewing angle. When the atmospheric transmission is high, only small amounts of water vapor, gases, and aerosols are present in the atmosphere. Under these conditions the self-emission of the atmosphere is low. Viewing the sky would show a uniform, cold background. There is, however, a strong angular dependence on the viewing angle as it is lowered from the zenith. At smaller elevation angles more aerosols and gases at ambient temperature are encountered along the line of sight, resulting in an average temperature close to that of the earth's surface. The atmospheric constituents degrade the thermal resolution by both scattering and absorption in the line of sight and reduce the magnitude of the thermal signature. On the other hand, scattering of radiation into the line of sight and self-emission from the atmosphere in the line of sight both serve to reduce the thermal contrast ΔT between the object and the background. Since nearly all fieldwork involving surveying or censusing will involve relatively long path lengths, the self-emission and scattering by the atmosphere (radiance emanating along the path) could be significant and must be taken into account. The directional dependence and the amount of background radiation (both emitted and reflected) that the imager will encounter is another important factor that must be incorporated into the overall methodology. Other atmospheric effects are taken up in Section "Atmospheric Effects" of this chapter.

Angular Dependence on the Diurnal Cycle

As the sun rises, the east-facing surfaces of objects will be warmed and the west-facing side may still be giving up heat. By flying circles around areas of interest the scene will look much different. The warmed side will contain much more clutter. See the example provided in Section "Diurnal Effects and the Background" of this chapter for a discussion and photographs of thermal images to illustrate the problems with solar loading from incident sunlight at large angles of incidence with respect to the normal to the surface of the earth. The angle and direction of solar loading changes continuously throughout the day, providing many opportunities for surveying or monitoring with minimum interference from background clutter.

EMISSIVITY

The Stefan–Boltzmann law was introduced in Chapter 5. This basic law of nature defines the relationship between the total emission of radiant energy of a blackbody and its absolute temperature. Since this relationship is wavelength independent and since there are no true blackbodies (perfect absorbers) in nature, we saw that the Stefan–Boltzmann law could be written in a very useful form, as was shown in (5.4) as $P_t = \varepsilon\sigma T^4$, where ε is called the emissivity of the surface. The emissivity of a surface is defined as the ratio of the energy absorbed by the surface to the energy absorbed by a blackbody ($\varepsilon = 1$). As a result, all surfaces have an emissivity between 0 and 1. This simply defined quantity is very complex in reality. We could use (5.4) to compute the power per unit area radiated into a hemisphere provided that we know the values of the emissivity and the true value of T.

In most imagery collected during aerial surveys the cross-section of animals will appear to be approximately rectangular in shape from a down-looking angle with respect to the nadir. In other words, to the first approximation, most animals can be considered as spherical (when the sides of the rectangle are not so different) or cylindrical in shape, depending on the viewing angle, the resolution of the imager, and the altitude of the aircraft. As a result of Kirchhoff's law, the relationship $(\varepsilon + \rho) = 1$ for opaque objects, there is a strong angular dependence on the emitted and reflected heat from an object. Flat objects in the FOV of the detector will show the angular dependence of the reflectivity, ρ, that reduces the emissivity from unity to zero as the angle of incidence increases from normal incidence to 90°. See Section "Viewing Angle" of Chapter 6 regarding the discussion of the dependence of the emissivity on the viewing angle. Additionally, the surface of the object may be diffuse (rough with high emissivity) or specular (smoothly polished with low emissivity).

If we were to examine the side view of a cylinder with a thermal imager, it would appear as a rectangular shape with blurred edges along the long axis as a result of the emissivity angular dependence on the viewing angle. The end-on view would be circular in shape, and the edges of the image would appear sharp with no blurring since the angular dependence of the emissivity would not play a role in the sharpness of the image contrast. Neither one of these two images viewed alone without the other would allow us to identify the object producing the images as a cylinder; however, the object would certainly be detected. The object would also be recognized as a nonbiotic object when the methodology outlined in Chapter 9 is utilized. In general, we have only the shape of the thermal signature to use for identification; however, as pointed out in Chapter 9, we can also make accessible other sources of information to assist in mitigating the uncertainty in viewing just the thermal signature of an object. This example suggests that a certain amount of training is needed to successfully identify the sources of thermal images collected in the field. This coupled with training in the proper use of thermal imaging cameras will benefit the field researcher with distinguishing thermal signatures in the collected imagery.

When selecting the angle, θ_C, a few other factors that may influence the imagery must be considered. In general, the emissivity of an object is highest when viewed at normal incidence, and for most surfaces the emissivity approaches zero as viewing angles approach the grazing angle. As a result, the emissivity, and therefore the apparent temperature of the object, will vary across the system's FOV, appearing higher for angles closer to the nadir.

This same angular effect is true on a more localized scale and has been observed (McCafferty et al., 1998) during laboratory thermography measurements to determine the radiative temperature of barn owls (*Tyto alba*). They reported that an examination of recorded thermal images showed that there was an apparent decrease in radiative temperature toward the outer edges of the image. This is commensurate with a reduction in emissivity where the surface of the animal becomes curved relative to the viewing angle of the imager. The thermal imager sees all radiation emanating from an object within the FOV of the imager. This includes the radiation reflected from the object's surface from all other sources radiating toward the object. In a laboratory setting strong contributions from reflected radiation can be minimized, and the changes in the emissivity caused by the curved surfaces of an object become evident. In fieldwork at longer ranges, this loss of information at the edges of the thermal signatures due to the angular dependence of the emissivity is not readily observable.

Reflectivity and Heat Transfer (Through the Reflectivity)

If the situation calls for a survey to be conducted during the day and there is direct solar radiation illuminating the scene then great care must be taken sorting through the imagery with regard to reflected solar radiation (solar glint) and absorbed solar radiation (solar loading). Solar glint is a reflection of solar radiation that depends directly on the viewing angle. Intermittent solar glint can occur from moving surfaces such as waves on water or snow fields with a rough surface. Scanning the imager in a northerly direction can frequently minimize these momentary reflections. For surfaces with high emissivity, there is very little solar reflection to worry about. Since animals have high emissivities, $\varepsilon > 0.9$, they reflect very little solar radiation; however, there is a problem with reflected solar radiation from the background and from specific objects in the background. Hot spots in the background need to be sorted out (are they signatures of animals or are they "solar glint"?). This can be accomplished by slightly changing the viewing angle while monitoring the suspected signature. If the signature changes or disappears then it is reflected solar radiation (i.e., it is not radiating). If it does not, then it may be due to a real source of heat (an animal that is emitting or an object that is reradiating previously absorbed solar radiation).

We have already noted that, in general, transmittance, absorptance, and reflectance are wavelength dependent. Water has relatively high transmittance in the visible and nearly zero in the infrared. The low transmittance in the infrared is due to both absorption and reflection (see, e.g., Figure 8.3). The 3–5 mm radiation emitted from the eyes of the beaver is reflected from the surface of the pond. If there were ripples on the surface of the pond then the reflected radiation would not be stationary and steady but would appear to be flickering as the radiation is reflected into and out of the camera's FOV as the surface changes in the vicinity of the reflections (i.e., becomes a mirror with periodic distortions). In general, the absorptance of materials is wavelength dependent and so is the emissivity. However, for opaque objects, such as animals, the transmittance is zero, the emissivity is high and nearly wavelength independent (gray body, $\varepsilon > 0.9$), and reflectivity is low.

Apparent Temperature Difference

There are many factors that contribute to the observed apparent temperature difference, and these are discussed in great detail elsewhere in this book. In the most basic analysis, the viewing angle can significantly change the apparent temperature difference ΔT_A. This is because the object of interest can be compared with a variety of backgrounds by merely changing the viewing angle. This concept must be utilized when the window of opportunity is developed for any particular field study. For example, when viewing an avian species, a change in the viewing angle could change the background from a perfect cold, clear sky to any number of other possibilities: clouds, tree canopy, water, grass fields, marsh, or any other natural background. These changes in background are far more significant with regard to their influence on the quality of the imagery than are the changes in the true temperature of the species being studied. Providing a near perfect background for the species of interest is the single most important consideration in setting up a field study. We need to control the viewing angle during any survey to provide maximum ΔT_A. An example is provided in (Figure 10.2), which shows a thermal image of small birds perching in the crown of a tree during leaf-off in coastal Virginia. This image was collected on a sunny morning at 10:30 am, and the air temperature was ~5°C. An optimized viewing angle was selected to provide the best possible contrast under the existing conditions at the time by imaging birds against the backdrop of the cold sky. The 3–5μm thermal imaging camera is handheld and the operator is on the ground with a viewing angle $\theta_C \sim 55°$ with respect to the azimuth.

For most surveys we would be either on the surface (ground or body of water) or in the air. In these situations there are many ways to configure the survey geometry or viewing angle. For example, the viewing platform could be arranged to accommodate surface-to-surface, surface-to-air, as discussed earlier, air-to-air, and air-to-surface (typical aerial survey), as will be the configuration for many field surveys. Other opportunities might be in caves, trees, buildings, or perhaps lairs or dens that would present an entirely different array of possible backgrounds. It is essential to select a set of viewing angles that will maintain the maximum ΔT_A for any of these configurations used during a particular experiment or survey.

We saw in Section "Apparent Temperature of the Object" of Chapter 5 that a thermal imager cannot distinguish what portion of the total radiation emanating from an object is reflected and what portion is due to self-emission. For all practical purposes we can minimize the effects of the change in apparent temperature across the FOV by keeping the maximum viewing angle below 45°. That is, we should try to keep (θ_C + ½ FOV) ≤ 45° in Figure 6.7 since for viewing angles greater than 45° the reflectivity begins to increase markedly while a concomitant decease occurs in the emissivity (see Section "Viewing Angle" of Chapter 6). In Figure 11.16 it should be noted that at angles approaching the grazing angle (90° angle of incidence or large viewing angles with respect to the nadir) $\rho \sim 1$ and $\varepsilon \sim 0$, and that animal hides or even rough black surfaces tend to become good reflectors, so care should be taken to avoid these situations.

When these extreme geometries are used for surveys along transects, which are parallel to the flight line, as shown in Figure 11.4 (transects B, C, and D), the animals closer to the flight line will appear to have a higher apparent temperature than animals situated in the transect farther from the flight line. See Section "Apparent Temperature Versus Viewing Angle" of Chapter 6 for a more detailed description of the angular dependence of the emissivity on the apparent temperature.

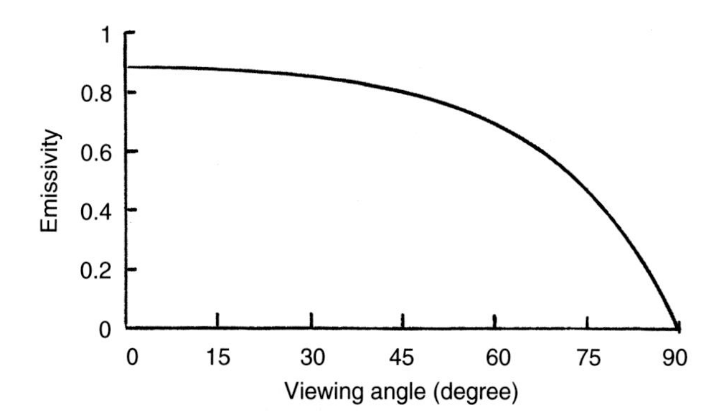

FIGURE 11.16 **A plot of the emissivity as a function of the viewing angle.** As the viewing angle increases with respect to the nadir or as the angle of incidence deviates from normal at the animal being viewed the emissivity decreases, and at 90° or grazing angles it approaches zero.

We still, however, must concern ourselves with the background, which can reflect significant amounts of radiation, both from other objects in the background and from the environment (i.e., sun, clouds, sky, and the earth).

BACKGROUND CLUTTER

What is "background clutter"? We refer to "unwanted" thermal signals in a scene of interest as "clutter" because these images are mixed in with the thermal images of animals (signatures) that we are trying to detect, thus making it very difficult to extract the desired signals from the background. Ideal thermal imagery for counting animals in the field would consist of a uniformly darkened background containing the thermal signatures of the animals and nothing else. These signatures would generally appear as white silhouettes with features such as shape and size (symmetry) commensurate with the particular animals being observed, and in the case of unsaturated signatures there is information divulged by the particulars of the plumage of the pelage of the animal (see Section "Signature Saturation" of Chapter 9). It is possible to achieve this type of imagery for surveys and censusing if care is taken to eliminate all "extraneous" infrared signatures, regardless of their origin.

We have already seen that a thermal imager can only detect infrared radiation, but it cannot distinguish between emitted, scattered, or reflected radiation (see Chapter 5). When a scene is cluttered, the background is no longer thermally uniform but contains a few to many various shaped and sized thermal signals or images, which are created by thermal radiation emanating from objects (biotic or nonbiotic) contained in the scene. The radiation may be from a number of possible sources: reflected or scattered solar radiation, solar radiation that has been absorbed and is now being re-emitted from rocks, bare soil, water bodies, and deadfall. Thermal radiation emanating from manmade devices with heat sources such as motor vehicles, electrical equipment (transformers and generators), and outdoor lighting can all be detrimental to gaining quality thermal signatures. Infrared signals emanating from manmade devices are usually strong and cannot be brought to equilibrium with the background so they are problematic for nighttime observations. Anthropogenic background clutter is more of a problem near cities or industrial complexes.

Background clutter has probably slowed the development of infrared thermal cameras as a survey tool more than any other single factor, yet the problem of "clutter" is one of the easiest to overcome. We saw earlier that a thermal imager cannot measure the absolute temperature of the object or the absolute temperature of the background. What it can measure is the radiation that appears to emanate from the objects and their backgrounds. We also determined that the radiation we measure includes the self-emission of the object and the self-emission of the atmosphere along the line of sight as well as the reflected ambient radiation from these sources. In Section "Background

Temperature" of Chapter 5 we found the emitted power differential (5.8) for an object in the field to be

$$\Delta P \sim 4\varepsilon\sigma T_{\mathrm{B}}^{3}\Delta T.$$

Once the background temperature, T_{B}, reaches thermal equilibrium, only objects that generate heat (animals) will exhibit a temperature difference with respect to the background, $\Delta T = (T_{\mathrm{O}} - T_{\mathrm{B}})$. During a survey, we need to ensure that all other ΔT originating from unwanted objects in the scene are kept near zero, or we want the temperature of unwanted objects to equal the background temperature ($T_{\mathrm{UO}} = T_{\mathrm{B}}$). That is, we want a uniform background. Note that this task can be reduced to that of keeping inanimate objects in the scene at a temperature equal to that of the background temperature. This merely means that we wait until the inanimate objects that were preheated by solar radiation (rocks, trees, deadfall, bare soil, etc.) and the background are in thermal equilibrium with one another.

To physically achieve the situation where the background is in thermal equilibrium is next to impossible; however, it is relatively easy to arrive at a situation where the background appears to be a uniform temperature. It is only necessary to arrive at the state when the power differential of the background and inanimate objects in the scene can be reduced far below the power differential presented by the animals and the background. That is, we can approximate the situation of a thermally uniform background, and since the cameras are equipped with gain and light level controls we are still able to see features of the background, which is important. Note that achieving the ideal conditions for detectabilities near 100% requires these conditions, and many times all it takes is allowing sufficient time for thermal stabilization to occur. Waiting for the window of opportunity to open will be dictated by the thermal situation and not by other factors such as convenience. Again, it cannot be overemphasized that familiarity with the operation of the thermal camera is a fundamental task that must be mastered before quality imagery can be obtained.

The background that we typically use when conducting a survey is, to a great extent, the animal's habitat, and this can vary significantly with species and time of year. Urban, rural, or wilderness areas can all be considered as background, depending on our search. Typical aerial surveys will have a wide range of backgrounds such as bare earth, sky, open range, marshy or wetlands, grass land, roads and other manmade structures, desert, plains, and forested mountains. In fact, the background is just as diverse as the habitat. A single tree branch could easily serve as the background when taking imagery of cavity-dwelling birds or insects. Even when the background has come to thermal equilibrium, the gain (determines the voltage output range of the detector for a given temperature input range) and light level (determines the minimum temperature to be detected) of the imager can be adjusted such that the details of the background (habitat) can still be viewed and recorded for later evaluation (see Section "Verifying Performance" of Chapter 8).

The following photos give an idea of what one can expect to see if the background temperature has reached a state of uniformity that is within the range of the gain and light level controls of the imaging camera. When the background reaches this state the apparent temperature difference ΔT_A between a biotic object and the background is sufficient to obtain uncluttered images. This does not mean that it is safe to assume that this situation can be obtained and managed as a general rule; in fact, it is generally not true. The window of opportunity is best achieved by allowing sufficient time and favorable micrometeorological conditions to prevail.

The difference between Figure 11.17A and B is produced by adjusting the gain and light level to create better quality imagery for the two deer depicted in the photograph. In this case the intensity or brightness of the deer signatures in Figure 11.17A is nearly the same as the brightness of the background near the center of the image. By increasing the gain and reducing the light level of the imager the detectability of the deer was significantly enhanced by effectively reducing the impact of the background clutter on the imagery. If just the light level or just the gain were adjusted then the resulting deer signatures, as seen in Figure 11.17B, would not be as distinct.

The photographs shown in Figure 11.18A and B illustrate the improvement in the detectability by reducing the light level when the brightness of the animal's signature, as compared to that of the background, is high enough so that the brightness of the entire image can be reduced, as shown in Figure 11.18B.

Thermal imagery collected during the daylight hours is more apt to have problems arising from cluttered backgrounds due to solar loading, but clutter-free imagery can still be obtained when proper care is taken. See the images in Figure 9.10A and B or Figure. 10.1A and B for more examples of how to

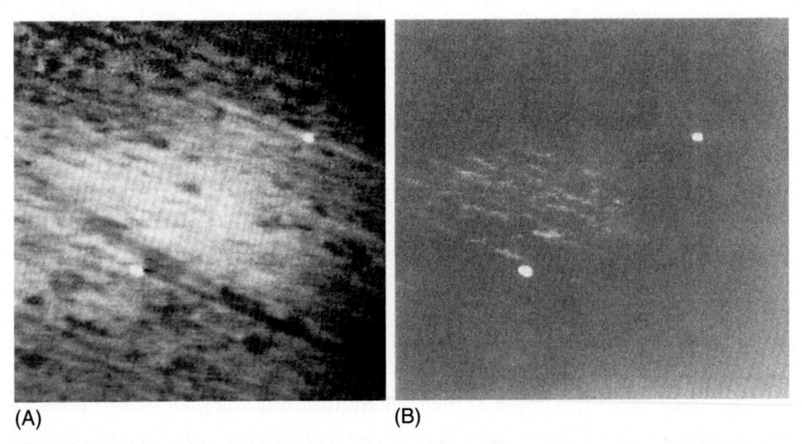

FIGURE 11.17 (A) Thermal imagery of deer taken at 200 m agl from a helicopter flying at ~93 km/hr. This image is cluttered with reflected solar radiation from the background. (B) Thermal image of the same deer as in Figure 11.17A except the light level has been reduced and the gain adjusted so that the clutter in the image is removed.

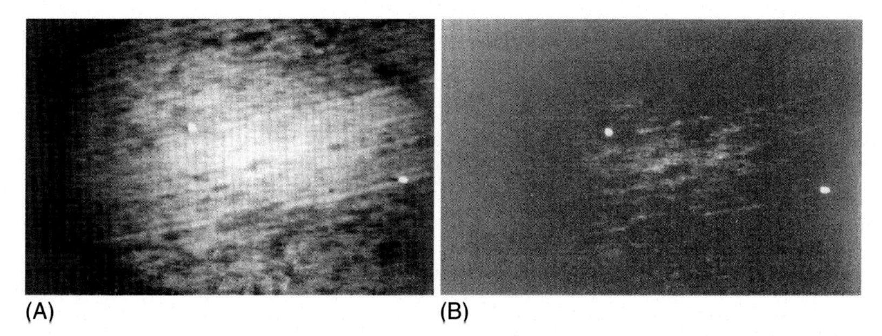

FIGURE 11.18 (A) Thermal image of deer taken at ~225 m agl from a helicopter flying at ~93 km/hr. This image is cluttered with reflected solar radiation from the background. (B) Thermal image of the same deer as in Figure 11.18A except the light level was reduced to reduce the background clutter.

mitigate background clutter in daytime thermal imagery. There are three things that can contribute to background clutter: (1) unwanted reflected radiation, (2) previously warmed objects (absorption and re-emission) not of interest, and (3) scattered radiation resulting in snow-like effects in imagery. Convection, conduction, and radiation all serve to moderate the temperature difference between objects and their backgrounds.

Reflected Solar Radiation

For most daytime field studies the chief component of the power reflected from the background will come from reflected solar radiation; however, other heat sources include reradiated solar energy from objects in the scene that are near the animal. Thermal loading and heat transfer processes occurring in the background are complex (see Figure 5.7), but the dominant source of radiative power reaching the imaging camera on clear days is reflected sunlight. These reflections occur from everything in the background but not uniformly. The angular arrangement of objects such as leaves, branches, grasses, soil types, rocks and their surfaces, and water all reflect solar radiation in the wavelength regions of the infrared detectors. Since this component can be quite large (generally much larger than the emitted component of the background), special care must be taken to minimize its effect. Reflected solar radiation is much more of a problem for 3–5 μm imagers than it is for 8–14 μm imagers because of the spectral shift in the blackbody curves to shorter wavelengths at higher source temperatures (see Figure 5.1). As seen in this figure, the peak of the blackbody curve shifts to shorter wavelengths as the temperature of the blackbody increases such that there is 16 times more solar radiation at the surface of the earth in the MWIR than there is in the LWIR. There are a number of ways to mitigate this major component of thermal nonuniformities in the background. By far, the best approach is to collect thermal imagery in the dark. If because of the particular animal ecology being studied the imagery must be acquired during daylight hours then collect imagery when the sun is masked by the prevailing atmospheric conditions, such as overcast skies, passing cloud cover, or in shady areas such as under vegetative canopies or those shielded by tall features in the terrain. Select the best conditions possible to provide the maximum contrast between the species of interest and the background.

Re-emitted Solar Radiation

If the effects of solar reflections within the background are mitigated, yet there are still thermal anomalies present in the background that are confounding the detection of thermal signatures associated with the species of interest, then they are probably coming from objects with high heat capacities that are present in the background or they could also be thermal signatures of a different species that shares the immediate survey area. In any event, the best remedy for this scenario is to collect imagery in the early morning hours to separate inanimate objects in the scene from biotic objects. If the prevailing processes of conduction, radiation, convection, and phase changes are sufficient to bring about a more thermally uniform background then the early morning imagery will allow the detection of biotic objects. The next step would be to examine the features such as shape and size (symmetry) commensurate with the particular animals being observed to determine if more than one species is present.

The time to reach a suitable state of thermal equilibrium in the background can be shortened significantly by changes in the atmospheric conditions. Wind, rain, and clouds for example all can assist the radiation and conduction processes in achieving equilibrium. Since biotic objects will be radiating energy as a result of metabolic processes they will be the only signatures of interest in our imagery once the background temperature becomes uniform. The task of getting an entire landscape to a state of thermal equilibrium requires a favorable combination of atmospheric conditions for a relatively long period of time (typically on the order of 24 h) to reach stable conditions suitable for collecting thermal imagery. These conditions might be a favorable combination of a steady rainfall for several hours followed by cloudy and breezy wind conditions until nightfall. The following predawn hours will be optimal for collecting thermal imagery on this landscape. For more arid regions where rainfall is rare overnight conditions with clear skies followed by a day of cloud cover would be optimal.

It should not be difficult to detect animals once a proper background has been established. If there is difficulty in picking biotic signatures out of the imagery then the problem is with the background. This is not a bad thing since it can usually be managed with a little patience and timing. It is very difficult to change the radiative properties of the animals, but it is relatively easy to adjust to the radiative properties of the background. If the surveys or observations do not have to be made during the cold months of the year then the thermoregulatory properties of many animals can be used to advantage during warmer months.

Scattered Radiation

It is difficult to collect thermal imagery in a highly scattering medium just as it is to collect visual imagery (but much less so). All imagery suffers from large-size scattering particles commensurate with snow, sleet, dust storms, or heavy rain. Visual imagery is also poor for small particles or aerosol scattering as well, which is quite common in many areas, both dry and moist. One of the greatest advantages of thermal imaging stems from the fact that scattering is a strongly wavelength dependent phenomenon, so that visible light is scattered for very small to very large particulates, but thermal wavelengths are only scattered by large particulates (see Section "Wavelength Selection" of Chapter 8).

Complete Snow Cover and Thermal Contrast

It is not necessary for the background to be snow covered to obtain conditions for achieving optimal apparent temperature differences and, in fact, the opposite is true. Parker (1972) reported that large ΔT_As were not encountered and would not be expected in a cold winter environment according to radiometric data and that the maximum apparent temperature differences between deer and the background occur in summer, despite warmer background temperatures. Nonetheless, many researchers still use complete snow cover to provide a thermal background for thermal imaging fieldwork. While it is true that complete snow cover creates the ideal background for visual surveys, it is far from the ideal background when using infrared thermal imagers for surveys. The thermoregulatory processes that come into play for animals in cold climates prevent the large anticipated ΔTs when the temperatures are below freezing even when the background is snow. In cases where surveys must be made on winter ranges for example it may not be possible to avoid snow cover or incomplete snow cover during fieldwork. Incomplete snow cover creates a situation where all the inanimate objects in the background that are subjected to solar radiation will present larger ΔTs than those of animals that are using natural thermoregulatory process in the cold environment to conserve heat. As a result, incomplete snow cover (particularly during daytime surveys) can be considered to be one of the worst thermal backgrounds possible and can actually create an extremely confounding background for thermal surveys since objects or areas that are not snow covered will appear as beacons (due to the reradiation of previously absorbed solar radiation) in a thermal image, which makes them very difficult, if not impossible, to distinguish from the signatures resulting from an animal (Wyatt et al., 1980). See Chapter 5 and the discussion of (5.8) along with Figure 5.5 for an explanation. Note further that the thermal contrast is related to the atmospheric conditions through the signal-to-noise ratio of the system and the NEDT of the thermal imager (see (7.3)). Additionally, Table 7.1 shows the strong dependence of the range on the atmospheric conditions for uniform contrast.

Wyatt et al. (1980) determined that relying on the thermal contrast (DT) between an object and its background, by itself, was inadequate for conducting effective aerial surveys of large animals. They reached this conclusion after subjecting the data of Marble (1967) and Parker (1972) taken with a thermal imager during daylight on snow-covered terrain to a Bayesian decision model. They suggested that deer could be successfully detected against a snow background but when the occurrence of snow-free objects was more frequent than that of deer, thermal imagery was inadequate. This should have been expected since the images were collected during daylight on partially snow-covered terrain. Under these conditions, detection is difficult, or may be impossible because any exposed objects (e.g., rocks, brush, or deadfall) will scatter, reflect, or reradiate previously absorbed solar energy. These unwanted sources of thermal energy in a scene are collectively called *background clutter*. When there is significant background clutter, the differential in apparent temperature between biotic objects and other objects in the background are minimized or even reversed.

Wyatt et al. (1985) determined from emissivity measurements that most biotic objects have emission spectra that do not have appreciable wavelength dependent features, whereas reflectance measurements indicated the presence of unique spectral signatures for mule deer (*Odocoileus hemionus*) and some commonly occurring backgrounds like snow and evergreen. The results of these studies formed the basis for the design of a multispectral classification approach for the remote detection of deer (Trivedi et al., 1982; Trivedi et al., 1984). It is unfortunate that their underlying premise for rejecting thermal imaging as a versatile and powerful tool for wildlife surveys and field studies was established in error. As pointed out earlier, a thermal imager will detect all objects whether they are biotic or not if they have a ΔT that is within the sensitivity range of the imager. That the imagers perform in this way is not a reason to dismiss the technology as inadequate for censusing deer populations. Nonbiotic objects can only radiate energy that they have absorbed from the sun.

Clutter from Species Mixing and Bedding Sites

During a survey of a particular species the signatures of other species occupying the same habitat could turn out to be an annoying source of clutter. They are not something that can be removed from the imagery, but they must be taken into account if specific population dynamics and/or data on only one of the species are required. In these situations the survey methods must be adapted to acquire imagery that will allow the two species in the imagery to be separated (identified), which may require shorter survey ranges, higher spatial resolution, narrower FOVs, or a combination of these parameters to accomplish the task. The price paid for these adaptations and subsequent results would most likely come in the form of time and effort. See the work of Wiggers and Beckerman (1993) for an example of ways to deal with thermal imaging surveys where mixed species, sex, and age were determined.

The possibility of mistaking the thermal signatures of bedding sites for actual animal signatures has been noted by Marble (1967) and McCullough et al. (1969). These unoccupied and recently vacated sites show up on thermal imagery because they are reradiating thermal energy that was transferred to the ground via conduction where the animal was bedded. The signatures associated with these sites are usually weak and indistinct (blob-like) when compared to the signature of a standing animal. Bedding sites could be more problematic in colder weather since for a short period of time they may appear as warmer than the apparent temperature associated with an animal conserving heat. The time constant associated with the signature presented by the bedding site is short and will quickly fade as the ΔT becomes smaller due to cooling of the site. Bedding sites were also noted by Wiggers and Beckerman (1993) but did not pose a problem with detecting animals (100%), identifying species correctly (93%), or identifying sex correctly (87%). The bedding sites were indistinct blobs and considerably weaker than the thermal signatures associated with the animals. Both McCullough et al. (1969, p. 145) and Marble (1967) point out that the temperature drops fast (<1 min) once a bedding site is vacated. Note that the apparent temperature of the bed was 1–2° warmer than the animal but dropped below the apparent temperature of the animal because the emissivity of the background and deer are different. The emissivity of the vacated bed and the background are the same, so the loss in the received signal is due to cooling through radiation, convection, and primarily conduction. Vacated bedding sites are easily detected from a rotary-winged aircraft and are recognized as vacated sites because the thermal loss from these areas causes the thermal signature associated with them to rapidly fade. If the flight speed is slow and the gain and light level of the camera are optimized then the animals in the scene will be many times brighter than quickly cooling vacated bedding sites and did not pose a problem during deer surveys in the Florida Panther National Wildlife Refuge and Big Cypress National Preserve (Havens and Sharp, 1998).

McCullough et al. (1969) point out that if one is preparing a survey for deer as the target species for example and there is a chance or likelihood of detecting medium-sized animals as well such as raccoons, coyotes, foxes, badgers, etc., then it might be wise to do the surveys during the daylight hours when these smaller animals tend to den. This is a very good suggestion regarding the possibility of removing signatures of other animals from the imagery, which to our knowledge has not yet been advanced by other workers. If care were taken to minimize background clutter then signature identification might be improved during these daytime counts by removing unwanted species.

Longevity of Signatures

The observable lifetime of signatures created by the species of interest by contact of the animal with some part of its background through heat transfer, which usually occurs by conduction, can provide information to the thermographer because there is an easily observed temporal decay of infrared signatures. The intensity of the image from such things as fresh scat, footprints, bedding sites, nests, etc. degrades over time. The signatures associated with some bedding sites and nest cavities lose a significant amount of intensity in seconds (see Section "Signature Dependence on Viewing Angles" of this chapter) after they are vacated, and for other sites, as discussed previously, the thermal decay of the signature may take less than a minute. The bright signature in the following photograph (Figure 11.19) beside that of the heron is a recently deposited pile of metabolic waste or excrement. The mixture of solid and liquid waste is at the same temperature as the bird when deposited and thermally decays on the order of minutes due to the localized processes of conduction, radiation, and convection (Figure 11.19).

Heat transferred by an animal to its surroundings can be significant under certain conditions (see, e.g., Figure 4.2B). Even though the signature associated with the transferred heat in this image is very bright it fades quickly when the processes of conduction, radiation, and convection begin to dissipate the heat to the surroundings. The signature of the heat transferred by an animal to its surroundings is not always blob-like in appearance and can contain information about the animal or its recent presence in the scene. See Figure 11.20A and B

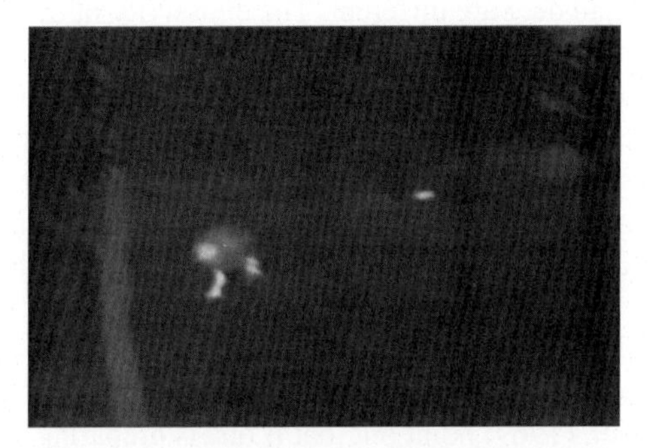

FIGURE 11.19 A thermal image of a heron and a recently deposited pile of metabolic waste or excrement in the background to the right of the bird.

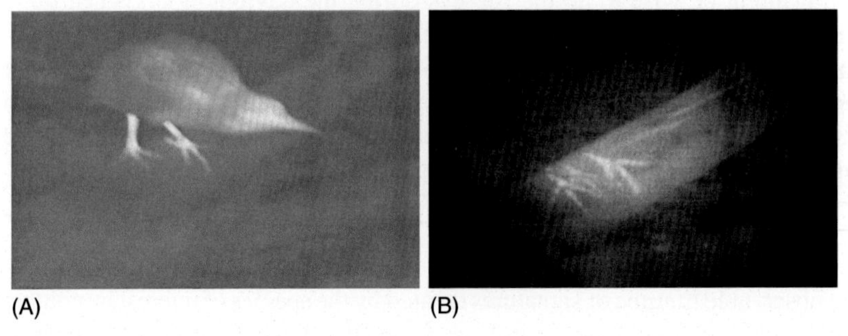

FIGURE 11.20 (A) A thermal image showing the signature of a heron standing on a damp log. (B) An image of the thermal footprints left by the heron on the damp log.

to see the result of a heron leaving its footprints on a log. The thermal signature associated with the footprint was not long-lived and faded away in less than a minute.

Some heat transfer processes are not clutter but may be the signature of choice, depending on the study goals. The transfer processes (conduction and convection) take place on timescales, which are very slow with respect to the radiative component detected by the thermal imager, so the decay of the signature can easily be recorded as a function of time.

Another example of this process is shown in Figure 11.21A and B. Here, an MWIR thermal imaging camera is being used in the reverse polarity mode where darker portions of the image are warmer than the lighter portions of the image. In Figure 11.21A, a hand is being held against a wall for a few seconds, and in Figure 11.21B the resulting thermal image of the heat transferred to the wall during the process can be visualized. The thermal signature of the handprint remained discernable for approximately a minute.

Small-Scale Backgrounds

Even very small survey plots, such as tree trunks with a nesting cavity, small ponds, caves, buildings, small wood lots, and other areas that might be amenable to small-scale investigations, can exhibit problems due to background clutter. Many outdoor studies and observations of animals can be conducted while using a viewing area that can be quite small—even small enough to be shaded by an umbrella to provide a uniform thermal background for the work. On a slightly larger scale, a section of woodland with a nearly full canopy may be suitable for any number of monitoring or surveying applications such as nests or den sites. The background of small study areas can be cleared of thermal clutter by using available shading and/or cloud cover, but it may require a little patience and planning to get there. It is instructive to engage in setting up and interrogating various small-scale landscapes to test out the imager and the methods needed to get good images for a wide variety of situations. This is good

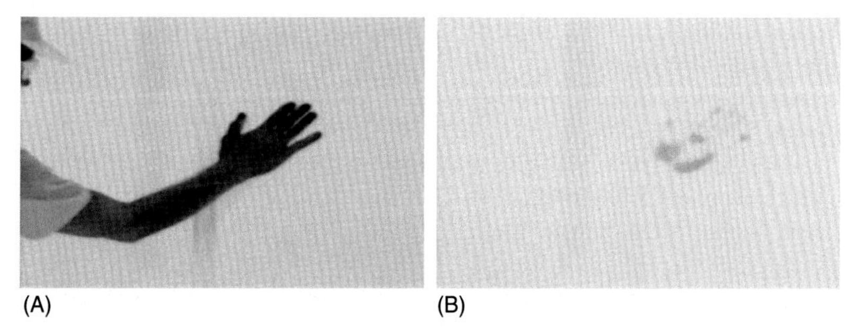

FIGURE 11.21 (A) A thermal image of a man with his hand pressed against a wall for a few seconds. (B) The image of the thermal handprint left on the wall.

hands-on training that will allow the thermographer to go through a complete process of constructing a controllable and changeable set of conditions to see how they affect the actual imagery.

DIURNAL CYCLE

The changes of environmental conditions produced by the cyclic rotation of the earth are primarily due to the periodic episodes of solar radiation impinging on the earth's surface. The light and heat, which accompany the cyclic bouts of solar loading, trigger the start or end periods of animal activity, depending on whether they are nocturnal or are active during the daylight hours. Additionally, the thermographer is faced with widely varying thermal conditions across the landscape that can make collecting thermal imagery difficult during the day, and to collect imagery of nocturnal animals he or she must wait until they become active within their habitats. These problems and how to address them are discussed in the next few sections.

Diurnal Effects and the Background

Image quality is always important in a thermal image, but the actual temperatures of the objects in the image are not so relevant. Particular attention should be given to establishing an apparent temperature difference between the background and the animal that allows for camera adjustments to assist in the detection, recognition, and identification processes. Survey designs and the subsequent image interpretation must take into consideration the effects of diurnal temperature variations. The proper use of these variations is very important when determining the proper timing for collecting thermal images. One of the greatest advantages of a thermal imaging camera is that it can be used to collect imagery at any time during the diurnal cycle.

There are two aspects of the diurnal cycle that we must consider when conducting studies in the field. First, the daily activity patterns of the species must be considered since habitat use throughout a 24-h period is in general not uniform. Second, the changes in the environment produced during a diurnal cycle will greatly influence the methodology selected for obtaining optimal imagery for any given survey. Field studies, and particularly population surveys, need to be conducted when animals are using preferred habitat and when the imager can be used to its fullest capacity. We are always looking for the optimum ΔT_A for the best imagery; that is, we want the apparent background temperature to be as uniform as possible to compare with the apparent temperature of the animal being photographed. Note that imagery collected during the daytime will have an exacting set of criterion that must be satisfied for optimizing the window of opportunity, just as imagery collected nocturnally (maybe more so). It doesn't matter when imagery is collected as long as the background clutter has been eliminated.

If we examine the general diurnal radiant temperature variation of bare soil, rocks, woody debris, and vegetation versus water we can get an idea of how important it is to manage these continually changing parameters to optimize the apparent temperature difference between the animal and its background. The diurnal variations in the radiant temperature of water are relatively small when compared to those of rock and bare soil, and water reaches its maximum temperature several hours after rock and soil. Soil and rock are generally much warmer during the day than water and are cooler during the night. Naturally occurring objects such as trees, grass, rocks, and soils are heated passively through the absorption of solar energy. Daily heating starts at sunrise. After mid-day the solar loading begins to taper, and these natural objects in the background begin to cool. After sunset the background temperature approaches the air temperature. The extreme rates of temperature variation of materials on the earth's surface are determined by the thermal conductivity of the materials, heat capacity, and inertia. Thermal conductivity is a measure of the rate at which heat passes through materials (see Chapter 4). Thermal capacity is a measure of how well a material stores heat. Water has a high thermal capacity since it cools slowly after a session of thermal loading. Thermal inertia is a measure of the response of a material to temperature changes and is higher in materials with a high conductivity, high heat capacity, and high density such as soils, rocks, and deadfall, which retain their heat well into the dark hours after sunset, cooling very slowly. These items become competitors with the thermal signatures of animals and are referred to as background clutter when their radiant temperature is significantly warmer than the background. There are a number of good references that discuss the details of thermal loading. Lillesand et al. (2008); Barrett and Curtis (1992); Campbell and Wynne (2011); Gates (1980); and Monteith and Unsworth (2008) all have detailed treatments of the effects of the diurnal cycle on the thermal behavior of the surface of the planet.

Field studies, and particularly population surveys, need to be conducted when animals are using their preferred habitat. For any given population, surveying animals when they are using their preferred habitat maximizes the number of animals available for detection. This is always desirable, and depending on the species this may mean that the fieldwork must then be conducted exclusively during the diurnal or nocturnal hours or during a combination of the two. In the past, the accuracy of many field studies was compromised by collecting data (counting animals) when it was convenient to do so because a species could be easily observed (diurnal) and not when they were actively using their preferred habitat (nocturnal). Beyer and Haufler (1994) examined the suitability of sampling designs relative to the daily activity patterns of elk (*Cervus elaphus*) for 56 separate studies and illustrated how sampling only a portion of an animal's daily activity can lead to biased data and erroneous management recommendations. The time of day best suited for collecting the thermal imagery in any sampling design must be weighed carefully against the part of the diurnal cycle when the species of interest are using the habitat being monitored or surveyed. Once this is determined it is then a simple matter to select the micrometeorological conditions to obtain the best imagery.

In Chapters 5 and 8 we saw that the operation and performance of a 3–5 μm thermal imager used in the field can be significantly influenced by extremes in the ambient temperature and by changes in local climatic conditions. For any given field study there is an optimal time for collecting imagery (a window of opportunity) that is governed by the prevailing local environmental and atmospheric conditions. As a result, this window of opportunity is strongly coupled to the diurnal cycle. In this section we will outline procedures for obtaining good imagery by selecting a window that will provide us with high-contrast thermal signatures in an environment that will allow us to use the imager to its full capability.

Radiation from the sun heats the earth on a cyclic basis. The amount of solar radiation reaching the surface of the earth (solar loading) depends upon the time of the year (season), atmospheric effects, and airborne constituents that can limit or moderate the intensity and duration of the exposure. Maximum exposure occurs during the summer solstice (June 21) and minimum exposure occurs during the winter solstice (December 21) in northern latitudes. Cloud cover can locally shield the surface of the earth and reduce the exposure. Concentrations of gases and aerosols in the atmosphere can absorb and scatter solar radiation that locally modifies the intensity and spatial distribution of heat at the earth's surface. The amount of heat actually transferred to the earth and the subsequent temperature rise of an object exposed to solar radiation is a complex process and depends upon absorption, convection (which depends upon wind speed), conduction to other objects in contact with it, and reradiation.

Daily heating begins at sunrise, and as the sun rises it preferentially heats up the east-facing parts of any objects in the scene. For example, the trunks of trees will be warming on their east-facing side while still cooling on the west-facing side. During the late afternoon the sun is warming the west-facing side of the trees and the east-facing side begins to cool. The directionality associated with the daily solar loading and its effect on cluttering the imagery with unwanted signatures can be avoided in some situations. If it is necessary to survey during these periods of oblique solar loading, it is possible to avoid the large number of unwanted thermal signature presented by the heated vegetation of the understory, trunks, and canopies of trees by flying circles around areas of interest. The scene will look much different as the direction of the camera changes from east to west, as we see in these photographs showing a Florida panther (*Puma concolor coryi*) bedded down in a stand of saw palmetto (*Serenoa repens*) in the Big Cypress National Preserve in southwest Florida.

The thermal images shown in Figure 11.22A and B were taken within minutes of one another, and the background clutter is considerably less confounding in the image showing the view of the camera while looking west to east. In Figure 11.22A the camera is viewing the scene from east toward the west, which is exposed to the morning solar radiation. The tree and shrub canopies are already reflecting and reradiating solar energy, which appears as clutter in the image.

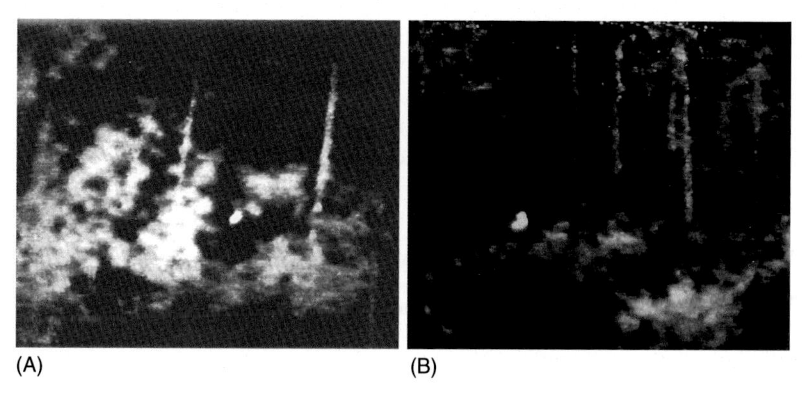

FIGURE 11.22 (A) A thermal image showing a westerly looking view of a female Florida panther bedded down in saw palmetto about an hour after sunrise. (B) A thermal image showing the easterly looking view of the same female panther.

After the mid-day solar loading begins to decline and natural objects begin to cool because they are reradiating previously absorbed solar energy, and after sunset, the background temperature begins to approach the air temperature. Objects with low thermal inertia (low heat capacity) will cool quickly, so leaves, grasses, and loose bare soils respond quickly to solar warming and cool rapidly in its absence.

A landscape that has considerable periods of solar exposure during the day and is littered with dark (near blackbody) objects such as deadfall, charred logs, rocks, pools of standing water, and patches of bare earth will be difficult to use as a background for an aerial survey. The thermal signatures of these objects as the result of reradiation of absorbed solar energy can be the size of large mammals and can easily be confused with the thermal signatures of survey animals. In fact, when there is significant thermal loading of the background (background clutter) animals as large as moose (*Alces alces*) cannot be counted accurately, Garner et al. (1995, p. 236). While these landscapes present a difficult task it is not an impossible one. If all is prepared for a survey then the observers only need to wait for the right time to conduct the count. If after an overnight cool down the thermal signatures of nonbiotic objects are still comparable to those of the survey animals then a suitable strategy would be to wait for an all-day rain followed by a night of drying prior to surveying in the predawn hours. This will bring into play the multiple effects of conductive, convective, and evaporative cooling of the background coupled with a longer time to reach equilibrium. Now the only signatures to be observed would be those that generate and radiate their own heat.

Solar Loading and Thermal Shading

The thermal shading provided by passing clouds is sufficient to produce rapid cooling in objects having low thermal inertia such as grasses, vegetation, and tree canopies. These objects also cool quickly after sunset. On the other hand, the apparent temperature of objects or mediums with high thermal inertia (i.e., water, rocks, dense bare soils, the trunks of trees, and deadfall) are relatively unaffected by the thermal shielding provided by passing clouds, and they cool slowly after sunset. Under normal environmental conditions (no precipitation) the temperatures of these objects with high thermal inertia typically approach that of the background after 8–10 h of radiative cooling or in the predawn hours of the diurnal cycle. The diurnal changes affecting surface temperatures and the retention of heat by objects with large thermal mass can interfere with images collected well after dark. The rate of heating and cooling experienced by an object during the diurnal cycle depends on its emissivity, absorption coefficient, and heat capacity (density). In Figure 11.23, the peak of each daily thermal oscillation shown in the curve occurs near sunset. At this point the temperature of the water begins to drop as the river reradiates heat to the cold sky until approximately mid-morning the next day, when it reaches a local minimum. Sometime before noon thermal loading begins again, and as heat is stored in the water the temperature rises again. The amount of solar loading depends on the location and season of the year. As the curve in Figure 11.23 indicates, the solar loading is sufficiently long to raise the true temperature of the water during the month of April.

The nearly perfect oscillatory nature of this curve is the result of stable weather conditions during the April 15–22, 2004, timeframe when this data was recorded. Note that an overcast day or periods of precipitation can significantly alter the shape of the curve. For example, the effect of daily solar loading on the river immediately following this time period is shown in Figure 11.24 where the true temperature and the daily temperature fluctuations are both modified by 36 h of cloud cover and precipitation.

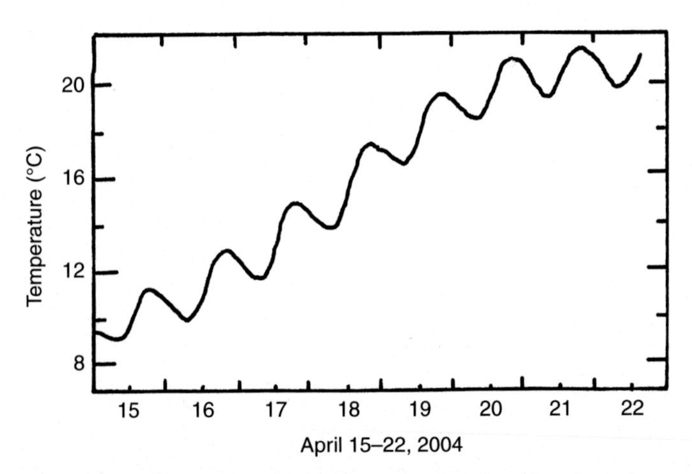

FIGURE 11.23 **The daily cyclic heating associated with the diurnal cycle on the water temperature of the Rappahannock River near Fredericksburg, Virginia.** The true temperature of the water from 15–22 April, 2004, increased approximately 11.5°C due to solar heating and the temperature cycled between 2 and 3 degrees daily during the diurnal cycle.

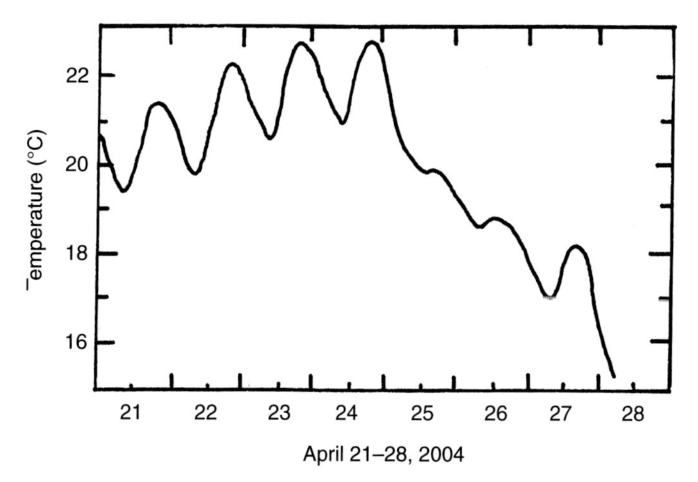

FIGURE 11.24 **The effect of solar heating associated with the diurnal cycle on the water temperature of the Rappahannock River near Fredericksburg, Virginia.** The temperature drop and the washout of the effects of solar loading produced by heavy cloud cover and rainfall are evident in the data for a 36-h period during April 25–26, 2004.

There are also significant variations in the thermal inertia of different soil types, and as a result there are significant differences in their responses to the diurnal cycle. The thermal inertia of unsaturated (dry) soils is influenced greatly by soil porosity. That is, any soil that is loose or sandy (high porosity) in nature will have low thermal inertia and will exhibit a wide temperature fluctuation during the diurnal cycle. Other soils, for example, wet clay soils that are dense or compacted, have high thermal inertias, and the observed diurnal temperature ranges for these materials are relatively low (Barrett and Curtis, 1992, p. 291).

Any changes in the environmental conditions induced by the diurnal cycle must be accounted for in the methodology selected so that adequate imagery is obtained. Figure 11.25 is a schematic representation of a typical diurnal cycle where the spherical object has an emissivity, ε_O, and it rests on a surface with emissivity ε_B, and furthermore, ($\varepsilon_O > \varepsilon_B$) such that the sphere and background exhibit a ΔT_A throughout the cycle. When ΔT approaches zero thermal crossover occurs, and the temperatures of the sphere and the background are the same.

At this point in time the thermal contrast is zero. Recall that ΔT (the object background differential) is given by $\Delta T = (T_O - T_B)$. At crossover, $T_O = T_B$, and according to (5.17), the apparent temperature difference, which was derived in Section "Apparent Temperature Difference, ΔT" of Chapter 5.

$$\Delta T_A = \left\{\varepsilon_O\left(T_O^4 - T_E^4\right) - \varepsilon_B\left(T_B^4 - T_E^4\right)\right\}^{1/4}$$

then becomes

$$\Delta T_A = \left\{\left(\varepsilon_O - \varepsilon_B\right)\left(T_B^4 - T_E^4\right)\right\}^{1/4}.$$

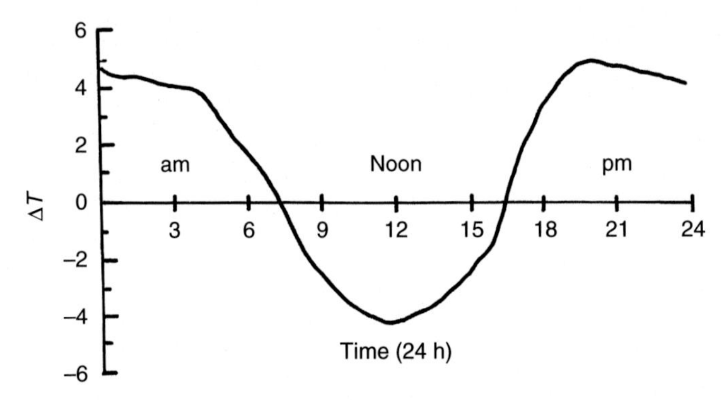

FIGURE 11.25 **A typical diurnal cycle for a spherically-shaped passive (nonbiological) object exposed to full sunlight (i.e., a stone in a grassy field).** The actual details of the cycle will depend on the location, existing atmospheric conditions, and season.

As a result of the different emissivities in this expression, there is an apparent temperature difference even when the object temperature and the background temperature are equivalent; however, the scene contrast is too low to be detected by the imager. Another consequence of (5.17) is that the contribution from the environment, T_E^4, is subtracted from both T_O^4 and T_B^4 so that its role is noticed in ΔT_A only through the difference in the emissivities of the stone and its background. The emissivity and the apparent temperature of the stone determine if sufficient thermal contrast exists between the stone and the background so that the thermal imager can detect it. This depends on a number of factors, the local atmospheric conditions and the signal-to-noise (S/N) ratio of the thermal imager being the most important (see Section "Signal-to-Noise Ratio" of Chapter 7). Thermal images can be collected at any time during the diurnal cycle, but to obtain quality imagery they should be collected when the temperature is relatively stable. It doesn't matter if it is day or night, but the S/N ratio of the imager must be above a threshold value to obtain quality images. From S/N ratio = $\Delta T\,(\beta^R)$/NETD (7.3) we can see that at thermal crossover the range R goes to zero, and the time during the diurnal cycle that the S/N ratio is too low to be measured actually begins before crossover and remains too low for a time after crossover. The length of this time depends on the NEDT, ΔT, and the local atmospheric attenuation.

If the threshold for the S/N ratio is taken as unity and the relative S/N ratio is estimated for the plot in Figure 11.25 then the window of opportunity for image collection is quite large. See Figure 11.26, which shows an example of the relative S/N ratio for a complete diurnal cycle. The three regions above the threshold (dashed line at $\Delta T = 1$) are all suitable for collecting imagery and include both the negative and positive ΔTs shown for the plot of Figure 11.25 that are above the threshold value. While there is still an apparent temperature difference associated with the portions of the curve below the threshold value, the detection capability is limited by the S/N ratio of the imager.

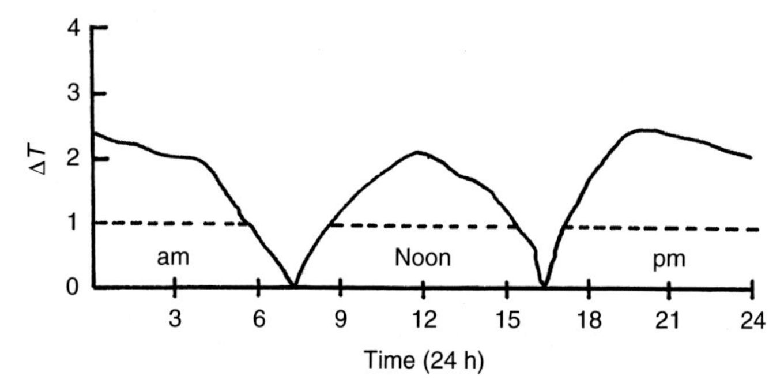

FIGURE 11.26 **The relative S/N ratio for a complete diurnal cycle.** The timespans on this curve that are above the dashed line (threshold S/N) are suitable windows of opportunity if all other imaging criterion are satisfactory.

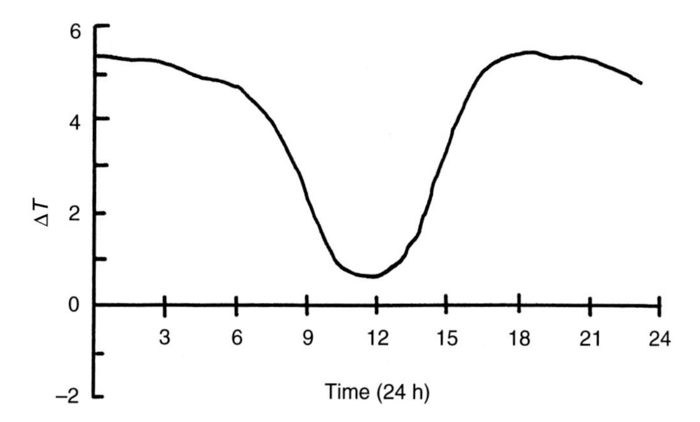

FIGURE 11.27 **The relative S/N ratio of a complete diurnal cycle for a homothermic animal does not exhibit thermal crossover; because of its metabolic processes it is always radiating heat to the background.**

Note that this figure is representative for one that does not consider the effects of the prevailing atmospheric conditions, which can considerably reduce the timespan in the three windows to collect imagery. There can also be an adverse effect on the S/N ratio due to extremely cold temperatures, as discussed in Section "Signal-to-Noise Ratio" of Chapter 7.

If the situation is such that the object of interest is a homothermic animal then the influence of the diurnal cycle on ΔT would typically be skewed to reflect the fact that the animal continues to radiate throughout the cycle while the background depends entirely on the solar loading. Figure 11.27 is a schematic representation of this situation. Interestingly, there is no thermal crossover feature since the apparent temperature of the animal is always higher than the apparent temperature of the background for homothermic animals. Nonetheless, if the thermal contrast diminishes during strong solar loading (i.e., $\Delta T \sim 0$), the S/N ratio of the imager may not be adequate to form good quality images.

In all of the curves representing the temperature changes during the diurnal cycle, the optimal windows will occur when the temperature changes are minimal as a function of time (i.e., the most stable).

Example of Solar Loading

The following provides an example of solar loading on image quality where the chief cause of image degradation is the loss of contrast between the animal and the background. The apparent temperature difference is minimized to the point that the thermal sensitivity of the imager is less than the S/N ratio. Note from (5.17) that when the thermal contrast ΔT is reduced to zero, or when $T_O = T_B$, there is still an apparent temperature difference between the animal and its background because of the difference in their emissivities, but when this ΔT_A is less than the NETD of the system, it is not detected.

An experiment was conducted during warm fall conditions in coastal Virginia in September, 1997, between 5:00 am and 9:30 pm to illustrate the effects of solar radiation (loading) on the image quality obtainable without compensating for these effects with the features available on a 3–5 μm imager. The cat "Sniper" was confined to a wooded area, which we observed with a tripod-mounted Inframetrics InfraCam imager to observe the image quality as a function of thermal loading across the FOV of the scene. A series of five photographs taken at different times shows the degradation of the thermal contrast as a function of solar loading. The first photograph was taken before sunrise, and Figure 11.28, taken at 5:30 am, shows a nice image of Sniper on a warm fall morning (23.5°C). The background is very uniform in this image, and the gain of the imager was set to keep the signature of Sniper from being saturated (a completely white signature) so that changes in the spatial distribution due to his pelage are discernable in the image.

Figure 11.28B, the photograph taken at 8:30 am (after sunrise), shows that thermal loading is beginning to reduce the contrast, ΔT, in the imagery between the cat and its background. In fact, there are several places in the scene that are emitting more radiation than Sniper. The following two photographs, shown in Figure 11.28C and D, were taken during the timespan of peak solar loading and show that the thermal contrast between the cat and its background is completely lost, providing no information of the scene or its contents. In some cases it is possible to get low-contrast images and still be able to extract a signature of interest from the imagery, but the thermal sensitivity of the camera is operating at is limit and the imagery is always inferior.

These photos were taken without adjusting the camera to optimize the imagery in order to provide a visual illustration of the effects of solar loading on the image quality. An improvement in the image quality is possible for these photographs by making adjustments to the camera settings for gain and light level as the solar loading progresses during the day; however, the imagery obtained during the peaking loading time will not be optimized and is generally not useful. Vegetation cools quickly (low thermal inertia) and more dense (solid) objects in

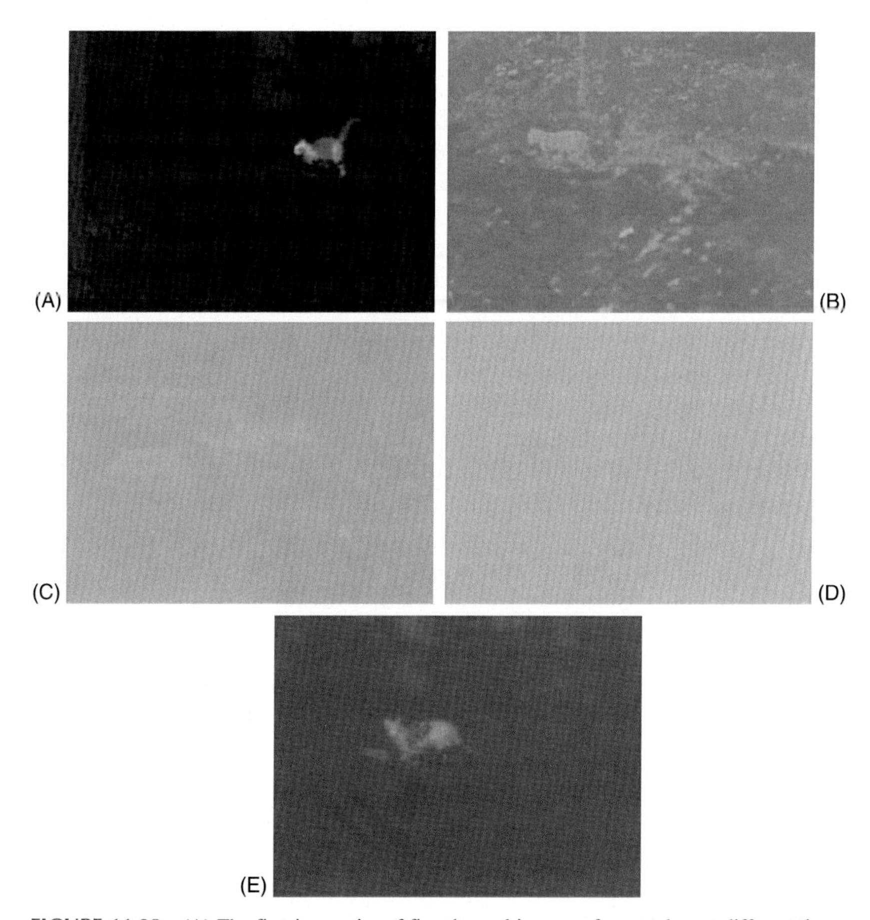

FIGURE 11.28 (A) The first in a series of five thermal images of a cat taken at different times during a diurnal cycle to illustrate the effect of the cycle on thermal contrast between the cat and its background. Cat at 5:00 am. (B) The imagery at 8:30 am is beginning to lose contrast. (C) The imagery at 10:00 am has lost all contrast and the cat is not detectable. (D) The imagery at 2:00 pm is still without contrast. (E) Contrast is restored to the imagery at 9:00 pm and the window of opportunity will be open until 5:00 am the next day, provided the microclimatology remains stable.

the background cool more slowly (high thermal inertia). Figure 11.28E was taken at 9:00 pm, and the background was already fairly uniform because it was comprised of small shrubs and grasses, which cooled quickly, and fairly good thermal contrast was restored to the imagery a few hours after dark as a result. In many cases it will take a full 12 h for the background to return to a state that allows quality imagery to be collected. It is obvious that, in this example, a strong window of opportunity exists for collecting good imagery between 9:00 pm and 5:00 am or for the rest of the night if there are no significant meteorological changes during this period.

Poikilothermic Animals and the Diurnal Cycle

If the animal of interest is poikilothermic, then the plot will have the general features shown in Figure 11.25. Poikilothermic animals will be mobile during the diurnal cycle to accommodate their required thermoregulation, and it is during this time that they will be at the greatest temperature difference with their backgrounds. Imagery should be collected when the animal is best contrasted with the background. If the animal is warm and trying to lose heat in order to regulate body temperature then the best opportunity would be when it moves into a cooler background and, if it is seeking to raise its body temperature, the best opportunity will be when it moves into a warmer background. Note that in the former case the signatures associated with the poikilotherm will be white (warm) against a dark (cool) background, and for the latter case the poikilotherm will be dark (cool) against a light (warm) background.

When we set about to collect data in the field we need to optimize the climatic conditions and the operation of the thermal imager to fit in a specific window of opportunity. That is, we are going to choose the best possible combination of conditions (climatic and equipment) that will allow us to fully utilize a given window of opportunity. We need to select a time during the diurnal cycle when the animals of interest are using the habitat being surveyed, and a time when the atmospheric conditions are favorable. For example, we want a high atmospheric transmission ($\tau \sim 1$) and moderate ambient temperature ($T \sim 25°C$). We need to select a time when the thermal contrast (ΔT) between the animals of interest and the background is optimized. We need to select an imager that has a good thermal sensitivity. The combination of these factors can be summarized by way of the S/N ratio, expressed as (7.3). Here the signal is just the thermal contrast $\Delta T = (T_O - T_B)$ between an object and its background that is attenuated by the atmospheric transmission β at some range R. The noise is given by NETD, a function of background temperature.

Survey conditions will vary considerably, depending upon when during the diurnal cycle data is collected. A rule of thumb is to try and ascertain when the local survey area is as close to thermal equilibrium as possible and as stable as possible. Gathering data on a perfect spring day (sunny, breezy, after a previous day of spring rain) is perhaps one of the worst possible situations. Under these conditions we could expect the apparent temperature difference between the background and animals of interest to be in a state of random fluctuation. The scene of interest will be exposed to a continual source of solar radiation that is being partially absorbed, reflected, and reradiated by all the objects in the scene and their backgrounds. Furthermore, the apparent temperature of the objects and the background will be modified continually in a random fashion by radiation, conduction, convection, and evaporation. Likewise, the apparent temperature difference between the objects and background will randomly fluctuate.

The observations made by Marble (1967); McCullough et al. (1969); and Parker (1972) that caution about collecting data during the daylight hours when there is direct sunlight or passing cloud cover have mostly been neglected in subsequent work with thermal imagers, even though it clearly creates a background whose true temperature and the apparent temperature of all objects (both biotic and nonbiotic) are continually fluctuating. Two separate considerations must be reconciled during any 24-h period to determine if it is suitable and beneficial to collect imagery. First, when are the animals using their habitats? And second, how do we overcome prevailing environmental conditions when animals are using their habitats so that we can collect quality imagery?

Homothermic Animals and the Diurnal Cycle

To determine the best time of the diurnal cycle to survey or observe homothermic animals we can examine the following four sets of conditions that conveniently follow from the four periods of change during the cycle. We provide images for comparison (Figure 11.29A and B and Figure 11.30A and B) when it was possible to obtain visible images in the field.

Daytime conditions (any season): The apparent temperature can depend on the animal's metabolic heat production, solar loading, reflection of direct solar radiation, convective cooling from wind, and passing cloud cover. These effects are acting collectively (and randomly) to produce the observed thermal contrast (ΔT) between the animal and the background at any given time.

Dusk conditions (any season): The apparent temperature can depend on the animal's metabolic heat production, a cooling background temperature, and perhaps some convective cooling from wind. Choosing windless days can minimize the effects of convective cooling.

Nighttime conditions (any season): The apparent temperature can depend on the animal's metabolic heat production, ambient temperature, and convective cooling from wind. Choosing windless nights can minimize the effects of convective cooling.

Predawn conditions (any season): The apparent temperature can depend on the animal's metabolic heat production, ambient temperature, and convective cooling from wind. Choosing windless mornings can minimize the effects of convective cooling.

If there is precipitation (rain or snow) then the thermal scene becomes one of very low contrast or "washed out" and ΔT approaches zero for all of the previously mentioned conditions. If the diurnal cycle is separated into these four periods of time then we see that daytime conditions pose a serious problem without even considering the thermal stability and uniformity of the background. It is difficult to collect data for an animal whose true temperature and apparent temperature are fluctuating randomly in time. Parker (1972) reported that $\sim 18°$ temperature fluctuations were obtained by exposing or shading a tanned, furred mule deer hide from direct sun. These temperature swings were occur-

FIGURE 11.29 **A standard visual photo and an infrared image of the same scene of a cat taken at ~4:00 pm on a warm fall day.**

ring in a time interval of 2 min, and perhaps more interesting is the fact that the initial change of ~10° occurred in a few seconds, indicating that there was a significant contribution to the observed radiant temperature from reflected sunlight. The build-up and decay rates of the measured radiant temperature for the exposed and shaded hides were remarkably similar and stabilized at approximately the same temperature during repeated experiments.

If we examine the three remaining sets of conditions, we see that they are essentially the same except that the ambient temperature will generally decrease from the period of dusk to dawn. In order to select a preferred time to establish a window of opportunity to collect field data, we must also consider the effects of the diurnal cycle on the background. The effects of solar loading were discussed earlier, and objects with high thermal inertia would require hours to dissipate the heat stored from the previous day's thermal loading cycle to the

FIGURE 11.30 **Bobcat images taken at dusk in the visible and in thermal infrared on a warm fall evening.**

background and environment through conduction, convection, radiation, and perhaps evaporation. This fact alone reduces the selection of the window of opportunity to the predawn hours. During this period of the diurnal cycle the background temperature is close to the ambient temperature and will provide the most uniform background possible without the modifying effects of precipitation. Excellent data can still be collected after sunrise if the background against which the biotic objects are contrasted is not exposed to direct sunlight. Many objects comprising the background have a high thermal inertia and can be heated very quickly (minutes), which will create unwanted thermal signatures (isolated objects) or serve to reduce the thermal contrast between animals and the backgrounds they occupy (grassy fields, watered areas, or expanses of bare soil). A solid methodology will expect these potential problems and provide a way to negate them.

A number of past efforts pointed out the problems associated with choosing the window of opportunity incorrectly. Parker (1972) correctly made the observation that "during periods of no direct solar irradiation of the deer surface, metabolic energy production was the major heat source and it was relatively constant." Large radiant temperature fluctuations (~20°C) of deer were reported by Marble (1967) when collected on clear days even with a complete snow background. McCullough et al. (1969) analyzed these data (both nighttime and daytime) and concluded that the daytime data were highly erratic and should be avoided.

ATMOSPHERIC EFFECTS

The transmission, absorption, reflection, emission, and refraction of light by the atmosphere are the most widely observed of all optical phenomena (Killinger et al., 1995). The atmosphere is made up of a long list of gases and aerosols, which absorb and scatter radiation. When collecting thermal images the imaging system and the animals are both surrounded by the atmosphere, and the radiation reaching the detector is the sum of the radiation collected by the camera optics emitted by both the animal and any radiation scattered or reemitted from particulates along the sight line of the camera. As a result the atmosphere exerts a significant influence on the quality of the imagery. In the extreme, when the magnitude of the path radiance is on the order of the attenuated radiation from the animal, the contrast can be very poor.

Atmospheric Transmission

The atmospheric transmission is high when there are only small amounts of water vapor, gases, and aerosols present in the atmosphere. At these times, the sky will appear cold to the thermal imager since deep space is very cold. Under these conditions (viewing the zenith), the self-emission of the atmosphere is low, and the average temperature of the background viewed by the imager is low. As the imager is lowered from the zenith to smaller elevation angles, more aerosols and gases at ambient temperature are encountered along the line of sight, resulting in an average temperature close to that of the earth's surface. For most of the detection and survey work encountered in the field small elevation angles are the norm. Down-looking angles from aircraft will experience the same high average temperatures associated with atmospheric layers close to the surface of the earth. Under these circumstances, atmospheric constituents degrade the thermal resolution. Radiation, scattering, and absorption in the line of sight reduce the magnitude of the thermal signature of the object to be detected, located, recognized, or identified. On the other hand, scattering of radiation from earth shine and sunshine into the line of sight along with the self-emission from the atmosphere in the line of sight both serve to reduce the thermal contrast between the object and the background.

Radiation

Since nearly all fieldwork involving censusing will utilize relatively long path lengths, the self-emission and scattering by the atmosphere in the path could be significant and must be taken into consideration. For short path lengths (<20 m), the transmittance for both the LWIR and MWIR windows can be considered as unity (see Figure 5.3). The directional dependence and the amount of background radiation (both emitted and reflected) that the imager encounters are other important factors that must be incorporated into the overall methodology. In the real world, the combinations and permutations of changes in the atmospheric conditions continually affect the apparent temperature difference between an object and the background. Thus, ΔT_A is a continually changing value that loosely follows a diurnal cycle. The atmospheric transmittance is not constant and can change in minutes. These changes in turn affect the window of opportunity. Absorption and scattering are the two primary mechanisms that reduce the atmospheric transmission. See Section "Signal-to-Noise Ratio" of Chapter 7 for a discussion of the atmospheric effects of absorption and scattering on the range.

Absorption

The main constituents of the atmosphere are oxygen and nitrogen, and they are generally transparent to visible and infrared wavelengths. Minor gaseous constituents are water vapor, carbon dioxide, and ozone, all of which absorb strongly in the infrared. The absorption process is a two-way process, so to speak, in that solar radiation is absorbed by these compounds on its way to the surface of the earth, and in turn the terrestrial radiation emitted by the earth's surface is also absorbed by these same compounds. It is a dynamic process in that as the earth's emitted infrared radiation (earth shine) is absorbed by the atmosphere, it is reradiated in all directions. About half of this radiation returns to the surface of the earth, and the rest is radiated out to space.

Scattering

Before solar radiation reaches the earth's surface, it is also attenuated via scattering caused by atmospheric aerosols, gases, and dust. This reduction is due primarily to the absorption by water vapor or the absolute humidity (water in g/m^3) in the atmosphere. Molecular, or Rayleigh, scattering occurs when the scattering particulates, such as air molecules, are small as compared with the wavelength of the radiation. Large-particle scattering (Mie scattering) is caused by aerosols, dust, and water droplets of a size comparable with the wavelength of the radiation (see Section "Wavelength Selection" of Chapter 8 for a discussion on infrared scattering in the atmosphere). If there is ground fog present in the predawn hours, care must be taken to collect data prior to any scattering of direct sunlight. Note that scattering can be significant prior to sunrise and after sunrise. Ground fog (water vapor) scatter will dominate the scene to such an extent that any detail in the scene is completely lost. If there is no direct sunlight during predawn and immediately after sunrise due to cloud cover then the range and the thickness and density of the water vapor will primarily determine the image quality.

Water Vapor

The amount of water vapor present in a unit volume of air, usually expressed in grams per cubic meter, is called the absolute humidity. It is a major absorber in the infrared and it can have a significant range of concentrations from a few tenths of a g/m^3 to 50 g/m^3, depending on location (arid and desert climates to tropical and seaside climates). The absolute humidity can be treated as an ideal gas (Gates, 1980), provided that condensation is not taking place. As a result, the density of the water vapor depends on its pressure, temperature, and molecular weight at any given time, which can change daily and with the season. The absolute humidity is a quantity calculated from the temperature and relative humidity. As the absolute humidity increases, the atmospheric transmission of infrared radiation in both the MWIR and LWIR decreases. Note also that for a given temperature the absolute humidity increases with the relative humidity. The relative humidity is expressed as a percentage. It is the ratio of the actual amount of water vapor present in the air at a given temperature to the maximum amount that the air could hold at that temperature.

Cloud Cover

A cloud is a visible aggregate of minute water droplets or ice particles in the atmosphere above the earth's surface. Cloudy conditions exist when the cloud cover obscures 5/8 (cloudy) to 8/8 (overcast) of the sky. Since clouds block solar radiation and also reduce the transfer of heat to the sky, the window of opportunity can be greatly improved by passing cloud cover. Passing clouds can significantly moderate the thermal signature of the background (inanimate objects). A heavy overcast lasting several days will nearly obliterate any thermal signature present in the background as the earth and all nonbiotic objects reach thermal equilibrium. A nearly idealistic situation exists for observing or locating homothermic animals after several days of complete overcast. In this situation there is a 24-h window of opportunity, and surveys can be conducted any time during the diurnal cycle. For some surveys it may be necessary to wait for these conditions.

Since clouds are aggregates of water droplets or ice crystals, they can scatter and absorb radiation received from both the sun and the earth. When the cloud is very thick, back-scattering is the dominant scattering mechanism, and a thick stratus cloud can reflect up to 70% of the solar light incident on the cloud. If you have ever been on an aircraft that is flying above the clouds and have looked down at the clouds you have probably witnessed the brightness of these reflections. Another 20% of the incident radiation from the sun may be absorbed by these clouds, leaving only ~10% that is transmitted to the surface of the earth. If you look up at the bottom of these clouds they appear very dark and grey. For thinner cumulus clouds, the situation is different and there is significant forward-scattering, making the clouds appear very bright and white.

The shading effects provided by passing clouds can be used to advantage in certain imaging applications (see, e.g., Section "Invertebrates (Arthropoda)" of Chapter 10.) We had the opportunity to examine orb weaving spiders (*Argiope aurantia*) on webs in a coastal Virginia marsh with a 3–5 μm thermal imager, and we observed that they were clearly visible in their webs when in direct sunlight. This was due to the reflected term in the equation, giving the apparent temperature difference ΔT_A. When the spider was shaded, it was still visible but quickly faded, becoming indistinguishable from its background as it came to thermal equilibrium with its background. That is, by blocking the solar radiation, the reflected terms in (5.16)

$$\Delta T_A = \left\{ \left(\varepsilon_O T_O^4 + \rho_o T_E^4 \right) - \left(\varepsilon_B T_B^4 + \rho_B T_E^4 \right) \right\}^{1/4}$$

are forced to zero (i.e., ρ_O and ρ_B both are zero) and (5.16) reduces to

$$\Delta T_A = \left\{ \left(\varepsilon_O T_O^4 \right) - \left(\varepsilon_B T_B^4 \right) \right\}^{1/4}.$$

When the spider and its surroundings are shaded $T_O \sim T_B$, which leaves the apparent temperature difference depending only on the difference of the emissivities. The atmospheric conditions can radically modify the background day-to-day or even hour-to-hour. On a very local scale, the shade of a large tree or a north-facing hill or rock may prove to be shaded enough to provide a suitable background for collecting thermal mages.

Solar Loading

We discussed the effects of solar loading on the background and animals in the preceding section on the diurnal cycle. By examining the US Geological Survey daily water temperatures on the Rappahannock River in Figure 11.24 it is easy to see the difference a rainy day can make with regard to solar loading. The extreme rates of temperature variation of materials on the earth's surface are determined by the materials' properties and the degree of solar loading they experience during the diurnal cycle. The thermal conductivity, heat capacity, and thermal inertia determine the amount and rate of thermal change possible for a specific amount of loading, just as they control the rate of cooling once loading ceases.

Precipitation

Short periods of heavy precipitation followed by overcast conditions can also quickly expand the window of opportunity. When it is raining or snowing there is no solar loading, and all objects in the background are rapidly coming to thermal equilibrium because conduction and convection can play a more significant role in the heat transfer processes. The high thermal conductivity of water serves to bring objects in the background to a uniform temperature through the dissipation of heat. After the precipitation ends, drying begins and evaporative cooling will remove additional heat from the background. Convection can play a major role in reducing the background temperature, especially in warmer climates. Once the temperature has stabilized in the background (i.e., has come to equilibrium), then homothermic animals are the only source of radiative heat that is significantly different from the background. Again, under these conditions, surveys can be conducted at any time during the diurnal cycle.

Wind

It is extremely important to collect thermal imagery when the wind speed is relatively low (calm to ~8 m/s), and it is advisable to avoid gusty situations. Wind is essentially forced convective cooling, which is a very efficient way to cool. The adverse effects of wind speed have been observed by many researchers and generally are related to a noticeable decrease in the apparent temperature difference. This is because the animal's body heat is being carried away on the wind by convection, and in an effort to maintain its thermal equilibrium it slows down the release of radiative energy. The Beaufort wind chart is a good way to recognize the wind conditions while working in the field. Table 11.1 shows the

TABLE 11.1 Beaufort Wind Chart

Beaufort		Wind Speed		
Number	Condition	Miles/h	Kilometers/h	Description
0	Calm	<1	<1.6	Smoke rises vertically
1	Light air	1–3	1.6–4.8	Direction of wind shown by smoke but not wind vanes
2	Light breeze	4–7	6.4–11	Wind felt on face, leaves rustle, wind vane moves
3	Gentle breeze	8–12	13–19	Leaves and twigs in constant motion and wind extends small flag
4	Moderate breeze	13–18	2–29	Wind raises dust, loose paper, and small branches move
5	Fresh breeze	19–24	31–39	Small-leafed trees begin to sway, crested wavelets form on inland waters

Beaufort number for a few conditions that should be easily recognized in the field. It is best to collect data when the Beaufort number is in the range of 0–5.

Patience

A landscape that has considerable periods of solar exposure during the day and is littered with dark (near blackbody) objects such as deadfall, charred logs, rocks, pools of standing water, and patches of bare earth will be difficult to use as a background for an aerial survey. The thermal signatures of these objects, as the result of reradiation of absorbed solar energy, can be the size of large mammals and can easily be confused with the thermal signatures of survey animals. In fact, when significant thermal loading of the background occurs (background clutter), animals as large as moose (*Alces alces*) cannot be counted accurately (Garner et al., 1995, p. 236). While these landscapes present a difficult task, it is not an impossible one. If all is prepared properly for a survey, then the observers only need to wait for the right time to conduct the count. If after an overnight cooldown the thermal signatures of nonbiotic objects are still comparable to those of the survey animals then a suitable strategy would be to wait for an all-day rain followed by a night of drying prior to surveying in the predawn hours. Alternatively, in dry or arid climates, it may be advisable to wait for a predawn survey following a day of cloud cover. This will allow the multiple effects of conductive, convective, and evaporative cooling to moderate the thermal background and give it ample time to reach equilibrium. Subsequently, the only signatures to be observed would be those that generate and radiate their own heat.

The earliest notion that thermal contrast is greater during daylight hours than at night was suggested by Marble (1967), who measured the radiant temperatures of big game animals, including white-tailed deer and mule deer. This observation was correct for those measurements not because it was daylight as opposed to nighttime but because of the large temperature changes that occurred during the diurnal cycle. Marble (1967) further suggested, and correctly so, that high overcast conditions during the daytime offered the best opportunity for high detectivities. The large temperature swings in the background and ambient overnight cause the thermoregulatory processes of the deer to conserve heat, which results in a very small apparent temperature differential between the deer and its background. During the day it is warmer, and the deer radiate more heat, so the differential between the deer and the background increases but is erratic in the sunlit conditions. When the day was overcast, both the problems of solar loading and cold background temperatures were mitigated. There are two primary reasons for these observations. Homothermic animals use thermoregulatory processes to maintain and conserve body temperature during colder periods, and during warmer periods they maintain their body temperature by losing heat. In addition, the emitted power differential of the signal that produces a response in the thermal imager is proportional to the cube of the background temperature (T_B^3) for a given thermal contrast ΔT (see (5.8)). It is better to be patient and wait until the conditions for data collection are optimal than to struggle with imagery data that suffers from inconsistency and is difficult to interpret.

Note that the atmospheric conditions at the time of a thermal imaging survey are of the utmost importance. The necessary atmospheric and meteorological conditions for collecting good thermal imagery cannot be ignored for the sake of comfort or convenience of the thermographer. For example, visual surveys from the air are enhanced by complete snow cover because the contrast between a deer and a typical forest setting provides a brown-on-brown contrast, while on snow it is a brown-on-white contrast. Visual images are formed by reflected sunlight, and when it gets dark the images are no longer visible or are unavailable. This is the reason why field scientists don't do visual surveys at night, even though it may be more convenient because of a busy daily schedule. The same rationale should carry over to field scientists doing thermal imaging surveys.

When looking for an object with a thermal imager, it is important to understand what exactly is responsible for forming the thermal images you are trying to collect. Thermal differences in a scene are invisible to the human eye because we can't see things in the infrared region of the spectrum, so this tells you that sunlight is not needed to form thermal images. In fact, you do not want sunlight, which is mostly visible light. A typical 8–12 μm thermal imager does not detect visible light, but it responds to radiation in the form of heat in the 8–14 μm spectral range. This occurs regardless of the source of that radiation (see Section "Spectral Domain" of Chapter 9). Nonetheless, to obtain a quality thermal image, you need to maximize the thermal contrast between the animal and its background. To do this correctly, you need to adjust the apparent temperature difference between the deer and its background, which is intimately tied to the local meteorological and atmospheric conditions, just as it is in a visual survey (i.e., lots of light and complete snow cover).

It is important to grasp the significance of the word "apparent" since the true temperature of the deer and/or its background is not important unless it gets to very low temperatures (see Section "Thermal Sensitivity"of Chapter 7). The temperature is called apparent because the thermal imager cannot discern the origin of the radiation emanating from the deer or its background but only the difference between them. For example, the deer has many different components contributing to the total radiation emanating from its surface that are detected by the thermal imager. Typically, they originate from the deer's basic metabolic activity and nearly everything that can be identified in the immediate vicinity of the deer, including reradiation of previously absorbed solar radiation, scattered radiation emanating from the environment, reflected radiation from other radiating objects in the background, and directly reflected solar radiation if the survey is conducted in the daylight. Most of the time it is advantageous to have the apparent temperature difference between the deer and its background be maximized, and this is easier to accomplish in warm weather because the apparent temperature of a deer is larger in warmer backgrounds (see Section "Background Temperature"of Chapter 5).

AUTOMATED DETECTION

One of the first attempts to look at the possibility of using the thermal imagery of an infrared scanning system to automatically detect mule deer was investigated by Wyatt et al. (1980). The results of that study led them to the determination that relying on the thermal contrast (ΔT) between an object and its background, by itself, was inadequate for conducting effective aerial surveys of large animals. They reached this conclusion after subjecting the data of Marble (1967) and Parker (1972), taken with a thermal imager during daylight on snow-covered terrain, to a Bayesian decision model. They suggested that deer could be successfully detected against a snow background, but when the occurrence of snow-free objects was more frequent than that of deer thermal imagery was inadequate. Under these conditions, detection is difficult, or may be impossible, because any exposed objects (e.g., rocks, brush, or deadfall) will scatter, reflect, or reradiate previously absorbed solar energy (see more on these effects in Section "Apparent Temperature Versus Viewing Angle" of Chapter 6). As pointed out earlier, a thermal imager will detect all objects whether they are biotic or not if they have a ΔT that is within the sensitivity range of the imager. It is up to the thermographer to remove these unwanted sources of thermal energy in a scene emanating from nonbiotic objects. When there is significant background clutter, the differential in apparent temperature between biotic objects and other objects in the background is minimized or even reversed. The temperatures of nonbiotic objects become elevated upon the absorption of solar radiation, so to exclude or reduce their effect on survey results, thermal imagery should be collected only after the nonbiotic objects have reached thermal equilibrium with the background. In this way the thermal imager will record only animals because of their metabolic activity.

Thermal Imaging and Automation

Several problems must be addressed when considering the design of automated thermal imaging detection applications. Thermal imagers collect radiation from animals and their background, and if there are sufficient differences in the apparent temperatures between the two then quality imagery can be obtained. This imagery can contain sufficient information to count and identify a large number of species and, in many cases, the user is able to make accurate evaluations regarding the activity, age, sex, and physical condition of the animal. These evaluations are not automatically determined and require the detailed examination of the thermographer. This being the case, we assume that there is adequate information obtained from the imagery, but the data processing was carried out in a human brain. The complexity and difficulty of processing thermal images so that robots can see well enough to make decisions in outdoor surroundings is truly a difficult thing to even imagine.

Going from the raw imagery to just counting the number of individual animals in the imagery has taken a significant effort on the part of many researchers and field scientists. It has been realized that the only way to count large numbers of animals during a short period of time is through automating the data extraction from the imagery. The basic problem that needs to be solved is how the number of animals in the collected imagery can be counted so that none are missed. We reviewed a number of efforts in Chapter 10 that used digital image processing, computer vision analysis, superposition of detection and tracking algorithms, and automated motion detection to solve some of these problems.

Automated Detection of Animals

Automated detection of animals using thermal imaging is complicated by a number of factors. Assuming that the thermal images are good enough to unambiguously separate animal signatures from the background and that they are easily visualized in the imagery then detection is easily accomplished. Is there more than one species present? The more species there are the more complex the automation of the detection task becomes because now the individual signatures must be identified as belonging to the species of interest. This in turn means that an algorithm of some sort based on shape, size, or intensity must be developed to perform the identification to complete the task of detection.

A number of efforts that do not require identification of species but instead require the detection of any and all animals present such as in agricultural mowing operations have seen marked improvements. Steen et al. (2012) used imagery collected with an uncooled bolometer in the LWIR with a 640 × 480 pixel FPA mounted on a tractor. The tractor was driven at different speeds to test an algorithm for the automatic detection of caged animals (rabbits, chickens), which were placed in grass ready for mowing. They concluded that the method has potential for automated detection of animals during mowing operations. They reported that under most circumstances detectivities were close to 100%, with the caution that dense crops may hamper the detection of animals.

When the work requires the automated detection and identification or classification of animals collected in the thermal images, the problem becomes more difficult. Nonetheless, Christiansen et al. (2014) have developed methods to detect and classify via a new thermal feature extraction algorithm used on imagery collected from a lift to simulate the platform of an unmanned aerial vehicle (UAV)-based detection and recognition system. The thermal signatures of detected objects are calculated using morphological operations, which are partly invariant to rotation, size, and posture. UAVs are an emerging technology (see Chapter 12), and in modern agriculture, they can be used for many purposes. UAVs can be preprogrammed to navigate flight paths commensurate with mowing operations, and they can be equipped with the required thermal imagers and data processing equipment to detect and enumerate animals in the path of agricultural machinery, thereby reducing wildlife mortality and promoting wildlife-friendly farming.

A system has been demonstrated by Israel (2011) that can detect roe deer fawns (*Capreolus capreolus*) in meadows prior to or during mowing operations. A UAV-mounted thermal imaging camera was tested in the field to demonstrate the application. Thermal images were collected from the UAV and transmitted as an analog video stream to a ground station, where the user followed the camera live stream on a monitor for manual animal detection in real time. In this system the camera was flown approximately 10 m agl with a view angle of $\theta_C = 0°$ with respect to the nadir. The thermal imager was a FLIR Tau 640® using 32° × 26° FOV optics with an IFOV = 0.89 mrad. The success demonstrated by the manual operation of this detection system provides the impetus to go forward with automation of the sensor platform. Sensor data fusion between the thermal and the visual camera should extend the utility of the system to detect fawns in a wider variety of weather situations, including sunny days.

Detecting cryptic and nocturnal species places more stringent requirements on the data-gathering efforts. Brawata et al. (2013) designed an automated thermal video recording system to monitor cryptic mammalian predators—dingo (*Canis lupus dingo*), red fox (*Vulpes vulpes*), and the feral cat (*Felis catus*)—at food and water resources in Australia. The thermal camera used was an FLIR ThermaCAM S45® with a 320 × 240 FPA microbolometer sensor operating in the LWIR. The 35 mm lens afforded a 24° × 18° FOV and a spatial resolution of 1.3 mrad. Since the automatic video capture system monitored wild carnivores, it was left unattended for extended periods of time to minimize the impact of human presence. The system remained on at all times but only recorded video when target species were identified in the thermal video frame when large temperature changes were detected between portions of an incoming frame and the average "background" frame. Observers using binoculars and image intensifiers monitored sites and correlated animal sightings with the recorded video. By using their equipment at focal lures (food and water) they were able to monitor three target predator species and determined that the optimal sampling distance for detection, identification, and for collecting behavioral data was 30–40 m. This range limitation may be due to the performance of the imager more than anything else, although there is very little information provided on the local environmental conditions at the time of the data collection, which could also have played a role.

Automated Enumeration of Animals

Again, if there is only a single species present and this is known before the imagery is collected, the problem is reduced to detection if they can be distinguished from one another in the imagery. If the grouping of the animals is too dense or too closely packed to count individuals within the group then a statistically-based algorithm is required to assist with the counting. This problem has been addressed by a number of workers who used thermal imagery to count avian species and bats. Significant advancements were made for animals present in large numbers in the dark when there is a reduced probability of interference from excessive background clutter. Advanced infrared detection and image processing for automated bat and bird censusing has been the subject of much work. Sabol and Hudson (1995) collected thermal imagery on videotape and subjected it to digital image processing routines to extract bat numbers from the imagery. Frank et al. (2003) demonstrated a real-time automated censusing system to make accurate and repeatable estimates of the number of Brazilian free-tailed bats (*Tadariada brasiliensis*) present independent of colony size, ambient light, or weather conditions and without causing disturbance to the colony. Melton et al. (2005) developed a thermal infrared detection and tracking system for bats in flight. Desholm et al. (2006) used LWIR thermal imagers to construct an automatic system for detecting avian collisions at offshore wind-energy facilities in Denmark. This system could be controled remotely and triggered automatically when a collision was eminent. Betke et al. (2008) studied the evening emergence of Brazilian free-tailed bats using thermal imaging and computer vision analysis. Their image analysis method allowed them to conduct censuses with an accurate and reproducible counting methodology, which was based on the total temporal record of colony emergences. Further, they suggested that similar image analysis methods could be developed to expand census capability to other crepuscular or nocturnal species of bats and birds. Hristov et al. (2008) present new areas of aeroecology research that highlight the use of thermal imaging and computer vision analysis, including population estimates, behavioral observations, thermoregulatory behavior, and bioenergetics (metabolic cost of flight, awaking from torpor, and foraging activities). Hristov et al. (2010) used a combination of thermal infrared imaging and computer vision analysis to provide an effective method for estimating colony size and emergence behavior of Brazilian free-tailed bats. Many of these papers are reviewed in Chapter 10 and are mentioned here because of the effort to bring automation to the detection and/or counting tasks.

Robotic Vision

Thermal infrared imaging provides the ability to identify targets during conditions of limited visibility, and it eliminates the need for a light source that could disclose the robot's location. A particular interest for a number of applications arises in a situation where the robot has detected an object but now needs to classify it. This type of problem is very similar to those faced by biologists trying to extract information from images acquired in the field; however, the imagery the robot needs to interpret is passive in nature and only emits the thermal energy absorbed during the daylight hours. The level of difficulty the robot faces in the classification process is much more difficult than the thermographer in the field. In these systems the robot will have a fusion of both active (sonar) and passive sensors (thermal imaging) to enable autonomous mobile robot operations in a wide variety of unstructured outdoor environments. Fehlman and Hinders (2010) describe their work on this type of robotic classifier that will ultimately afford robots with the intelligence to automatically interpret the information in signal data to make decisions without the need for an interpretation by humans. Many of the relationships between the performance of a thermal imager and the environment that thermographers and robots must manage to classify objects and animals in the field are the same.

Large Area Surveys

An improved procedure for the detection and enumeration of Pacific walrus (*Odobenus rosmarus divergens*) signatures in airborne thermal imagery was demonstrated by Burn et al. (2009). They determined that digital photographs concomitantly taken with collected thermal images at spatial resolutions of 1–4 m per pixel indicated a linear relationship between the number of walruses in a group and the amount of heat that they produced. This linear relationship existed for all spatial resolutions tested, indicating that the number of walruses in a group hauled out on sea ice could be estimated using their thermal signatures.

This important work demonstrates that the use of airborne thermal imagery to survey marine mammal populations has the potential to sample considerably larger areas per unit of time than visual photographic surveys and is applicable to a number of other marine mammals. Here the thermal camera is used to cover vast areas of open sea and ice to locate (detect) hauled out walruses and followed with visual high-resolution photography and data processing to count walrus.

Multispecies abundance estimations were made by Conn et al. (2013) using automated detection systems. They present a hierarchical modeling framework for jointly analyzing automated detection and double sampling data obtained during population surveys of ice-associated seals in the Bering Sea. Note that when the thermal signatures of animals are different from their backgrounds, the imagery can be used to enumerate animal populations in conjunction with high-resolution digital photography to provide information about species identity and classification (sex and age) within species. They demonstrated their approach on simulated data and also on aerial survey data of ice-associated seals. In both cases, automatic detection data consisted of thermal imagery, and double sampling consisted of automated high-resolution digital photography. With this method, the thermal imagery finds hot spots when thermal signatures are compared with the background, and the digital photographs are studied to get information on the species composition of each thermal signature as well as the number of animals present. The technique and model presented by combining multiple datasets within a hierarchical modeling framework provide a powerful approach for analyzing animal abundance over vast spatial regions. Most modern airborne thermal imaging systems are equipped with high-resolution digital imagery capability; in fact, image fusion and multispectral imaging is becoming the norm for new thermal imagers of all kinds.

Image Fusion

There are efforts to develop integrated sensor technologies to move image fusion into the mainstream of remote sensing (Jiang et al., 2013; Sahu and Parsai, 2012). The process of image fusion is one where two or more images are combined such that the combined images reveal more information than is contained in the individual images. Many electro-optic/infrared multispectral systems for intelligence, surveillance, and reconnaissance applications combine at least one reflective band electro-optic sensor (visible) with one thermal infrared sensor (Coffey, 2012). Even though new imagers are becoming more versatile in their function and capabilities with features such as multispectral imaging and image fusion, it is unlikely that all automated detection systems will become a universal tool. In fact, the problem of automating systems gets more difficult with even the slightest loss of specificity. If the goal is to automatically detect any homothermic animal observed within the scan area, the problem is manageable as long as nonbiotic signatures are reduced below some predetermined background threshold. If a single known species is to be detected (e.g., counted) the problem becomes easier since recognition of the species based on observing specific ΔTs is no longer a part of the problem. It becomes a matter of minimizing background clutter. It is extremely unlikely that a universal system can be advanced that will detect with a high degree of certainty a wide range of animals in a particular scan area based solely on the magnitude of the thermal contrast ΔT. More likely, specific systems will be advanced for each species because of their unique physiology and thermoregulatory processes, which vary from season to season, daily during the diurnal cycle, and by activity and habitat use. The point is that the thermal signature of an animal is continually changing (more or less) as is the thermal background of the animal.

THERMOGRAPHY AND THERMOREGULATION

So far in this chapter we have been discussing the difficulties of conducting thermal imaging surveys and observations in the field. Our main concerns have been the local environmental changes brought about by absorption and scattering in the atmosphere due to particulates, the weather, and the diurnal cycle changes that modify the local climate. On occasion these modifications can make the collection of quality thermal images moderately difficult. The applications discussed so far did not involve the measurement of the true temperature of either the animals or the temperature of the background to get quality imagery. Such is not the case when collecting imagery to determine if an organism is functioning properly or why an organism is malfunctioning. In these cases the thermographer must make accurate temperature measurements when mapping out the spatial distribution of the heat producing the thermal image. Note that these applications are quite different than applications that primarily require high detectivities, that is, surveillance applications.

In many of these applications the background and the target of the investigation are one and the same, and they can be relatively small (a region of the human body or of an animal that is suspected of disease) or large (e.g., the thermoregulatory regions of an elephant). The thermographer is essentially mapping out the spatial distribution of the apparent temperature difference between the anomaly and the rest of the background. It is an apparent temperature difference because both the background and the anomaly will be reflecting radiation from the surroundings, and the imager cannot distinguish what portion of the temperature distribution is due to emitted radiation and what portion is due to radiation reflected from the surroundings.

In Chapter 10 we reviewed and discussed thermographic work such as observing working beehives to determine the heat management problems bee colonies must overcome to survive large temperature swings due to seasonal variations and thermal loading due to direct sunlight (Bonoan et al., 2014; Shaw et al., 2011). We also discussed a number of studies treating the thermoregulatory processes in animals in Chapters 6 and 10 (e.g., the work of McCafferty et al., 1998; Hammel, 1956; and Simonis et al., 2014).

There are a number of different studies on animals in zoos and laboratories that avoid the problems dealing with adverse climatic conditions. Many of these studies are devoted to examining the thermoregulatory processes by using information gathered with thermal imagery of the surface temperatures of skin, fur, or feathers under various stimuli and conditions. Cilulko et al. (2013) present a review of the use of infrared thermal imaging in studies of wild animals. They consider a number of applications that used thermal imagers to advance studies of wild animal populations, including disease diagnosis, control of reproductive processes, thermoregulation, analyses of animal behavior, and the detection of animals and estimation of population size. A survey of thermal cameras and applications has been complied by Gade and Moeslund (2014). The survey includes 205 references to dealing with infrared imaging cameras and their current applications, which include animals (thermographic or thermoregulatory as opposed to surveillance applications), agriculture, buildings, gas detection, industrial and military applications, and the detection, tracking, and recognition of humans in addition to a section on medical analysis. Sensor fusion advancements are also reviewed as well as several robotic vision applications. McCafferty (2007) discusses the value of using infrared thermography to study the thermal physiology of mammals. He points out that using infrared thermography provides a noninvasive technique for measuring radiative surface temperature, and therefore it can either be used to infer underlying circulation that is related to physiology, behavior, disease, or simply to detect a warm body against a cool background. McCafferty (2013) reviews the utility of thermal infrared imaging for applications in avian science. In this review he includes a table that summarizes the advantages, limitations, and constraints in study design for thermal imaging that is well worth consulting before heading into the field.

Energy Budget of Target Species

The energy budget of animals is important for the thermographer to know in detail because if he is trying to understand the physiological adaptations necessary for a species to adjust to changes in habitat conditions, and the behavioral modifications needed to make it happen, then he or she needs a reliable model of the species energy budget to predict results. Good detailed discussions of the energy budgets of animals and the energy exchange processes to maintain their body temperatures within a tolerable range are provided by Gates (1980) and Monteith and Unsworth (2008).

Homothermic Animals

The metabolic rate scales as the mass of the organism to the ¾ power. This scaling is observed in many fundamental biological variables, including metabolic rate, lifespan, growth rate, heart rate, and a number of others. Typically, the scaling is a simple power law expressed as $Y = Y_O M^b$ where Y is the observable, Y_O a constant, M the mass of the organism, and the exponent b approximates a simple multiple of ¼. There has been an interesting debate as to the details regarding the value of b and the interpretation of the scaling for the metabolic rate (see Gillooly et al., 2006; Clarke and Fraser, 2004; and Clarke, 2004, 2006). This debate is not going to be a real important factor for us since the temperature dependence of the metabolic rate is a first-order approximation and adequately describes the across-species relationship of the effect of size and temperature on the resting metabolic rate. Clarke points out that it is implicitly assumed that the within-species and across-species relationships between the temperature and metabolic rate are identical. Therefore, it does not provide for cases of laboratory acclimation, seasonal acclimation, or evolutionary adaptation without evoking a change in the activation energy (i.e., making it a variable in the rate equation). For our purposes we don't really care what the temperature of the species is as long as it is different from its background.

Consider placing all animals (shrews to elephants) in the same thermally uniform background. They all will be observable because of the ¾ scaling law of the Standard Metabolic Rate if all other factors are equal (conductive, convective, and evaporative). Any thermal shielding due to the thermal insulative properties (hide, fur, feathers, blubber, etc.) of a particular species will affect the observed signatures since only the radiation emitted by the animal will be collected. In addition, observation of the smallest animals at longer ranges will be difficult to observe because they will be on the order of a single pixel with the optics presently being used.

Once the background clutter has been minimized by constructing an optimized window of opportunity we can easily determine if the signatures captured by the infrared imager are biotic or inanimate. The window will be set for a warm yet uniform background to compare the biotic objects against. In the case of homothermic animals the apparent temperature difference between the bright signatures and the background will be due to heat generated by the animal through metabolic activity. In the case of poikilothermic animals the apparent temperature difference between the dark signatures and the background will be due to the lack of heat generated through metabolic activity. Note that poikilothermic animals can also be warmer than their backgrounds. For example, if a tortoise is warming in sunlight and then moves off to a shaded area it will present a bright signature against the cooler shaded background.

Thermographers have studied homotherms and poikilotherms in both large and small sizes, ranging from butterflies (Lavers et al., 2009) to polar bears (Simonis et al., 2014) and some in between such as starlings (Speakman and Ward, 1998) and peacocks (Simonis et al., 2014).

Chapter 12

On the Horizon

Chapter Outline

DRONES

Unmanned aerial systems and unmanned aerial vehicles (UAVs), commonly referred to as drones, are aircrafts without on-board pilots that carry a payload and are connected to a remote operator. The definition of drones includes fixed-wing, rotary-wing, and airship platforms (Budiyono, 2007). Depending on the model and application, drones can cost significantly less to operate than helicopters and manned fixed-wing aircrafts, which require special piloting skills. Drone unit costs can vary from several hundred dollars to $4.5 million for a Predator® (Odido and Madara, 2013). A modified Predator drone has been used for ecological studies by the National Aeronautics and Space Administration for activities such as real-time monitoring of wildfires (Anderson and Gaston, 2013). Drones are expected to play an increasing role in remote sensing applications with the global drone industry predicted to reach US$15 billion by 2020 (Coffey, 2015). In the past, thermal imagery obtained from drones was mostly limited to military or government agency activities. Advances in imager technology, particularly in imager size (miniaturization), weight, and cost coupled with the rapid advance in drone platform technology have opened up opportunities for civilian applications. Some of the potential civilian applications include disaster management, safety and rescue, environmental monitoring, agricultural pest detection, animal detection during farm harvesting, fire prevention, real-estate marketing, film making, and wildlife surveys (Israel, 2011; Anderson and Gaston, 2013; Odido and Madara, 2013). The focus of this chapter is on the rapid evolution and use of lightweight drones, which are garnering publicity and widespread civilian use (Figure 12.1) as well as advances in thermal optics.

The use of drones in wildlife survey applications has been shown to be effective. Vas et al. (2015) showed that birds, common to wetland habitats, could be approached to within 4 meters without undue disturbance if approached horizontally rather than vertically, possibly due to the semblance of a predator

Thermal Imaging Techniques to Survey and Monitor Animals in the Wild: A Methodology
http://dx.doi.org/10.1016/B978-0-12-803384-5.00012-9

FIGURE 12.1 **Quadcopter drone.**

attack. Israel (2011) demonstrated the use of drones in the detection of newborn deer in farmland during harvesting operations. Newborn deer mortality during pasture mowing causes concern both from an animal welfare perspective and human health due to bacteria from undetected carcasses accidentally being incorporated into the fodder for livestock. Drones outfitted with thermal imagers were able to detect and identify newborn deer as well as smaller animals in an agricultural setting (Israel, 2011).

Lightweight multipropeller drones are limited by payload weight and subsequently flight time duration. Thermal imagers used in drone applications are primarily uncooled (8–14 μm) units due to the relatively light weight when compared to cooled (3–5 μm) units. Thermal imager weight is important since for most drone applications weight is the limiting factor. Thermal imagers of different resolutions can vary in weight from around 100 g to just under a kilogram (Stark et al., 2014).

The use of thermal imagers with drones enable a number of primary requirements for a successful wildlife survey, as presented in Chapter 11. Four-propeller (quadcopter) and eight-propeller (octocopter) drones provide a stable flight with precise control. Altitude, speed, and direction are controled by the regulation of each propeller's speed, and positioning can be maintained with GPS, Inertial Measurement Unit, barometric height sensors, and magnetic field sensors (Vasterling and Meyer, 2013). The thermal imager camera can be mounted to avoid undesirable objects in the FOV and can allow for panning, an essential component when using a thermal imager for survey work.

In most cases it is possible to not only fly the unit manually with real-time Wi-Fi signals transmitted to a viewing monitor or tablet but also to set a planned flight mission. This is particularly valuable in setting up transects, but flights are generally limited to a few kilometers or line of sight and battery life of about 20 min. Most units are outfitted with an automatic "return to home" feature, which is triggered when the battery power reaches a certain level in order to provide for a safe return to the launch site.

Video output to a remote control and or laptop/tablet is recommended for visual flight control. The recording of the infrared video stream via downlink to a remote control or connected computer, respectively, allows for extracting single frames in postprocessing, which helps to optimize the spatial coverage and identification of target objects. In addition (when conducted during daylight), a visual picture or video should be recorded at the same time as the thermal image.

As with a fixed-wing or rotary-wing aircraft platform, the purpose is to get a georeferenced thermal image survey of the target species undistorted by background clutter. A real advantage of drones is the ability to access areas or terrains and altitudes inaccessible to standard aerial survey platforms. It is also necessary to know which height is actually recorded (i.e., height above ground calibrated at the launching site or continuously calculated with absolute GPS). As discussed in Chapter 11, the thermal image footprint depends on the angle and distance to the target.

As part of the survey operation, extra batteries or the ability to recharge batteries is essential due to the limited battery life of present models. Model manuals for operation should be consulted in regard to wind and weather condition parameters. Optimizing the thermal image acquisition preflight preparation, as discussed in Chapter 10, is essential. Prior to conducting a survey with a drone, it is necessary to check with the appropriate authorities in regard to various permissions or restrictions.

MINIATURIZED THERMAL CAMERAS

Hemispherical Drone Optics

In a report by Coffey (2015), several new bioinspired photonic devices still under development were highlighted. Their innovators described how they were able to create new and improved optoelectronic systems based on the natural design and structure that living organisms have produced through billions of years of evolution. Two devices of particular interest to the advancement of future thermal imaging surveillance applications are briefly described in the report: (1) a tunable hemispherical electronic eye camera system that combines rigid device elements (photodiodes) with soft elastomers and (2) a high-speed, uncooled MWIR detector with a thermal sensitivity or NETD = 0.02°C. Both of these devices (based on projected performance, size, and weight) are on the horizon as viable candidates for inclusion in modern surveillance drones.

The sophisticated eyesight of arthropods such as ants, beetles, arachnids, and crustaceans is due to hemispherical eyes that provide vision with a wide FOV (180°), high acuity to motion, low aberration, and an infinite depth of field. These optical features are only possible with hemispherical imaging systems and this makes them ideal devices for drone surveillance, endoscopic imaging, and other areas that require compact cameras with simple zoom optics and wide FOVs. An arthropod vision system similar to the eyes of fire ants and bark beetles was used as a basis for fabricating a digital camera comprising an array of 180 imaging elements (Jung et al., 2011). The numerous small, simple eye elements (ommatidium) comprising the compound eye of an arthropod were mimicked by the 180 silicon photodiodes, each topped with a single convex elastomeric microlens. The photodiodes were electronically interconnected to form a mesh that was embedded in a thin elastomeric membrane, which provides tunable access (via hydraulic actuation) to a range of hemispherical shapes in the order of a centimeter in diameter. When the detector/microlens array is combined with a tunable lens, which is also controled by a hydraulic actuator, one obtains a hemispherical camera capable of adjustable zoom and excellent imaging characteristics. The analytic scaling relationships between the overall deformation and applied pressure for tunable hemispherical imaging systems that offer adjustable zoom capabilities were established by Lu et al. (2013), and they have important implications in real applications. In the tunable hemispherical imaging system, operation of the tunable electronic eye camera system requires coordinated concomitant tuning of geometries of both the lens and detector array. When the system was tested using the 180-element compound eye, the camera system successfully retrieved high-contrast images and demonstrated 160° FOV. The images were recognizable but "blurry," as could be expected for a system with only 180 elements, and the resolution was comparable to the vision of many arthropods and mammals.

Blue *Morpho* Butterfly Detector

Infrared thermal imagers based on microbolometer technology (see Chapter 8) make up a great part of the commercial market for surveillance applications. It is not unreasonable to assume that because of their excellent performance, small size, and relatively low cost they will become the imagers most suitable to be fielded with UAVs. The blue morpho butterfly (*Morpho peleides*) is shown in Figure 12.2 at rest with its bright blue wings extended. The vivid, iridescent blue coloring is due to microscopic scales on the backs of their wings, which reflect light. The undersides of the morpho's wings are dull brown with many eye spots and provide camouflage against predators such as birds and insects when its wings are folded. When in flight this coloring scheme results in a contrast-

FIGURE 12.2 **The blue morpho butterfly (*Morpho peleides*).** *(Photo credit: Kade Havens)*

ing display of bright blue and dull brown with each wing beat making the blue morpho look like it is disappearing and reappearing as it flies.

The bright blue iridescence of blue morpho butterfly wings is caused by multilayer interference of light on the horizontal lamellae and from light diffraction on the vertical ridges of the scales (Vukusic et al., 1999). The scales of *Morpho* butterflies are formed by chitin from cuticle, which is secreted by epidermal cells. The top side of the scale has a complex hierarchical structure consisting of longitudinal ridges that support a series of alternating and opposed lamellae folds. The ridges are joined at intervals by transverse cross ribs. Pris et al. (2012) showed that the complex geometry of the Morpho butterfly wing structure exhibits an observable optical response to temperature primarily because of its thermal expansion and to a lesser extent because of a change in the refractive index. These changes result in the "wavelength conversion" of MWIR radiation (3–8 mm) into visible iridescence changes with a spatial resolution potentially localized to individual ridges on a butterfly scale.

Briefly, the thermal response of Morpho butterfly scales originates from the thermal expansion of the complex hierarchical structure, which causes an increase in spacing between the ridges, an expansion in the lamellar structure, and a thermally induced reduction in the refractive index of the structure following periodic (chopped) MWIR illumination. This leads to modulation in the multilayer interference and diffraction pattern by a "wavelength conversion" process to visible radiation. The extremely low thermal mass of the air-filled nanoarchitectures of the morpho butterfly scale structure coupled with the infrared absorption properties of natural insect cuticle (enhanced by using single-walled carbon nanotubes as a dopant in the scale structure) was used to explore an innovative design of bioinspired MWIR detector. They demonstrated experimentally and theoretically that absorption of MWIR photons by the morpho scale nanostructure modulates visible light iridescence upon illumination with white light.

The demonstration by Pris et al. (2012) of a high-speed (35–40 Hz), uncooled MWIR detector with a thermal sensitivity or NETD = 0.02°C based on this structure (a significant improvement over existing bolometers) bodes well for its potential use in low-cost, high-speed thermal imaging without the added weight and size of bulky heat sinks. Commercial thermal sensing products could reach the market in the next few years for applications such as before-and-after medical diagnosis of patient inflammation, public safety and homeland security surveillance, and handheld thermal imagers for firefighters (Coffey, 2015). Their small size and weight makes them ideally suited for equipping surveillance drones. Drones integrated with multiple sensors such as visual color cameras, thermal imagers, gas sensors, particulate detectors, and hyper-spectral imaging will soon be ubiquitous (Coffey, 2015).

References

Adams, K.P., Pekins, P.J., Gustafson, K.A., Bontaites, K., 1997. Evaluation of infrared technology for aerial moose surveys in New Hampshire. Alces 33, 129–139.

Allison, N.L., Destefano, S., 2006. Equipment and techniques for nocturnal wildlife studies. Wildl. Soc. Bull. 34 (4), 1036–1044.

Ammerman, L.K., McDonough, M., Hristov, N.I., Kunz, T.J., 2009. Census of the endangered Mexican long-nosed bat *Leptonycteris nivalis* in Texas, USA, using thermal imaging. Endangered Species Res. 8, 87–92.

Amstrup, S.C., York, G., McDonald, T.L., Nielson, R., Simac, K., 2004. Detecting denning polar bears with forward looking infra-red (FLIR) imagery. Bioscience 54 (4), 337–344.

Anderson Jr., C.R., Moody, D.S., Smith, B.L., Lindzey, F.G., Lanka, R.P., 1998. Development and evaluation of sightability models for summer elk surveys. J. Wildl. Manag. 63 (3), 1055–1066.

Anderson, D.R., Burnham, K.P., Lubow, B.C., Thomas, L., Corn, P.S., Medcia, P.A., Marlow, R.W., 2001. Field trials of line transect methods applied to estimation of desert tortoise abundance. J. Wildl. Manag. 65 (3), 583–597.

Anderson, K., Gaston, K.J., 2013. Lightweight unmanned aerial vehicles will revolutionize spatial ecology. Front. Ecol. Environ. 11 (3), 138–146.

Anderson, R.J., Sharp, E.J., Wood, G.L., Clark III, W.W., Vuong, Q., Salamo, G.J., Neurgaonkar, R.R., 1993. Mutually pumped phase conjugator as a moving object correlator. Opt. Lett. 18, 986–988.

Anderson, S.H., Gutzwiller, K.J., 1996. Habitat evaluation methods. In: Bookhout, T.A. (Ed.), Research and Management Techniques for Wildlife and Habitats, fifth ed., rev. The Wildlife Society, Bethesda, MD, pp. 592–606.

Anderson, S.S., 1978. Day and night activity of Grey bull seals. Mammal Rev. 8 (1–2), 43–46.

Atkinson, D., Morley, S.A., Hughes, R.N., 2006. From cells to colonies: at what levels of body organization does the 'temperature-size rule' apply? Evol. Dev. 8 (2), 202–214.

Barber, D.G., Richard, P.R., Hochheim, K.P., 1989. Thermal remote sensing for walrus population assessment in the Canadian arctic. International Geoscience and Remote Sensing Sumposium and Twelfth Canadian Symposium on Remote Sensing, Vancouver, BC, Canada 10–14 July.

Barber, D.G., Richard, P.R., Hochheim, K.P., 1991. Calibration of aerial thermal infrared imagery for walrus population assessment. Artic 44, 58–65.

Barrett, E.C., Curtis, L.F., 1992. Introduction to Environmental Remote Sensing, third ed. Chapman & Hall, London, UK, p. 426.

Bart, J., Droege, S., Geissler, P., Peterjohn, B., Ralph, C.J., 2004. Density estimation of wildlife surveys. Wildl. Soc. Bull. 32 (4), 1242–1247.

Beasom, S.L., 1979. Precision in helicopter censusing of white-tailed deer. J. Wildl. Manag. 43 (3), 777–780.

Beasom, S.L., Leon III, F.G., Synatzske, D.R., 1986. Accuracy and precision of counting white-tailed deer with helicopters at different sampling intensities. Wildl. Soc. Bull. 14, 364–368.

Beasom, S.L., Hood, J.C., Cain, J.R., 1981. The effect of strip width on helicopter censusing of deer. J. Range Manag. 34 (1), 36–37.

Beaver, J.T., Harper, C.A., Kissell Jr., R.E., Muller, L., Basinger, P.S., Goode, M.J., Van Manen, F.T., Winton, W., Kennedy, M.L., 2014. Aerial vertical-looking infrared imagery to evaluate bias of distance sampling techniques for white-tailed deer. Wildl. Soc. Bull. 38 (2), 419–427.

Beavers, S.C., Ramsey, F.L., 1998. Detectability analysis in transect surveys. J. Wildl. Manag. 63 (3), 948–957.

Beier, P., 1995. Dispersal of juvenile cougars in fragmented habitat. J. Wildl. Manag. 59 (2), 228–237.

Beier, P., Cunningham, S.C., 1996. Power of track surveys to detect changes in cougar populations. Wildl. Soc. Bull. 24 (3), 540–546.

Beier, P., Vaughan, M.R., Conroy, M.J., Quigley, H., 2006. Evaluating scientific inferences about the Flordia panther. J. Wildl. Manag. 70 (1), 236–245.

Becker, E.F., 1991. A terrestrial furbearer estimator based on probability sampling. J. Wildl. Manag. 55, 730–737.

Belant, J.L., Seamans, T.W., 2000. Comparison of 3 devices to observe white-tailed deer at night. Wildl. Soc. Bull. 28 (1), 154–158.

Bell, T.E., 1995. Remote sensing. IEEE Spectrum. March. pp. 24–30.

Berger, J., 1999. Anthropogenic extinction of top carnivores and interspecific animal behavior: implications of the rapid decoupling of a web involving wolves, bears, moose and ravens. Proc. R. Soc. Lond. 266, 2261–2267.

Berger, J., 2008. National Wildlife Magazine, Feb–Mar. Moose Moms hit the Road, p. 10, Pregnant Moose get cozy with humans, www.npr.org.

Bergerud, A.T., Dalton, W.J., Butler, H., Camps, L., Ferguson, R., 2007. Woodland caribou persistence and extirpation in relicpopulations on Lake Superior. Rangifer Spec. Issue 17, 57–78.

Bernatas, S., Nelson, L., 2004. Sightability model for California bighorn sheep in canyonlands using forward-looking infrared (FLIR). Wildl. Soc. Bull. 32 (3), 638–647.

Bernatas, S., 2010. Thermal Infrared survey for white-tailed deer. Vision Air report submitted to Fort Thomas, KY. 10 pp.

Beringer, J., Hanson, L.P., Sexton, O., 1998. Detection rates of white-tailed deer with a helicopter over snow. Wildl. Soc. Bull. 26 (1), 24–28.

Best, R.G., Fowler, R., 1981. Infrared emissivity and radiant surface temperatures of Canada and snow geese. J. Wildl. Manag. 45 (4), 1026–1029.

Best, R.G., Fowler, R., Hause, D., Wehde, M., 1982. Aerial thermal infrared census of Canada geese in South Dakota. Photogram. Eng. Remote Sens. 48 (12), 1869–1877.

Betke, M., Hirsh, D.E., Makris, N.C., McCraken, G.F., Procopio, M., Hristov, N.I., Tang, S., Bagghi, A., Reichard, J.D., Horn, J.W., Crampton, S., Cleveland, C.J., Kunz, T.H., 2008. Thermal imaging reveals significantly smaller Brazilian free-tailed bat colonies than previously estimated. J. Mammol. 89 (1), 18–24.

Betke, M., Hirsh, D.E., Bagchi, A., Hristov, N.I., Makris, N.C., Kunz, T.H., 2007. Tracking large variable numbers of objects in clutter. In: Proceedings of the IEEE Computer Society Washington, DC, June 2007, pp. 1–8.

Beyer Jr., D.E., Haufler, J.B., 1994. Diurnal versus 24-hour sampling of habitat use. J. Wildl. Manag. 58 (1), 178–180.

Biass, E.H., Gourley, S., 2001. Night vision technology update. Armada Int. 5, 28–36.

Bilkovic, D.M., Havens, K., Stanhope, D., Angstadt, K., 2014. Derelict fishing gear in Chesapeake Bay, Virginia: spatial patterns and implications for marine fauna. Mar. Pollut. Bull. 80, 114–123.

Birkebak, R.C., 1966. Heat transfer in biological systems. Int. Rev. Gen. Exp. Zool. 2, 269–344.
Black, B.B., Collopy, M.W., 1982. Nocturnal activity of great blue herons in a north Florida salt marsh. J. Field Ornithol. 55 (4), 403–406.
Bokma, F., 2004. Evidence against universal metabolic allometry. Funct. Ecol. 18 (2), 184–187.
Bonoan, R.E., Goldman, R.R., Wong, P.Y., Starks, P.T., 2014. Vasculature of the hive: heat dissipation in the honey bee (Apis mellifera) hive. Naturwissenschaften 101, 459–465.
Bontaites, K.M., Gustafson, K.A., Makin, R., 2000. A Gasaway-type moose survey in New Hampshire using infrared thermal imagery: preliminary results. Alces 36, 69–75.
Bookhout, T.A. (Ed.), 1996. Research and Management Techniques for Wildlife and Habitats. The Wildlife Society, Bethesda, MD, p. 740.
Boonstra, R., Krebs, C.J., Boutin, S., Eadie, J.M., 1994. Finding mammals using far-infrared thermal imaging. J. Mammal. 75 (4), 1063–1068.
Boonstra, R., Eadie, J.M., Krebs, C.J., Boutin, S., 1995. Limitations of far infrared thermal imaging in locating birds. J. Field Ornithol. 66 (2), 192–198.
Borchers, D.L., Buckland, S.T., Zucchini, W., 2004. Estimating Animal Abundance: Closed Populations. Springer-Verlag, London, p. 314.
Brawata, R.L., Raupach, T.H., Neeman, T., 2013. Techniques for monitoring carnivore behavior using automatic thermal video. Wildl. Soc. Bull. 37 (4), 862–871.
Brodsky, L.M., 1988. Ornament size influences mating success in male Rock Ptarmigan. Anim. Behav. 36, 662–667.
Brooks, J.W., 1970. Infra-red scanning for polar bear. In: Bears and Their Management, IUCN Pub. No. 23, Calgary, Canada, pp. 138–141.
Brooks, R.P. (Ed.), 1985. Nocturnal Mammals: Techniques for Study. School of forest resources research paper No. 48. Pennsylvania State University, University Park, PA, p. 57.
Buckland, S.T., Anderson, D.R., Burnham, K.P., Laake, J.L., 1993. Distance Sampling: Estimating Abundance of Biological Populations. Chapman & Hall, London, UK.
Buckland, S.T., Anderson, D.R., Burnham, K.P., Laake, J.L., Borchers, D.L., Thomas, L., 2001. Introduction to Distance Sampling: Estimating Abundance of Biological Populations. Oxford University Press, New York, NY, p. 432.
Budzier, H., Gerlach, G., 2011. Thermal Infrared Sensors: Theory Optimization, and Practice. John Wiley & Sons, West Sussex, UK, p. 299.
Budiyono, A., 2007. Advances in unmanned aerial vehicles technologies. Chin. Sci. Bull. 52 (1), 1–13.
Buech, R.R., 1985. Methodologies for observing beavers (*Castor canadensis*) during the activity period. In: Brooks, R.P. (Ed.), Nocturnal Mammals: Techniques for Study. School of Forest Resources, Research Paper No. 48. Pennsylvania State University, University Park, pp. 29–33.
Burn, D.M., Udevitz, M.S., Speckman, S.G., Benter, R.B., 2009. An improved procedure for detection and enumeration of walrus signatures in airborne thermal imagery. Int. J. Appl. Earth Observ. Geoinform. 11, 324–333.
Burn, D.M., Webber, M.A., Udevitz, M.S., 2006. Applications of airborne thermal imagery to surveys of Pacific walrus. Wildl. Soc. Bull. 34, 51–58.
Burney, S.G., Williams, T.L., Jones, C.H.N. (Eds.), 1988. Applications of Thermal Imaging. Adam Hilger, Philadelphia, PA.
Butler, M.J., Ballard, W.B., Wallace, M.C., 2003. An evaluation of population estimation techniques for Rio Grande wild turkeys in the Texas Panhandle and southwest Kansas. Progress report to Texas Parks and Wildlife Department. Austin, TX, October.
Butler M.J., Ballard, W.B., Wallace, M.C., 2005. Estimating Rio-Grande wilde turkey populations in Texas. Progress report to Texas Parks and Wildlife Department. Austin, TX.

Butler, M.J., Wallace, M.C., Ballard, W.B., DeMaso, S.J., Applegate, R.D., 2005. From the field: the relationship of Rio Grande wild turkey distributions to roads. Wildl. Soc. Bull. 33 (2), 745–748.
Campbell, J.B., Wynne, R.H., 2011. Introduction to Remote Sensing, fifth ed. Guilford Press, New York, NY, p. 667.
Cano, J.M., Nicieza, A.G., 2006. Temperature, metabolic rate, and constraints on locomotor performance in ectotherm vertebrates. Funct. Ecol. 20, 464–470.
Carr, N.L., Rodgers, A.R., Kingston, S.R., Hettinga, P.N., Thompson, L.M., Renton, J.L., Watson, P.J., 2010. Comparative woodland caribou population surveys in Slate Islands Provincial Park, Ontario. Rangifer, Special Issue No. 20, pp. 205–217.
Cassey, P., McArdle, B.H., 1999. An assessment of distance sampling techniques for estimating animal abundance. Environmetrics 10, 261–278.
Caughley, G., 1974. Bias in aerial survey. J. Wildl. Manag. 38 (4), 921–933.
Caughley, G., Sinclair, R., Scott-Kemmis, D., 1976. Experiments in aerial survey. J. Wildl. Manag. 40 (2), 290–300.
Caughley, G., 1977. Analysis of Vertebrate Populations. John Wiley & Sons, New York, NY, p. 234.
Caughley, G., Grice, D., 1982. A correction factor for counting emus from the air, and its application to counts in western Australia. Aust. Wildl. Res. 9, 253–259.
Caughley, G., Goddard, J., 1972. Improving the estimates from inaccurate censuses. J. Wildl. Manag. 36 (1), 135–140.
Cena, K., Clark, J.A., 1972. Effect of solar radiation on temperatures of working honey bees. Nat. New Biol. 236, 222–223.
Cena, K., Clark, J.A., 1973. Thermographic measurements of the surface temperatures of animals. J. Mammol. 54 (4), 1003–1007.
Chadwick, D., 2013. Ghost cats. National Geographic. December 2013. pp. 61–83.
Charman, W.N., 1996. Optics of the eye. In: Bass, M., Van Stryland, E.W., Williams, D.R., Wolfe, W.L. (Eds.), Handbook of Optics, vol. 1, McGraw Hill, New York, pp. 24.3–24.54.
Chatten, J.E., 1952. Appraisal of California waterfowl concentrations by aerial photography. Trans. N. American Wild. Conf., Wildlife Management Inst., Washington, DC, pp. 421–426.
Chen, G., Haddadi, A., Hoang, A.-M., Chevallier, R., Razeghi, M., 2015. Demonstration of type-II superlattice MWIR minority carrier unipolar imager for high operation temperature application. Opt. Lett. 40 (1), 45–47.
Christiansen, P., Steen, K.A., Jorgensen, R.N., Karstoft, H., 2014. Automated detection and recognition of wildlife using thermal cameras. Sensors 14, 13778–13793.
Cilulko, J., Janiszewski, P., Bogdaszewski, M., Szczygielski, E., 2013. Infrared thermal imaging in studies of wild animals. Eur. J. Wildl. Res. 59, 17–23.
Ciupa, R., Rogalski, A., 1997. Performance limitations of photon and thermal infrared detectors. Opto-Electr. Rev. 5 (4), 257–266.
Clark, J.A., 1976. Effects of surface emissivity and viewing angle on errors in thermography. Acta Thermogr. 1, 138–141.
Clarke, A., 2004. Is there a universal temperature dependence of metabolism? Funct. Ecol. 18, 252–256.
Clarke, A., 2006. Temperature and the metabolic theory of ecology. Funct. Ecol. 20, 405–412.
Clarke, A., Fraser, K.P.P., 2004. Why does metabolism scale with temperature? Funct. Ecol. 18, 243–251.
Clarke, A., Johnston, N.M., 1999. Scaling of metabolic rate with body mass and temperature in teleost fish. J. Anim. Ecol. 68 (5), 893–905.
Cobb, D.T., Francis D.L., Etters, R.W., 1996. Validating a wild turkey population survey using cameras and infrared sensors. Proc. Natl. Wild Turkey Symp. no. 7, pp. 213–218.

Cobb, D.T., Fuller, R.S., Francis, D.L., Sprandel, G.L., 1997. Research priorities for monitoring wild turkeys using cameras and infrared sensors. Proc. Annu. Conf. Southeast Assoc. Fish Wildl. Agencies 51, 362–372.

Cochran, W.W., Warner, D.W., Tester, J.R., Kuechle, V.B., 1965. Automatic radio-tracking system for monitoring animal movements. BioScience 15, 98–100.

Cockburn, W., 2006. Common errors in medical thermal imaging. Inframation Proc. 7, 165–177.

Coffey, V.C., 2011. Seeing in the dark: defense applications of IR imaging. Opt. Photonics News 22 (4), 26–31.

Coffey, V.C., 2012. Multi-spectral imaging moves into the main stream. Opt. Photonics News 23 (4), 18–24.

Coffey, V.C., 2015. Global UAV market to grow to US $15 billion by 2020. Opt. Photonic News. March, p. 13.

Coffey, V.C., 2015. Bio-inspired photonics. Opt. Photonics News. April, p. 30.

Coffey, V.C., 2015. Applied Robotics: How robots are changing our world. Photonics Spectra. June, p. 50.

Collier, B.A., Ditchkoff, S.S., Raglin, J.B., Smith, J.M., 2007. Detection probability and sources of variation in white-tailed deer spotlight surveys. J. Wildl. Manag. 71 (1), 277–281.

Colwell, R.N., 1966. Uses and limitations of multispectral remote sensing. In: Proceedings of the Fourth Symposium on Remote Sensing of Environment. Ann Arbor: Institute of Science and Technology, University of Michigan, pp. 71–100.

Conn, P.B., Ver Hoef, J.M., McClintock, B.T., Moreland, E.E, London, J.M., Cameron, M.F., Dahle, S.P., Boveng, P.L., 2013. Estimating multispecies abundance using automated detection systems: ice associated seals in the Bering Sea. Methods Ecol. Evol. 5, 1280–1293, doi: 10.1111/2041-210X.12127.

Conner, L.M., Rutledge, J.C., Smith, L.L., 2010. Effects of mesopredators on nest survival of shrub-nesting songbirds. J. Wildl. Manag. 74 (1), 73–80.

Conroy, M.J., Beier, P., Quigley, H., Vaughan, M.R., 2006. Improving the use of science in conservation: lessons from the Florida panther. J. Wildl. Manag. 70 (1), 1–7.

Conroy, M.J., Carroll, J.P., 2009. Quantitative Conservation of Vertebrates. Wiley-Blackwell, West Susex, UK, p. 342.

Conroy, M.J., Goldsberry, J.R., Hines, J.E., Stotts, D.B., 1988. Evaluation of aerial transect surveys for wintering American black ducks. J. Wildl. Manag. 52 (4), 694–703.

Cooke, S.J., Hinch, S.G., Wikelski, M., Andrews, R.D., Kuchel, L.J., Wolcott, T.G., Butler, P.J., 2004. Biotelemetry: a mechanistic approach to ecology. Trends Ecol. Evol. 19 (6), 334–343.

Crabtree, G.W., Lewis, N.S., 2007. Solar energy conversion. Physics Today, March, pp. 37–42.

Croon, G.W., McCullough, D.R., Olsen Jr., C.E., Queal, L.M., 1968. Infrared scanning techniques for big game censusing. J. Wildl. Manag. 32, 751–759.

Daniels, M., 2006. Estimating red deer *Cervus elaphus* populations: an analysis of variation and cost-effectivness of counting methods. Mammal Rev. 36 (3), 235–247.

Danylyshyn, D., Humphreys, J., 1982. Binoculars. Wildl. Rev. 10, 13–14.

Davis, A.P., Lettington, A.H., 1988. Principals of thermal imaging. In: Burney, S.G., Williams, T.L., Jones, C.H.N. (Eds.), Applications of Thermal Imaging. Adam Hilger, Philadelphia, PA, pp. 1–34.

Dawson, T.J., Hulbert, A.J., 1970. Standard metabolism, body temperature, and surface areas of Australian marsupials. Am. J. Physiol. 218 (4), 1233–1238.

Desholm, M., Fox, A.D., Beasley, P.D.L., Kahlert, J., 2006. Remote techniques for counting and estimating the number of bird-wind turbine collisions at sea: a review. Ibis 148, 76–89.

DeYoung, C.A., 1985. Accuracy of helicopter surveys of deer in south Texas. Wildl. Soc. Bull. 13, 146–149.

Digby, P.S.B., 1955. Factors affecting the temperature excess of insects in sunshine. J. Exp. Biol. 32, 279.

Ditchkoff, S.S., Raglin, J.B., Smith, J.M., Collier, B.A., 2005. Capture of white-tailed deer fawns using thermal imaging technology. Wildl. Soc. Bull. 33 (3), 1164–1168.

Drake, D., Aquila, C., Huntington, G., 2005. Counting a surburban deer population using forward looking infrared radar and road counts. Wildl. Soc. Bull. 33 (2), 656–661.

Dunbar, M.R., MacCarthy, K.A., 2006. Use of infrared thermography to detect signs of rabies infection in raccoons (*Procyon lotor*). J. Zoo Wildl. Med. 37 (4), 518–523.

Dunn, W.C., Donnelly, J.P., Krausmann, W.J., 2002. Using thermal infrared sensing to count elk in the southwestern United States. Wildl. Soc. Bull. 30 (3), 963–967.

Dymond, J.R., Trotter, C.M., Shepard, J.D., Wilde, H., 2000. Optimizing the airborne thermal detection of possums. Int. J. Remote Sens. 21 (17), 3315–3326.

Eberhardt, L.L., Chapman, D.G., Gilbert, J.R., 1979. A review of marine mammal census methods. Wildl. Manage. 63, 1–46.

Eberhardt, L.L., Garrott, R.A., White, P.J., Grogan, P.J., 1998. Alternative approaches to aerial censusing of elk. J. Wildl. Manag. 62 (3), 1046–1055.

Edwards, G.P., Pople, A.R., Saalfeld, K., Caley, P., 2004. Introduced mammals in Australian rangelands: future threats and the role of monitoring programs in management strategies. Austral Ecol. 29 (1), 40–50.

Efron, U. (Ed.), 1995. Spatial Light Modulator Technology: Materials, Devices, and Applications. Marcel-Dekker, New York, p. 665.

Engeman, R.M., Sugihara, R.T., Pank, L.F., Dusenberry, W.E., 1994. A comparison of plotless density estimators using Monte Carlo simulation. Ecology 75 (6), 1769–1779.

Erickson, Bledsoe, L.J., Hanson, M.B., 1989. Bootstrap correction for diurnal activity cycle in census data for Antarctic seals. Mar. Mammal Sci. 5 (1), 29–56.

Evans, C.D., Troyer, W.A., Lensink, C.J., 1966. Aerial census of moose by quadrat sampling units. J. Wildl. Manag. 30 (4), 767–776.

Fafarman, K.R., DeYoung, C.A., 1986. Evaluation of spotlight counts of deer in south Texas. Wildl. Soc. Bull. 14, 180–185.

Farnsworth, P.B., Cox, P.A., 1988. A laser illuminator designed for pollination studies with a night vision device. Biotropica 20 (4), 334–335.

Fehlman, II, W.L., Hinders, M.K., 2010. Passive infrared thermographic imaging for mobile robot object identification. J. Field Robotics 27 (3), 281–310.

Feynman, R.P., Leighton, R.B., Sands, M., 1963. The Feynman Lectures on Physicsvol. IAddison-Wesley, Reading, MA.

Focardi, S., Franzetti, B., Ronchi, F., 2013. Nocturnal distance sampling of a Mediterranean population of fallow deer is consistent with population projections. Wildl. Res. 40, 437–446.

Focardi, S., De Marinis, A.M., Rizzotto, M., Pucci, A., 2001. Comparitive evaluation of thermal infrared imaging and spotlighting to survey wildlife. Wildl. Soc. Bull. 29 (1), 133–139.

Focardi, Isotti, S.R., Tinelli, A., 2002. Line transect estimates of ungulate populations in a Mediterranean forest. J. Wildl. Manag. 66, 48–58.

Fox, A.D., Desholm, M., Kahlert, J., Christensen, T.J., Petersen, I.K., 2006. Information needs to support environmental impact assessment of the effects of European marine offshore wind farms on birds. Ibis 148 (Suppl. 1), 129–144.

Fox, A.D., Desholm, M., Kahlert, J., Christensen, T.K., Peterson, I.K., 2006. Information needs to support environmental impact assessment of the effects of European marine offshore wind farms on birds. Ibis 148, 129–144.

Frank, J.D., Kunz, T.H., Horn, J., Cleveland, C., Petronio, S., 2003. Advanced infrared detection and image processing for automated bat censusing. In: Andresen B.F., Fulop G.F. (Eds.), Infrared Technology and Applications XXIX. Proc. SPIE, vol. 5074, pp. 261–271.

Franke, U., Goll, B., Hohmann, U., Heurich, M., 2012. Aerial ungulate surveys with a combination of infrared and high-resolution natural colour images. Anim. Biodivers. Conserv. 35 (2), 285–293.

Franzetti, B., Ronchi, F., Marini, F., Scacco, M., Calmanti, R., Calabrese, A., Paola, A., Paolo, M., Focardi, S., 2012. Nocturnal line transect sampling of wild boar (Sus scrofa) in a Mediterranean forest: long-term comparison with capture-mark-resight population estimates. Eur. J. Wildl. Res. 58, 385–402.

Frederickson, L.H., Laubhan, M.K., 1996. Managing wetlands for wildlife. In: Bookhout, T.A. (Ed.), Research and Management Techniques of Wildlife and Habitats. fifth ed., rev. The Wildlife Society, Bethesda, MD.

Freilich, J.E., LaRue Jr., E.L., 1998. Importance of observer experience in finding desert tortoises. J. Wildl. Manag. 62 (2), 590–596.

Gade, R., Moeslund, T.B., 2014. Thermal cameras and applications: a survey. Mach. Vis. Appl. 25, 245–262.

Galligan, E.W., Bakken, G.S., Lima, S.L., 2003. Using a thermographic imager to find nests of grassland birds. Wildl. Soc. Bull. 31 (3), 865–869.

Garn, L.E., Sharp, E.J., 1974. Pyroelectric vidicon target materials. IEEE Parts Hybrids Packg. 10 (4), 208–221.

Garner, D.L., Underwood, H.B., Porter, W.F., 1995. Use of modern infrared thermography for wildlife population surveys. Environ. Manag. 19 (2), 233–238.

Garton, E.O., Horne, J.S., Aycrigg, J.L., Ratti, J.T., 2012. Research and experimental design. In: Silvy, N.J. (Ed.), The Wildlife Techniques Manual (Research). John Hopkins University Press, Baltimore, MD, pp. 1–40.

Gasaway, W.C., DuBois, S.D., Harbo, S.J., 1985. Biases in aerial transect surveys for moose during May and June. J. Wildl. Manag. 49 (3), 777–784.

Gasaway, W.C., DuBois, S.D., Reed, D.J., Harbo, S.J., 1986. Estimating moose population parameters from aerial surveys. Biol. Pap. University of Alaska, No. 22. 89 pp.

Gates, D.M., 1980. Biophysical Ecology. Springer Verlag, New York.

Gauthreaux Jr., S.A., 1969. A portable ceilometer technique for studying low-level nocturnal migration. Bird Banding 40, 309–320.

Gauthreaux Jr., S.A., Livingston, J.W., 2006. Monitoring bird migration with a fixed-beam radar and a thermal-imaging camera. J. Field Ornithol. 77 (3), 319–328.

Gebremedhin, K.G., Wu, B., 2001. A model of evaporative cooling of wet skin surface and fur layer. J. Thermal Biol. 26, 537–545.

Gilbert, P.F., Grieb, J.R., 1957. Comparison of air and ground deer counts in Colorado. J. Wildl. Manag. 21 (1), 33–37.

Gill, R.M.A., Thomas, M.L., Stocker, D., 1997. The use of portable thermal imaging for estimating deer population density in forest habitats. J. Appl. Ecol. 34, 1273–1286.

Gillooly, J.F., Allen, A.P., Savage, V.M., Charnov, E.L., West, G.B., Brown, J.H., 2006. Response to Clarke and Fraser: effects of temperature on metabolic rate. Funct. Ecol. 20, pp. 400–404.

Glazier, D.S., 2005. Beyond the '3/4-power law': varation in the intra-and interspecific scaling of metabolic rate in animals. Biol. Rev. 80 (4), 611–622.

Goodwin, D.B., Cappiello, G., Coppeta, D., Govignon, J., 1991. Hybrid digital/optical ATR system. Proc. SPIE 1564, 536–549.

Gompper, M.E., Kays, R.W., Ray, J.C., Lapoint, S.D., Bogan, D.A., Cryan, J.R., 2006. A comparison of noninvasive techniques to survey carnivore communities in northeastern North America. Wildl. Soc. Bull. 34 (4), 1142–1151.

Graves, H.B., Bellis, E.D., Kunth, W.M., 1972. Censusing white-tailed deer by airborne thermal infrared imagery. J. Wildl. Manag. 36, 875–884.

Gregory, S.K., 2005. Comparison of density estimators for white-tailed deer using aerial thermal infrared videography. Masters thesis. University of Arkansas, Monticello, 92 pp.

Grulois, T., Druart, G., Guerineau, N., Crastes, A., Sauer, H., Chavel, P., 2014. Extra-thin infrared camera for low cost surveillance applications. Opt. Lett. 39 (11), 3169–3172.

Gutzwiller, K.J., Marcum, H.A., 1997. Bird reactions to observer clothing color: implications for distance sampling techniques. J. Wildl. Manag. 61, 935–947.

Hafsteinsson, M.T., Misund, O.A., 1995. Recording the migration behavior of fish schools by multi-beam sonar during conventional acoustic surveys. ICES J. Mar. Sci. 52 (6), 915–924.

Hammel, H.T., 1956. Infrared emissivities of some arctic fauna. J. Mammal. 37, 375–378.

Harder, B., 2004. Degraded darkness. Conserv. Pract. 5 (2), 21–27.

Haroldson, B.S., 1999. Evaluation of thermal infrared imaging for detection of white-tailed deer. Masters thesis. University of Missouri, Columbia, p. 49.

Haroldson, B.S., Wiggers, E.P., Berringer, J., Hansen, L.P., McAninch, J.B., 2003. Evaluation of aerial thermal imaging for detecting white-tailed deer in a deciduous forest environment. Wildl. Soc. Bull. 31 (4), 1188–1197.

Harris, M.P., Lloyd, C.S., 1977. Variation in counts of seabirds from photographs. Br. Birds 70, 200–205.

Hart, J.S., 1956. Seasonal changes in insulation of the fur. Can. J. Zool. 34, 53–57.

Havens, K.J., Bilkovic, D.M., Stanhope, D., Angstadt, K., Hershner, C., 2008. Derelict blue crab trap impacts on marine fisheries in the lower York River, Virginia. North Am. J. Fish. Manag. 28, 1194–1200.

Havens, K.J., Bilkovic, D.M., Stanhope, D., Angstadt, K., 2011. Fishery failure, unemployed commercial fishers, and lost crab pots: an unexpected success story. Environ. Sci. Policy 14, 445–450.

Havens, K.J., Priest, III, W.I., Jennings, A., 1995. The use of night-vision equipment to observe wildlife in forested wetlands, VA. J. Sci. 46 (4), 227–234.

Havens, K.J., Sharp, E.J., 1995. The use of thermal imagery in the aerial survey of panthers (and other animals) in the Florida Panther National Wildlife Refuge and the Big Cypress National Preserve. Final Report to U.S. Fish and Wildlife Service (Naples, Florida).

Havens, K.J., Sharp, E.J., 1998. Using thermal imagery in the aerial survey of animals. Wildl. Soc. Bull. 26, 17–23.

Hecht, E., Zajac, A., 1979. Optics. Addison-Wesley, Reading, MA.

Heilbrun, R.D., Silvy, N.J., Peterson, M.J., Tewes, M.E., 2006. Estimating bobcat abundance using automatically triggered cameras. Wildl. Soc. Bull. 34 (1) 69–73.

Heinrich, B., 2003. Winter World. Harper Collins, New York, NY.

Helmuth, B., 2002. How do we measure the environment? Linking intertidal thermal physiology and ecology through biophysics. Integr. Comp. Biol. 42, 837–845.

Hemami, M.R., Watkinson, A.R., Gill, R.M.A., Dolman, P.M., 2007. Estimating abundance of introduced Chinese muntjac *Muntiacus reevesi* and native roe deer *Capreolus capreolus* using portable thermal imaging equipment. Mammal Rev. 37, 246–254.

Hill, S.B., Clayton, D.H., 1985. Wildlife After Dark: A Review of Nocturnal Observation Techniques (Occasional paper; no. 17). James Ford Bell Museum of Natural History, University of Minnesota, Minneapolis, MN, pp. 1–23.

Hirons, G., Lindsley, M., 1986. Beating nature's camouflage: locating woodcock on the ground in woodland by the use of a thermal imager. IWRB Woodcock Snipe Res. Group Newsl. 12, 5–8.

Hogan, H., 2007. Infrared imaging: the long and the short of it. Photonics Spectra. April, pp. 49–56.

Holst, G.C., 2000. Common Sense Approach to Thermal Imaging. JCD Publishing, Winter Park, FL.

Hone, J., 1988. A test of the accuracy of line and strip transect estimators in aerial survey. Aust. Wildl. Res. 15, 493–497.

Horn, J.W., Arnett, E.B., Kunz, T.H., 2008. Behavioral responses of bats to operating wind turbines. J. Wildl. Manag. 72 (1), 123–132.

Hovis, J.A, 1997. Red-cockaded woodpecker surveys. Red-cockaded woodpecker status/Goethe Forest. Florida Game and Freshwater Fish Comm. Final Perf. Rep. Tallahassee. 18 pp., ii.

Hristov, N.L., Betke, M., Kunz, T.H., 2008. Applications of thermal infrared imaging for research in aeroecology. Integr. Comp. Biol. 48 (1), 50–59.

Hristov, N.L., Betke, M., Theriault, D.E.H., Bagchi, A., Kunz, T.H., 2010. Seasonal variation in colony size of Brazilian free-tailed bats at Carlsbad Cavern based on thermal imaging. J. Mammal. 91 (1), 183–192.

Hu, W., Ye, Z., Liao, L., Chen, H., Chen, L., Ding, R., He, L., Chen, X., Lu, W., 2014. 128 × 128 long-wavelength/mid-wavelength two-color HgCdTe infrared focal plane array detector with ultra-low spectral cross talk. Opt. Lett. 39 (17), 5184–5187.

Hubbs, A.H., Karels, T., Boonstra, R., 2000. Indices of population size for burrowing mammals. J. Wildl. Manag. 64 (1), 296–301.

Israel, M., 2011. A UAV-based roe deer fawn detection system. Int. Arch. Photogram. Remote Sens. Spatial Inform. Sci, vol. XXXVIII-1/C22, ISPRS Zurich 2011 Workshop, Sept. 14–16, 2011, Zurich, Switzerland.

Jakob, M., 1949. Heat Transfervols. I & IIJohn Wiley & Sons, New York, NY.

Jacobson, H.A., Kroll, J.C., Browning, R.W., Koerth, B.H., Conway, M.H., 1997. Infrared-triggered cameras for censusing white-tailed deer. Wildl. Soc. Bull. 25 (2), 547–556.

Jiang, D., Zhuang, D., Huang, Y., 2013. Investigation of image fusion for remote sensing application. INTECH Open Access Publisher. dx.doi.org/10.5772/56946.

Johnson, J., 1958. Analysis of imaging forming systems. In: Proceedings of the Image Intensifier Symposium, pp. 249–273. Warfare Electrical Engineering Dept. US Army Engineering Research and Development Laboratories, Ft. Belvoir, VA. This article is reprinted *in* Johnson, R.B., Wolfe, W. L. (Eds.), 1985. Selected Papers on Infrared Design. Proc. SPIE, vol. 513, pp. 761–781.

Johnson, C.B., Owen S L.D., 1995. Image tube intensified electronic imaging. In: Bass, M. (Ed.), Handbook of Optics, Vol. I, Fundamentals, Techniques, and Designs. McGraw-Hill, Inc, New York, pp. 21.1–21.32.

Johnson, D.H., 1996. Population analysis. In: Bookhout, T.A. (Ed.), Research and Management Techniques of Wildlife and Habitats, fifth ed., rev. The Wildlife Society, Bethesda, MD, pp. 419–444.

Jones, R.C., 1957. Quantum efficiency of photoconductors. Proc. IRIS 2, 9.

Jung, I.W., Xiao, J.L., Malyarchuk, V., Lu, C.F., Li, M., Liu, Z.J., Yoon, J., Huang, Y.G., Rogers, J.R., 2011. Dynamically tunable hemispherical electronic eye camera system with adjustable zoom capability. Proc. Natl. Acad. Sci. USA 108 (5), 1788–1793.

Kaplan, H., 1999. Practical Applications on Infrared Thermal Sensing and Imaging Equipment, second ed. SPIE Press, Bellingham, WA, vol. TT34.

Karanth, K.U., Nichols, J.D., 1998. Estimation of tiger densities in India using photographic captures and recaptures. Ecology 79, 2852–2862.

Karanth, K.U., Nichols, J.D., Kumar, N.S., 2004. Photographic sampling of elusive mammals in tropical forests. In: Thompson, W.L. (Ed.), Sampling Rare or Elusive Species. Island Press, Washington, DC, pp. 229–247.

Kats, M.A., Blanchard, R., Zhang, S., Genevet, P., Ko, C., Ramanathan, S., Capasso, F., 2013. Vanadium dioxide as a natural disordered metamaterial: perfect thermal emission and large broadband negative differential thermal emittance. Phys. Rev. 3, 041004 (1–7).

Kiana, K., Koenen, K.G., DeStefano, S., Krausman, P.R., 2002. Using distance sampling to estimate seasonal densities of desert mule deer in a semidesert grassland. Wildl. Soc. Bull. 30, 53–63.

Kie, J.G., Bleich, V.C., Medina, A.L., Yoakum, J.D., Thomas, J.W., 1996. Managing rangelands for wildlife. In: Bookhout, T.A. (Ed.), Research and Management Techniques for Wildlife and Habitats, fifth ed., rev. The Wildlife Society, Bethesda, MD.

Kidd, J.B., Kissell Jr., R.E., 2009. Relationships between groundwater level and furbearer abundance in the northern Arkansas Mississippi alluvial valley. Ecohydrology 2, 472–479.

Kilgo, J.C., Sargent, R.A., Miller, K.V., Chapman, B.R., 1997. Landscape influences on breeding bird communities in hardwood fragments in South Carolina. Wildl. Soc. Bull. 25 (4), 878–885.

Killinger, D.K., Churnside, J.H., Rothman, L.S., 1995. Atmospheric optics. In: Bass, M. (Ed.), Handbook of Optics, Vol I, Fundamentals, Techniques, and Designs. McGraw-Hill, Inc, New York, pp. 44.1–44.50.

King, J.O., King S D.T., 1994. Use of a long-distance night vision device for wildlife studies. Wildl. Soc. Bull. 22, 121–125.

Kingsley, M.C.S., Hammill, M.O., Kelly, B.P., 1990. Infrared sensing of the under-snow lairs of the ringed seal. Mar. Mammal Sci. 6 (4), 339–347.

Kinkel, L.K., 1989. Lasting effects of wing tags on ring-billed gulls. Auk 106, 619–624.

Kirkwood, J.J, Cartwright, A., 1991. Behavioral observations in thermal imaging of the big brown bat, *Eptesicus fuscus*. SPIE, vol. 1467, Thermosense XIII, pp. 369–371.

Kirkwood, J.J., Cartwright, A., 1993. Comparison of two systems for viewing bat behavior in the dark. Proc. Indiana Acad. Sci. 102, 133–137.

Kissell Jr., R.E., Tappe, P.A., 2004. An assessment of thermal infrared detection rates using white-tailed deer surrogates. J. Arkansas Acad. Sci. 58, 70–73.

Kissell Jr., R.E., Nimmo, S.K., 2011. A technique to estimate white-tailed deer Odocoileus virginianus density using vertical-looking infrared imagery. Wildl. Biol. 17, 85–92.

Kissell Jr., R. E., Tappe, P.A, Gregory, S.K., 2004. Assessment of population estimators using aerial thermal infrared videography data. University of Arkansas, Monticello. Report for Arkansas Fish and Game Commision. pp. 28.

Klir, J.J., Heath, J.E., 1992. An infrared thermographic study of surface temperature in relation to external thermal stress in three species of foxes: the red fox (*Vulpes vulpes*), arctic fox (*Alopex lagopus*) and kit fox (*Vulpes macrotis*). Physiol. Zool. 65 (5), 1011–1021.

Koenen, K.K.G., DeStefano, S., Krausman, P.R., 2002. Using distance sampling to estimate seasonal densities of desert mule deer in a semidesert grassland. Wildl. Soc. Bull. 30, 53–63.

Koerth, B.H., McKown, C.D., Kroll, J.C., 1997. Infrared-triggered camera versus helicopter counts of white-tailed deer. Wildl. Soc. Bull. 25 (2), 557–562.

Korhonen, H., Harri, M., 1986. Heat loss of farmed raccoon dogs and blue foxes as evaluated by infrared thermography and body cooling. Comp. Biochem. Physiol. 84A (2), 361–364.

Kozlowski, L.J., Kosonocky, W.F., 1995. Infrared detector arrays. In: Bass, M. (Ed.), Handbook of Optics, Vol I, Fundamentals, Techniques, and Designs. McGraw-Hill, Inc, New York, pp. 23.1–23.27.

Krebs, C.J., 1985. Ecology: The Experimental Analysis of Distribution and Abundance, third ed. Harper & Row Publ., New York, NY, p. 800.

Krebs, C.J., 1989. Ecological Methodology. Harper Collins, New York, NY, p. 645.

Kruse, P.W., McGlauchlin, L.D., McQuistan, R.B., 1962. Elements of Infrared Technology: Generation, Transmission, and Detection. John Wiley & Sons, Inc, New York, NY.

Kufeld, R.C., Olterman, J.H., Bowden, D.C., 1980. A helicopter quadrat census for mule deer on Uncompahgre Plateau, Colorado. J. Wildl. Manag. 44 (3), 632–639.

Kummer, S., 2004. All IR cameras are not created equal. Photonics Spectra. February, pp. 58–60.

Kuntz, T.H., 2001. Seeing in the dark: recent technology advances for the study of free ranging bats. Twelfth International Bat Research Conf., Bat Research News, 42 (3), 91.

Kuntz, T.H., Arnett, E.B., Cooper, B.M., Erickson, W.P., Larkin, R.P., Mabee, T., Morrison, M.L., Strickland, M.D., Szewczak, J.M., 2007. Assessing impacts of wind-energy development on nocturnally active birds and bats: a guidance document. J. Wildl. Manag. 71 (8), 2449–2486.

Kuzyakin, V.A., 1983. Results of modeling winter transect counts. In: Zabrodin, V.A. (Ed.), Scientific Proceedings: Winter Transect Counts of Game Animals. Central Research Laboratory of Glavokhota (in Russian), Moscow, pp. 193–229.

Lancaster, W.C., Thompson, S.C., Speakman, J.R., 1997. Wing temperature in flying bats measured by infrared thermography. J. Therm. Biol. 22 (2), 109–116.

Lancia, R.A., Nichols, J.D., Pollock, K.H., 1996. Estimating the number of animals in wildlife populations. In: Bookhout, T.A. (Ed.), Research and Management Techniques for Wildlife and Habitats, fifth ed., rev. The Wildlife Society, Bethesda, Md, pp. 215–253.

Larkin, R.P., Diehl, R.H., 2012. Radar techniques for wildlife research. In: Silvy, N.J. (Ed.), The Wildlife Techniques Manual-Research, seventh ed. John Hopkins University Press, Baltimore, MD, pp. 319–335.

Lavers, C., Franks, K., Floyd, M., Plowman, A., 2005. Application of remote thermal imaging and night vision technology to improve endangered wildlife resource management with minimal animal distress and hazard to humans. J. Phys. Conf. Ser. 15, 207–212.

Lavers, C., Franklin, P., Plowman, A., Sayers, G., Bol, J., Shepard, D., Fields, D., 2009. Non-distructive high-resolution thermal imaging techniques to evaluate wildlife and delicate biological samples. J. Phys. Conf. Ser. 178 (2009), 012040.

Lazarevic, L., 2009. Improving the efficiency and accuracy of nocturnal bird surveys through equipment selection and partial automation. PhD thesis. Brunel University, London, UK. p. 220.

Leckie, D.G., 1982. An error analysis of thermal infrared line-scan-data for quantitative studies. Photogram. Eng. Remote Sens. 48 (6), 945–954.

LeResche, R.E., Rausch, R.A., 1974. Accuracy and precision of aerial moose censusing. J. Wildl. Manag. 38 (2), 175–182.

Les, C.B., 2010. Uncooled ir cameras and detectors:costing less, scalingup. Photonics Spectra, November, pp. 22–23.

Liang, J., Williams, D.R., Miller, D.T., 1997. Supernormal vision and high resolution retinal imaging. J. Opt. Soc. Am. 14 (11), 2884.

Liang, J., Williams, D.R., 1997. Aberrations and retinal image quality of the normal human eye. J. Opt. Soc. Am. 14 (11), 2873.

Liechti, F., Bruderer, B., Paproth, H., 1995. Quantification of nocturnal bird migration by moonwatching: comparison with radar and infrared observations. J. Field Ornithol. 66 (4), 457–468.

Liechti, F., Deter, P., Komenda-Zehnder, S., 2003. Nocturnal bird migration in Mauritania – first records. J. Ornithol. 144, 445–450.

Lillesand, T.M., Kiefer, R.W., Chipman, J.W., 2008. Remote Sensing and Image Interpretation, sixth ed. John Wiley & Sons, Hoboken, NJ, p. 776.

Litvaitis, J.A., Titus, K., Anderson, E.M., 1996. Measuring vertebrate use of terrestrial habitats and foods. In: Bookhout, T.A. (Ed.), Research and Management Techniques for Wildlife and Habitats. fifth ed., rev. The Wildlife Society, Bethesda, MD.

Locke, S.L., Lopez, R.R., Peterson, M.J., Silvy, N.J., Schwertner, T.W., 2006. Evaluation of portable infrared cameras for detecting Rio Grande wild turkeys. Wildl. Soc. Bull. 34 (3), 839–844.

Locke, S.L., Parker, I.D., Lopez, R.R., 2012. Use of remote cameras in wildlife ecology. In: Silvy, N.J. (Ed.), The Wildlife Techniques Manual-Research, seventh ed. John Hopkins University Press, Baltimore, MD, pp. 311–318.

Loudon, R., 1983. The Quantum Theory of Light, second ed. Oxford University Press, Oxford, UK.

Lowery, G.H., Newman, R.J., 1966. A continent wide view of bird migration on four nights in October. Auk 83, 547–586.

Lu, C., Li, M., Xaio, J., Jung, I., Wu, J., Huang, Y., Hwang, K.-C., Rogers, J.A., 2013. Mechanics of tunable hemispherical electronic eye camera systems that combine rigid device elements with soft elastomers. J. Appl. Mech. 80, 061022:1–7.

Ludwig, J., 1981. Proportion of deer seen in aerial counts. Minnesota Wildl. Res. Quart. 41, 11–19.

Mackenzie, D., 2006. Modeling the probability of resource use, the effect of, and dealing with, detecting a species imperfectly. J. Wildl. Manag. 70 (2), 367–374.

Maehr, D.S., Land, E.D., Roelke, M.E., 1991. Mortality patterns of panthers in southwest Florida. Proc. Annu. Conf. Southeast. Fish Wildl. Agencies 45, 201–207.

Maier, T., Bruckl, H., 2009. Wavelength-tunable microbolometers with metamaterial absorbers. Opt. Lett. 34 (19), 3012–3014.

Maier, T., Bruckl, H., 2010. Multispectral microbolometers for the midinfrared. Opt. Lett. 35 (22), 3766–3768.

Mannan, R.W., Conner, R.N., Marcot, B., Peek, J.M., 1996. Managing forestlands for wildlife. In: Bookhout, T.A. (Ed.), Research and Management Techniques for Wildlife and Habitats. fifth ed., rev. The Wildlife Society, Bethesda, MD.

Maldague, X.P.V., 1992. Nondestructive Evaluation of Materials by Infrared Thermography. Springer-Verlag, New York, NY.

Marble, H.P., 1967. Radiation from big game and background: a control study for infrared scanner census. MS thesis. University of Montana, Missoula, 86 pp.

Marcos, S., 2001. Refractive surgery and optical aberrations. Opt. and Photonics News, January, pp. 22–25.

Marini, F., Franzetti, B., Calabrese, A., Cappellini, S., Focardi, S., 2009. Response to human presence during nocturnal line transect surveys in fallow deer (*Dama dama*) and wild boar (*Sus scrofa*). Eur. J. Wildl. Res. 55, 107–115.

Martin, C., 2009. Mammalian Survey Techniques for Level II Natural Resource Inventories on Corps of Engineers Projects (Part I). ERDC TN-EMRRP-SI-34.

Martorello, D.A., Eason, T.H., Pelton, M.R., 2001. A sighting technique using cameras to estimate population size of black bears. Wildl. Soc. Bull. 29, 260–267.

McCafferty, D.J., Moncrieff, J.B., Taylor, I.R., Boddie, G.F., 1998. The use of IR thermography to measure the radiative temperature and heat loss of a barn owl (*Tyto alba*). J. Therm. Biol. 23 (5), 311–318.

McCafferty, D.J., 2007. The value of infrared thermography for research on mammals: previous applications and future directions. Mammal Rev. 37 (3), 207–223.

McCafferty, D.J., Gilbert, C., Paterson, W., Pomeroy, P.P., Thompson, D., Currie, J.I., Ancel, A., 2010. Estimating metabolic heat loss in birds and mammals by combining infrared thermography with biophysical modeling. Comp. Biochem. Physiol 158, 337–345.

McCafferty, D.J., 2013. Applications of thermal imaging in avian science. Ibis 155, 4–15.

McCracken, G.F., Gustin, M.K., 1991. Nursing behavior in Mexican free-tailed bat maternity colonies. Ethology 89, 305–321.

McCullough, D.R., Olsen Jr., C.E., Queal, L.M., 1969. Progress in large animal census by thermal mapping. In: Johnson, P.L. (Ed.), Remote Sensing in Ecology. University of Georgia Press, Athens, pp. 138–147.

McCullough, D.R., 1982. Evaluation of night spotlighting as a deer study technique. J. Wildl. Manag. 46, 963–973.

McCullough, D.R., Hirth, D.H., 1988. Evaluation of the Peterson–Lincoln estimator for a white-tailed deer population. J. Wildl. Manag. 52, 534–544.

McCullough, D.R., Weckerly, F.W., Garcia, P.A., Evett, R.R., 1994. Sources of inaccuracy in black-tailed deer herd composition counts. J. Wildl. Manag. 58 (2), 319–329.

McDonald, L.L., 2004. Sampling rare populations. In: Thompson, W.L. (Ed.), Sampling Rare or Elusive Species. Island Press, Washington, DC, pp. 11–42.

McMahon, B.F., Evans, R.M., 1992. Nocturnal foraging in the American white pelican. Condor 94, 101–109.

Meehan, T.D., 2006. Mass and temperature dependence of metabolic rate in litter and soil invertebrates. Physiol. Biochem. Zool. 79 (5), 878–884.

Melton, R.E., Sabol, B.M., Sherman, A., 2005. Poor man's missile tracking technology: Thermal IR detection and tracking of bats in flight. In: Watkins, W.R., Clement, D., Reynolds, W.R. (Eds.), Targets and Backgrounds XI. Proc. SPIE, vol. 5811, pp. 24–33.

Miller, D.T., 2000. Retinal imaging and vision at the frontiers of adaptive optics. Physics Today, January, pp. 31–36.

Miller, C.A., Wilson, D.E., 1997. *Pteropus tonganus*. Mammalian Species 552, 1–6.

Moen, A.N., 1968. Surface temperatures and radiant heat loss from white-tailed deer. J. Wildl. Manag. 32 (2), 338–344.

Monteith, J., Unsworth, M., 2008. Principles of Environmental Physics. Elsevier, Academic Press, Boston, MA, p. 418.

Morelle, K., Bouche, P., Lehaire, F., Leeman, V., Lejeune, P., 2012. Game species monitoring using road-based distance sampling in association with thermal imagers: a covariate analysis. Anim. Biodivers. Conserv. 35 (2), 253–265.

Morrison, M.L., 2006. Bird movements and behaviors in the Gulf Coast region: relation to potential wind energy developments. Subcontract Report NREL/SR-500-39572. US Dept. of Energy. Contract No. DE-AC36-99-GO10337.

Martorello, D.A., Eason, T.H., Pelton, M.R., 2001. A sighting technique using cameras to estimate population size of black bears. Wildl. Soc. Bull. 29 (2), 560–567.

National Research Council [NRC], 2007. Environmental Impacts of Wind-Energy Projects. The National Academies Press, Washington, DC, USA.

Naugle, D.E., Jenks, J.A., Kernohan, B.J., 1996. Use of thermal infrared sensing to estimate density of white-tailed deer. Wildl. Soc. Bull. 24 (1), 37–43.

Nelson, M.E., Mech, L.D., Frame, P.F., 2004. Tracking of white-tailed deer migration by global positioning system. J. Mammal. 85 (3), 505–510.

Night Vision Thermal Imaging Systems. Performance Model. User's Manual & Reference Guide. March 12, 2001. Document Rev 5. US Army Night Vision and Electronic Sensors Directorate, Fort Belvoir, VA.

Nisbet, I.G.T., 1959. Calculations of flight directions of birds observed crossing the face of the moon. Wilson Bull. 71, 237–243.

Nowak, R.M., 1999. Arctic fox. In: Walker's Mammals of the World, sixth ed., vol. I. Johns Hopkins University Press, Baltimore, Maryland, USA, pp. 644–646.

Odido, D., Madara, D., 2013. Emerging technologies: use of unmanned aerial systems in the realization of Vision 2030 goals in the counties. Int. J. Appl. Sci. Technol. 3 (8), 107–127.

Ono, M., Igarashi, T., Ohno, E., Sasaki, M., 1995. Unusual thermal defense by a honeybee against mass attack by hornets. Lett. Nat., 337, 334–336.

Ovadia, O., Pinshow, B., Lotem, A., 2002. Thermal imaging of house sparrow nestlings: the effect of begging behavior and nestling rank. Condor 104, 837–842.

Overton, W.S., Davis, D.E., 1969. Estimating the number of animals in wildlife populations. In: Giles, R.H. (Ed.), Wildlife Management Techniques, third ed. The Wildlife Soc, Washington, DC, pp. 403–456.

Oxley, D.J., Fenton, M.B., Carmody, G.R., 1974. The effects of roads on populations of small mammals. J. Appl. Ecol. 11 (1), 51–59.

Packard, J.M., Summers, R.C., Barnes, L.B., 1985. Variation of visibility bias during aerial surveys of manatees. J. Wildl. Manage. 49, 347–351.

Page, V., Goudail, F., Refregier, P., 1999. Improved robustness of target location in nonhomogeneous backgrounds by use of the maximum-likelihood ratio test location algorithm. Opt. Lett. 24 (20), 1383–1385.

Palmer, J.M., 1995. The measurement of transmission, absorption, emmission, and reflection. In: Bass, M. (Ed.), Handbook of Optics, Vol II, Devices, Measurements, and Properties. McGraw-Hill, Inc, New York, pp. 25.1–25.25.

Parker Jr., H.D., Driscoll, R.S., 1972. An experiment in deer detection by thermal scanning. J. Range Manag. 25, 480–481.

Parker Jr., H.D., 1972. Airborne infrared detection of deer. PhD thesis. Colorado State University, Fort Collins, 186 pp.

Pauley, G.R., Crenshaw, J.G., 2006. Evaluation of Paintball, Mark-Resight surveys for estimating mountain goat abundance. Wildl. Soc. Bull. 34 (5), 1350–1355.

Pekola, J., Schoelkopf, R., Ullom, J., 2004. Cryogenics on a chip. Physics Today. May, pp. 41–47.

Peterson, J.T., Bayley, P.B., 2004. A bayesian approach to estimating presence when a species is undetected. In: Thompson, W.L. (Ed.), Sampling Rare or Elusive Species. Island Press, Washington, DC, pp. 173–188.

Pierce, B.L., Lopez, R.R., Silvy, N.J., 2012. Estimating animal abundance. In: Silvy, N.J. (Ed.), The Wildlife Techniques Manual (Research), seventh ed. John Hopkins University Press, Baltimore, MD, pp. 284–310.

Pojar, T.M., Bowden, D.C., Gill, R.B., 1995. Aerial counting experiments to estimate pronghorn density and herd structure. J. Wildl. Manag. 59 (1), 117–128.

Pollock, K.H., Kendall, W.L., 1987. Visibility bias in aerial surveys: a review of estimation procedures. J. Wildl. Manag. 51 (2), 502–510.

Pollock, K.H., Nichols, J.D., Simons, T.R., Farnsworth, G.L., Bailey, L.L., Sauer, J.R., 2002. Large scale wildlife monitoring studies: statistical methods for design and analysis. Environmetrics 13, 105–119.

Pollock, K.H., Marsh, H., Bailey, L.L., Farnsworth, G.L., Simmons, T.R., Alldreidge, M.W., 2004. Separating components of detection probability in abundance estimation: an overview with diverse examples. In: Thompson, W.L. (Ed.), Sampling Rare or Elusive Species. Island Press, Washington, DC, pp. 43–58.

Porter, W.P., Gates, D.M., 1969. Thermodynamic equilibria of animals with environment. Ecol. Monogr. 39 (3), 227–244.

Potvin, F., Breton, L., 2005. From the field: testing 2 aerial survey techniques on deer in fenced enclosures-visual double-counts and thermal infrared sensing. Wildl. Soc. Bull. 33 (1), 317–325.

Pris, A.D., Utturkar, Y., Surman, C., Morris, W.G., Vert, A., Zalyubovskiy, S., Deng, T., Ghiradella, H.T., Potyrailo, R.A., 2012. Towards high-speed imaging of infrared photons with bio-inspired nanoarchitectures. Nat. Photonics 6, 195–200.

Quattrochi, D.A., Luvall, J.C., 1999. Thermal infrared remote sensing for analysis of landscape ecological processes: methods and applications. Landscape Ecol. 14, 577–598.

Quiroga, R.Q., Reddy, L., Kreiman, G., Koch, C., Fried, I., 2005. Invarient visual representation by single neurons in the human brain. Nature 435, 1102–1107.

Ratches, J.A., Lawson, W.R., Obert, L.P., Bergmann, R.J., Cassidy, T.W., Swenson, J.W., 1975. Night Vision Laboratory static performance model for thermal imaging systems. US Army Electronic Command, Night Vision Laboratory, Fort Belvoir, VA. ECOM. 7043.

Ratti, J.T., Garton, E.O., 1996. Research and experimental design. In: Bookhout, T.A. (Ed.), Research and Management Techniques for Wildlife and Habitats. fifth ed., rev. The Wildlife Society, Bethesda, MD, pp. 1–23.

Reed, J.K., Shepard, A.N., Koenig, C.C., Scanlon, K.M., Gilmore Jr., R.G., 2005. Mapping, habitat characterization, and fish surveys of the deep water Oculina coral reef Marine Protected Area: a review of historical and current research. In: Freiwald, A., Roberts, J.M. (Eds.), Cold-Water Corals and Ecosystems. Springer-Verlag, Berlin Heidelberg, pp. 443–465.

Reichard, J.D., Prajapati, S.I., Austad, S.N., Keller, C., Kunz, T.H., Thermal windows on Brazilian free-tailed bats facilate thermogrgulation during prolonged flight. Integr. Comp. Biol. 50 (3), 358–370.

Reynolds, D.R., Riley, J.R., 2002. Remote-sensing, telemetric and computer-based technologies for investigating insect movement: a study of existing and potential techniques. Comput. Electron. Agric. 35, 271–307.

Reynolds, P., Duck, C., Youngson, D., Clem, D., 1993. An evaluation of airborne thermal imaging for the census of red deer *Cervus elaphus* populations in extensive open habitats in Scotland. Proc. Int. Union Game Biol. Congress 21 (2), 162–168.

Rice, W.R., Harder, J.D., 1977. Application of multiple aerial sampling to a mark-recapture census of white-tailed deer. J. Wildl. Manag. 41, 197–206.

Richards, A., Johnson, G., 2005. Radiometric calibration of infrared cameras accounting for atmospheric path effects. Proc. SPIE, vol. 5782, Thermo Sense XXVII, pp. 9–29.

Richardson, D.M., Bradford, J.W., Range, P.G., Christensen, J., 1999. A video probe system to inspect red-cockaded woodpecker cavities. Wildl. Soc. Bull. 27 (2), 353–356.

Roberts, C.W., Pierce, B.L., Braden, A.W., Lopez, R.R., Silvy, N.J., Frank, P.A., Ransom Jr., D., 2006. Comparison of camera and road survey estimates for white tailed deer. J. Wildl. Manag. 70 (1), 263–267.

Robinson, R., Smith, T.S., Larson, R.T., Kirschhoffer, B.J., 2014. Factors influencing the efficacy of forward-looking infrared in polar bear den detection. BioScience 64 (8), 735–742.

Rodda, G.H., 1992. Foraging behavior of the brown tree snake, Boiga irregularis. Herpetol. J. 2, 110–114.

Rodgers, A.R., Rempel, R.S., Abraham, K.F., 1996. A GPS-based telemetry system. Wildl. Soc. Bull. 24 (3), 559–566.

Rodgers Jr., J.A., Kubilis, P.S., Nesbitt, S.A., 2005. Accuracy of aerial surveys of waterbird colonies. Waterbirds 28 (2), 230–237.

Rodriguez-Duran, A., Soto-Centeno, J.A., 2003. Temperature selection by tropical bats roosting in caves. J. Thermal Biol. 28, 465–468.

Rogalski, A., 2003a. Infrared detectors: status and trends. Prog. Quant. Electron. 27, 59–210.

Rogalski, A., 2003b. Quantum well photoconductors in infrared detector technology. J. Appl. Phys. 93, 4355–4391.

Romesburg, H.C., 1981. Wildlife science: gaining reliable knowledge. J. Wildl. Manag. 45 (2), 293–313.

Root, A.I., 1972. The ABC and XYZ of Bee Culture, thirty-fourth ed. A.I. Root Co. Medina, Ohio.

Routledge, R.D., 1981. The unreliability of population estimates from repeated incomplete aerial surveys. J. Wildl. Manag. 45 (4), 997–1000.

Russel, R.W., Dunne, P., Sutton, C., Kerlinger, P., 1991. A visual study of migrating owls at Cape May Point, New Jersey. Condor 93, 55–61.

Sabol, B.M., Hudson, M.K., 1995. Technique using thermal infrared-imaging for estimating populations of gray bats. J. Mammal. 76 (4), 1242–1248.

Sahu, D.K., Parsai, M.P., 2012. Different image fusion techniques: a critical review. IJMER 2 (5), 4298–4301.

Sale, R., 2006. A Complete Guide to Arctic Wildlife. Firefly Books Inc, Buffalo, New York, USA, pp. 385–387.

Samuel, M.D., Pollock, K.H., 1981. Correction of visibility bias in aerial surveys where animals occur in groups. J. Wildl. Manag. 45 (4), 993–1000.

Samuel, M.D., Garten, E.O., Schlegel, M.W., Carson, R.G., 1987. Visibility bias during aerial surveys of elk in northcentral Idaho. J. Wildl. Manag. 51 (3), 622–630.

Samuel, M.D., Fuller, M.R., 1996. Wildlife radiotelemetry. In: Bookhout, T.A. (Ed.), Research and Management Techniques for Wildlife and Habitats. fifth ed., rev. The Wildlife Society, Bethesda, MD, pp. 370–418.

Savidge, J.A., Seibert, T.F., 1988. An infrared trigger and camera to identify predators at artificial nests. J. Wildl. Manag. 52 (2), 291–294.

Scheer, B.T., 1966. Animal Physiology. John Wiley & Sons, Inc, New York, NY, p. 407.

Schmidt-Nielsen, K., 1970. Animal Physiology. Prentice-Hall, Inc, Englewood Cliffs, NJ, p. 145.

Scholander, P.F., Walters, V., Hock, R., Irving, L., 1950. Body insulation of some arctic and tropical mammals. Biol. Bull. 99, 225–236.

Scholander, P.F., Hock, R., Walters, V., Johnson, F., Irving, L., 1950. Heat regulation in some arctic and tropical mammals and birds. Biol. Bull. 99, 237–258.

Scott, J.M., Temple, S.A., Harlow, D.L., Shaffer, M.L., 1996. Restoration and management of endangered species. In: Bookhout, T.A. (Ed.), Research and Management Techniques of Wildlife and Habitats. fifth ed., rev. The Wildlife Society, Bethesda, MD, pp. 531–539.

Seber, G.A.F., 1982. The Estimation of Animal Abundance and Related Parameters. Charles Griffin & Company LTD, London.

Seber, G.A.F., 1986. A review of estimating animal abundance. Biometrics 42, 267–292.

Seubert, J.L., 1948. A technique for nocturnal studies of birds and mammals by the use of infra-red projection and electronic reception. MS thesis. Ohio State University, Columbus, OH, 87 pp.

Seyrafi, K., 1973. Electro-Optical Systems Analysis. Electro-Optical Research Co, Los Angeles, CA.

Sharp, E.J., Wood, G.L., Salamo, G.J., Anderson, R.J., Yarrison-Rice, J.M., Neurgaonkar, R.R., 1994. Photorefractive image processing using mutually-pumped phase conjugators. Proc. SPIE Optical Pattern Recognition V 2237, 347–359.

Shaw, J.A., Nugent, P.W., Johnson, J., Bromenshenk, J.J., Henderson, C.B., Debnam, S., 2011. Long-wave infrared imaging for non-invasive beehive population assessment. Opt. Express 19 (1), 399–408.

Shenkenberg, D.L., 2009. Seize the night. Photonics Spectra. June, pp. 52–53.

Shirvaikar, M.V., Trivedi, M.M., 1995. A neural network filter to detect small targets in high clutter backgrounds. IEEE Trans. Neural Netw. 6 (1), 252–257.

Shupe, T.E., Beasom, S.L., 1987. Speed and altitude influences on helicopter surveys of mammals in brushland. Wildl. Soc. Bull. 15, 552–555.

Sidle, J.G., Nagle, H.G., Clark, R., Gilbert, C., Stuart, D., Willburn, K., Orr, M., 1993. Aerial thermal infrared imaging of sandhill cranes on the Platte River, Nebraska. Nebraska Remote Sens. Environ. 43, 333–341.

Silver, H., Colovos, N.F., Holter, J.B., Hayes, H.H., 1969. Fasting metabolism of white-tailed deer. J. Wildl. Manag. 33 (3), 490–498.

Silver, H., Holter, J.B., Colovos, N.F., Hayes, H.H., 1971. Effect of falling temperature on heat production in fasting white-tailed deer. J. Wildl. Manag. 35 (1), 37–46.

Silver, H., Colovos, N.F., Hayes, H.H., 1959. Basal metabolism of white-tailed deer-a pilot study. J. Wildl. Manag. 23 (4), 434–438.

Silvy, N.J. (Ed.), 2012. The Wildlife Techniques Manual, (Research), seventh ed. The Johns Hopkins University Press, Baltimore, MD, p. 686.

Simonis, P.M., Rattal, E.M., Qualim, Mouhse, A., Vigneron, J.P., 2014. Radiative contribution to thermal conductance in animal furs and other woolly insulators. Opt. Express 22, 1940–1951.

Siniff, D.B., Skoog, R.O., 1964. Aerial censusing of caribou using stratified random sampling. J. Wildl. Manag. 28 (2), 391–401.

Smart, J.C., Ward, A.I., White, P.C.L., 2004. Monitoring woodland deer populations in the UK: an imprecise science. Mammal Rev. 34, 99–114.

Soria, M., Freon, P., Gerlotto, F., 1996. Analysis of vessel influence on spatial behavior of fish schools using a multi-beam sonar and consequences for biomass estimates by echo-sounder. ICES J. Mar. Sci. 53, 453–458.

Sprafke, T., Beletic, J.W., 2008. High performance infrared focal plane arrays for space applications. Opt. Photonics News, June, pp. 22–27.

Speakman, J.R., Ward, S., 1998. Infrared thermography: principles and applications. Zoology 101, 224–232.

Stark, B., Smith, B., Chen, Y., 2014. Survey of thermal infrared remote sensing for unmanned aerial systems. IEEE International Conf. on Unmanned Aircraft Systems (ICUAS) May 27–30, pp. 1294–1299.

Stedman, R., Diefenbach, D.R., Swope, C.B., Finley, J.C., Luloff, A.E., Zinn, H.C., San Julian, G.J., Wang, G.A., 2004. Integrating wildlife and human-dimensions research methods to study hunters. J. Wildl. Manag. 68 (4), 762–773.

Steen, K.A., Villa-Henricksen, A., Therkildsen, O.R., Karstof, H., Green, O., 2012. Automatic detection of animals using thermal imaging. Sensors 12, 7587–7597.

Steinhorst, R.K., Samuel, M.D., 1989. Sightability adjustment methods for aerial surveys of animal populations. Biometrics 45, 415–425.

Stephens, P.A., Zaumyslova, O.Y., Miquelle, D.G., Myslenkov, A.I., Hayward, G.D., 2006. Estimating population density from indirect sign: track counts and the Formozov–Malyshev–Pereleshin formula. Anim. Conserv. 9, 339–348.

Stevens, B., Bony, S., 2013. Water in the atmosphere. Physics Today. June, pp. 29–34.

Stewart, K.H., 1988. Meteorological and land survey applications of thermography. In: Burnay, S.G., Williams, T.L., Jones, C.H. (Eds.), Applications of Thermal Imaging. Adam Hilger, Bristol, England, pp. 126–155.

Stirling, I., 1988. Polar Bears. University of Michigan Press, Michigan, p. 144.

Stoll Jr., R.J., McClain, M.W., Clem, J.C., Plageman, T., 1991. Accuracy of helicopter counts of white-tailed deer in western Ohio farmland. Wildl. Soc. Bull. 19, 309–314.

Storm, D.J., Samuel, M.D., Van Deelen, T.R., Malcolm, K.D., Rolley, R.E., Frost, N.A., Bates, D.P., Richards, B.J., 2011. Comparison of visual-based helicopter and fixed wing forward looking infrared surveys for counting white tailed deer *Odocoileus virginianus*. Wildl. Biol. 17, 431–440.

Storm, G.L., Cottam, D.F., Yahner, R.H., Nichols, J.D., 1992. A comparison of two techniques for estimating deer density. Wildl. Soc. Bull. 20, 197–203.

Swanson, G.A., Sargeant, A.B., 1972. Observation of nighttime feeding behavior of ducks. J. Wildl. Manag. 36 (3), 959–961.

Tappe, P.A.S., Kissell Jr., R.E., McCammon, E.E., 2003. Ground-based and aerial thermal infrared imaging for estimating white-tailed deer population densities. University of Arkansas-Monticello. Final Report for Arkansas Fish and Game Commision, pp. 29.

Temple, S.A., Cary, J.C., 1998. Modeling dynamics of habitat-interior bird populations in fragmented landscapes. Conserv. Biol. 2 (4), 340–347.

Thomas, D.C., 1967. Population estimates of barren-ground caribou March to May, 1967. Can. Wildl. Serv. Rep. 9, 44.

Thompson, B.C., Baker, B.W., 1981. Helicopter use by wildlife agencies in North America. Wildl. Soc. Bull. 9 (4), 319–323.

Thompson, W.L., 2002. Toward reliable bird surveys: accounting for individuals present but not detected. Auk 119 (1), 18–25.

Thompson, W.L., 2004. Future directions in estimating abundance of rare or elusive species. In: Thompson, W.L. (Ed.), Sampling Rare or Elusive Species. Island Press, Washington, DC, pp. 389–399.

Thompson, W.L. (Ed.), 2004. Sampling Rare or Elusive Species. Island Press, Washington, DC, p. 412.

Thompson, W.L., White, G.C., Gowan, C., 1998. Monitoring Vertebrate Populations. Academic Press, Inc, SanDeigo, CA, p. 359.

Tibbals, E.C., Carr, E.K., Gates, D.M., Krieth, F., 1964. Radiation and convection in conifers. Am. J. Bot. 51 (5), 529–538.

Trivedi, M.M., Wyatt, C.L., Anderson, D.R., 1982. A multispectral approach to remote detection of deer. Photogram. Eng. Remote Sens. 48 (12), 1879–1889.

Trivedi, M.M., Wyatt, C.L., Anderson, D.R., Voorheis, H.T., 1984. Designing a deer detection system using a multistage classification approach. Photogram. Eng. Remote Sens. 50 (4), 481–491.

Truett, J.C., Short, H.L., Williamson, S.C., 1996. Ecological impact assessment. In: Bookhout, T.A. (Ed.), Research and Management Techniques of Wildlife and Habitats, fifth ed., rev. The Wildlife Society, Bethesda, MD.

Tuttle, M.D., 1979. Status, causes of decline, and management of endangered gray bats. J. Wildl. Manag. 43, 1–17.

Unsworth, J.W., Kuck, L., Garton, E.O., 1990. Elk sightability model validation at the National Bison Range, Montana. Wildl. Soc. Bull. 19 (2), 113–115.

Unsworth, J.W., Leban, F.A., Garton, E., Leptich, D.J., Zager, P., 1994. Aerial Survey: User's manual. Second edition, Idaho Department of Fish and Game, Boise, Idaho, USA.

Vallese, F., 2010. Cooled ir detectors. Photonics Spectra. December, pp. 38–41.

Vas, E., Lescroël, A., Duriez, O., Boguszewski, G., Grémillet, D., 2015. Approaching birds with drones: first experiments and ethical guidelines. Biol. Lett. 11, 20140754. http://dx.doi.org/10.1098/rsbl.2014.0754.

Vasterling, M., Meyer, U., 2013. Challenges and opportunities for UAV-borne thermal imaging. In: Kuenzer, C., Dech, S. (Eds.), Thermal Infrared Remote Sensing. Springer, The Netherlands, pp. 69–92.

Vaudo, J.J., Lowe, C.G., 2006. Movement patterns of the round stingray *Urobatis halleri* (Cooper) near a thermal outfall. J. Fish Biol. 68, 1756–1766.

Vaughan, T.A., 1978. Mammalogy. W.B. Saunders Co, Philadelphia, PA, p. 522.

Venier, L.A., Fahrig, L., 1996. Habitat availability causes the species abundance-distribution relationship. OIKOS 76, 564–570.

Vollmer, M., Mollmann, K.-P., 2010. Infrared Thermal Imaging: Fundamentals, Research and Applications. Wiley-VCH, Verlag, Germany, p. 593.

Vukusic, V., Sambles, J.R., Lawrence, C.R., Wootton, R.J., 1999. Quantified interference and diffraction in single *Morpho* butterfly scales. Proc. R. Soc. Lond. B 266, 1403–1411.

Wakeling, B.F., Cagle, D.N., Witham, J.H., 1999. Performance of forward looking infrared surveys on cattle, elk, and turkey in northern Arizona. In: Research on the Colorado Plateau: proceedings; 1996, Fourth Biennial Conference, vol. 4, pp. 77–88.

Wakeling, B.F., Engel-Wilson, R.W., Rodgers, T.D., 2003. Reliability of infrared surveys for detecting and enumerating turkeys within forested habitats in north-central Arizona. In: Proceedings of the Biennial Conference of Research on the Colorado Plateau, vol. 6, pp. 187–192.

Watmough, D.J., Fowler, P.W., Oliver, R., 1970. The thermal scanning of a curved isothermal surface: implications for clinical thermography. Phys. Med. Biol. 15 (1), 1–8.

Watts, B.D., Bradshaw, D.S., Wilson, M.D., 2006. Investigation of red-cockaded woodpeckers in Virginia: year 2005 report. Center for Conservation Biology Technical Report Series., CCBTR-06-03. College of William and Mary, Williamsburg, VA. p. 18.

Weir, K., 2011. Operation knock on wood. Nat. Conservancy 4, 46–53.

West, G.B., Brown, J.H., 2004. Life's universal scaling laws. Physics Today. September, pp. 36–42.

Wiggers, E.P., Beckerman, S.F., 1993. Use of thermal infrared sensing to survey white-tailed deer populations. Wildl. Soc. Bull. 21, 263–268.

Wilde, R.H., 2000. Thermal infrared imaging for counting deer. Conserv. Sci. Newsl. 39, 11–13.

Williams, B.K., Nichols, J.D., Conroy, M.J., 2001. Analysis and Management of Animal Populations. Academic Press, San Diego, CA, p. 817.

Williams, S.C., DeNicola, A.J., Ortega, I.M., 2008. Behavioral responses of white-tailed deer subjected to lethal management. Can. J. Zool. 86, 1358–1366.

Williams, T.L., 2009. Thermal Imaging Cameras: Characteristics and Performance. Taylor & Francis Group, Boca Raton, FL, p. 218.

Willis, K., Horning, M., Rosen, D.A.S., Trites, A.W., 2005. Spatial variation of heat flux in Stellar sea lions: evidence for consistent avenues of heat exchange along the body trunk. J. Exp. Mar. Biol. Ecol. 315, 163–175.

Wilson, G.J., Delahay, R.J., 2001. A review of methods to estimate the abundance of terrestrial carnivores using field signs and observation. Wildl. Res. 28, 151–164.

Wilson, R.B., 1985. Use of thermal imagers for nocturnal field studies: a demonstration. In : Brooks, R.P. (Ed.), Nocturnal Mammals: Techniques for Study. School for Forest Resources, Pennsylvania State University. Research Paper #48, p. 57.

Wilton, M.L., Garner, D.L., Inglis, J.E., 1994. The use of infrared trail monitors to study moose movement patterns. Alces 30, 153–157.

Witmer, G.W., 2005. Wildlife population monitoring: some practical considerations. Wildl. Res. 32, 259–263.

Wolfe, W.L. (Ed.), 1965. Handbook of Military Infrared Technology. U.S. Government Printing Office, Washington, DC.

Wolfe, W.L., Kruse, P.W., 1995. Thermal Detectors. In: Bass, M. (Ed.), Handbook of Optics, Vol. I, Fundamentals, Techniques, and Designs. McGraw-Hill, Inc, New York, pp. 19.1–19.14.

Wood, G.L., Clark, W.W., Miller, M.J., Salamo, G.J., Sharp, E.J., Neuragonkar, R.R., Oliver, J.R., 1995. Photorefractive materials. In: Efron, U. (Ed.), Spatial Light Modulator Technology. Marcel-Dekker, New York, pp. 161–215, 665.

Wride, M.C., Baker, K., 1977. Thermal imagery for census of ungulates. In: Proceedings of Eleventh International Symposium of Remote Sensing of the Envionment. University of Michigan, Ann Arbor. pp. 1091–1099.

Wyatt, C.L., Trivedi, M.M., Anderson, D.R., 1980. Statistical evaluation of remotely sensed thermal data for deer census. J. Wildl. Manag. 44 (2), 397–402.
Wyatt, C.L., Trivedi, M.M., Anderson, D.R., Pate, M.C., 1985. Measurement techniques for spectral characterization for remote sensing. Program. Eng. Remote Sens. 51 (2), 245–251.
Zalewski, E.F., 1995. Radiometry and photometry. In: Bass, M. (Ed.), Handbook of Optics, Vol. II, Devices, Measurements, and Properties. McGraw-Hill, Inc, New York, pp. 24.3–24.51.

Subject Index

A

B

C

F

G

H

I

J

K

L

O

P

Q

R

S

T

U

V

W

Z

Printed in the United States
By Bookmasters